HANDBOOK OF BIOENERGY ECONOMICS AND POLICY

For other titles published in this series, go to
www.springer.com/series/6360

NATURAL RESOURCE MANAGEMENT AND POLICY

Editors:
David Zilberman
Department of Agricultural and Resource Economics
University of California, Berkeley
Berkeley, CA 94720

Renan Goetz
Department of Economics
University of Girona, Spain

Alberto Garrido
Department of Agricultural Economics and Social Sciences
E.T.S. Ingenieros Agrónomos, Madrid, Spain

EDITORIAL STATEMENT

There is a growing awareness to the role that natural resources such as water, land, forests and environmental amenities play in our lives. There are many competing uses for natural resources, and society is challenged to manage them for improving social well being. Furthermore, there may be dire consequences to natural resources mismanagement. Renewable resources such as water, land and the environment are linked, and decisions made with regard to one may affect the others. Policy and management of natural resources now require interdisciplinary approach including natural and social sciences to correctly address our society preferences.

This series provides a collection of works containing most recent findings on economics, management and policy of renewable biological resources such as water, land, crop protection, sustainable agriculture, technology, and environmental health. It incorporates modem thinking and techniques of economics and management, Books in this series will incorporate knowledge and models of natural phenomena with economics and managerial decision frameworks to assess alternative options for managing natural resources and environment.

The Series Editors

HANDBOOK OF BIOENERGY ECONOMICS AND POLICY

MADHU KHANNA
University of Illinois, Urbana-Champaign, USA

JÜRGEN SCHEFFRAN
University of Illinois, Urbana-Champaign, USA

DAVID ZILBERMAN
University of California, Berkeley, USA

 Springer

Editors

Madhu Khanna
Department of Agricultural and Consumer
 Economics
University of Illinois
Urbana-Champaign
1301 W. Gregory Dr.
Urbana IL 61801
301A Mumford Hall
USA
khanna1@illinois.edu

David Zilberman
Department of Agricultural and Resource
 Economics
College of Natural Resources
University of California
Berkeley
206 Giannini Hall
Berkeley CA 94720-3310
USA
zilber11@berkeley.edu

Jürgen Scheffran
ACDIS, CABER
University of Illinois
Urbana-Champaign
505 East Armory Ave.
Champaign IL 61820
USA
scheffra@illinois.edu
Insitute for Geography
Hamburg University
ZMAW, Bundesstr. 53
20146 Hamburg
Germany
juergen.scheffran@zmaw.de

ISBN 978-1-4419-0368-6   e-ISBN 978-1-4419-0369-3
DOI 10.1007/978-1-4419-0369-3
Springer New York Dordrecht Heidelberg London

Library of Congress Control Number: 2009936725

Printed on acid-free paper

Acknowledgments

This volume grew out of two conferences held in 2007 to address the opportunities and challenges of transition to a bio-economy. The first, an international symposium on "Fueling Change with Renewable Energy," was held at the University of Illinois at Urbana-Champaign in April 2007, while the second, "Intersection of Energy and Agriculture: Implications of Biofuels and the Search for a Fuel of the Future," was held at the University of California at Berkeley in October 2007. We gratefully acknowledge financial support provided by the Energy Biosciences Institute, University of California at Berkeley; Giannini Foundation of Agricultural Economics, University of California; the Environmental Council, University of Illinois at Urbana-Champaign; the Farm Foundation; and the Economics Research Service of the USDA. We thank the Program in Arms Control, Disarmament and International Security (ACDIS), the Center for Advanced BioEnergy Research (CABER), and the Department of Agricultural and Consumer Economics at the University of Illinois at Urbana-Champaign for their support. We also thank the reviewers of the chapters in this book for their thoughtful comments that helped to improve the book. Finally, our thanks go to Becky Heid and Amor Nolan for editorial assistance.

Contents

Contributors

Amy W. Ando Department of Agricultural and Consumer Economics, University of Illinois, Urbana-Champaign, IL, USA, amyando@illinois.edu

John C. Beghin Department of Economics, Iowa State University, Ames, IA, USA, beghin@iastate.edu

Hans P. Blaschek Center for Advanced BioEnergy Research, University of Illinois Urbana-Champaign, IL, USA, blaschek@illinois.edu

Jean-Christophe Bureau AgroParisTech, UMR Economie Publique, Paris, France, bureau@grignon.inra.fr

Xiaoguang Chen Department of Agricultural and Consumer Economics, University of Illinois, Urbana-Champaign, IL, USA, xchen29@illinois.edu

Harry de Gorter Department of Applied Economics and Management, Cornell University, Ithaca, CA, USA, hd15@cornell.edu

Frank G. Dohleman University of Illinois, Urbana, IL, USA; Monsanto Company, St. Louis, MO, USA, frank.g.dohleman@monsanto.com

Fengxia Dong Center for Agricultural and Rural Development, Iowa State University, Ames, IA, USA, fdong@iastate.edu

Amani Elobeid Center for Agricultural and Rural Development, Iowa State University, Ames, IA, USA, amani@iastate.edu

Mandy Ewing International Food Policy Research Institute, Washington, DC, USA, m.ewing@cgiar.org

Thaddeus Ezeji Department of Animal Sciences and Ohio State Agricultural Research and Development Center (OARDC), Ohio State University, Wooster, OH, USA, ezeji.1@osu.edu

Jacinto F. Fabiosa Center for Agricultural and Rural Development, Iowa State University, Ames, IA, USA, jfabiosa@iastate.edu

John (Jake) Ferris Department of Agricultural, Food, And Resource Economics, Michigan State University, East Lansing, MI, USA, jakemax33@comcast.net

P.W. Gassman Center for Agricultural and Rural Development, Iowa State University, Ames, IA, USA, pwgassma@iastate.edu

José M. Gil Centre de Recerca en Economia i Desenvolupament Agroalimentari (CREDA-UPC-IRTA), Castelldefels, Spain, chema.gil@upc.edu

Peter Goldsmith Department of Agricultural and Consumer Economics, University of Illinois, Urbana-Champaign, IL, USA, pgoldsmi@illnois.edu

Barry K. Goodwin Department of Agricultural and Resource Economics, North Carolina State University, Raleigh, NC, USA, barry_goodwin@ncsu.edu

Carolina Guimaraes Escola Superior de Agricultura Luiz de Queiroz, Department of Economics, Management and Sociology, University of Sao Paulo, Sao Paulo, Brazil, cpguimar@esalq.usp.br

Hervé Guyomard INRA, Paris, France, guyomard@roazhon.inra.fr

Emily A. Heaton Department of Agronomy, Iowa State University, IA, USA, heaton@iastate.edu

Gal Hochman Department of Agricultural and Resource Economics, University of California, Berkeley, CA, USA, galh@berkeley.edu

Haixiao Huang Energy Biosciences Institute, University of Illinois, Urbana-Champaign, IL, USA, hxhuang@illinois.edu

Florence Jacquet INRA, UMR Economie Publique, Paris, France, fjacquet@grignon.inra.fr

Satish Joshi Department of Agricultural, Food, And Resource Economics, Michigan State University, East Lansing, MI, USA, satish@msu.edu

David R. Just Department of Applied Economics and Management, Cornell University, Ithaca, CA, USA, drj3@cornell.edu

Fredrich Kahrl Department of Agricultural and Resource Economics, University of California, Berkeley, CA, USA, fkahrl@berkeley.edu

Seungmo Kang Energy Biosciences Institute, University of Illinois, Urbana-Champaign, IL USA, skang2@illinois.edu

Madhu Khanna Department of Agricultural and Consumer Economics, University of Illinois, Urbana-Champaign, IL, USA, khanna1@illinois.edu

L.A. Kurkalova North Carolina A & T State University, Greenboro, NC, USA, lakurkal@ncat.edu

Christine Lasco Department of Agricultural and Consumer Economics, University of Illinois, Urbana-Champaign, IL, USA, mlasco2@illinois.edu

Hyunok Lee Department of Agricultural and Resource Economics, University of California, Berkeley, CA, USA, hyunok@primal.ucdavis.edu

Stephen P. Long   Energy Biosciences Institute, University of Illinois, Urbana-Champaign, IL, USA, slong@uiuc.edu

Joao Martines   Escola Superior de Agricultura Luiz de Queiroz, Department of Economics, Management and Sociology, University of Sao Paulo, Sao Paulo, Brazil, martines@usp.br

Thein Maung   Department of Agribusiness and Applied Economics, North Dakota State University, Fargo, ND, thein.maung@ndsu.edu

Bruce A. McCarl   Department of Agricultural Economics, Texas A&M University, College Station, TX, USA, mccarl@tamu.edu

Seth Meyer   Food and Agricultural Policy Research Institute (FAPRI), University of Missouri, Columbia, MO, USA, meyerse@missouri.edu

Siwa Msangi   International Food Policy Research Institute, Washington, DC, USA, s.msangi@cgiar.org

Gerald C. Nelson   International Food Policy Research Institute, Washington, DC, USA, g.nelson@cgiar.org

Hayri Önal   Department of Agricultural and Consumer Economics, University of Illinois, Urbana-Champaign, IL, USA, h-onal@illinois.edu

Yanfeng Ouyang   Department of Civil and Environmental Engineering, University of Illinois, Urbana-Champaign, IL, USA, yfouyang@illinois.edu

Nathan D. Price   Department of Chemical and Biomolecular Engineering, University of Illinois, Urbana-Champaign, IL, USA, ndprice@illinois.edu

Deepak Rajagopal   Energy and Resources Group, University of California, Berkeley, CA, USA, deepak@berkeley.edu

Renato Rasmussen   Department of Agricultural and Consumer Economics, University of Illinois, Urbana-Champaign, IL USA, re.lima@gmail.com

David Roland-Holst   Department of Agricultural and Resource Economics, University of California, Berkeley, CA, USA, dwrh@are.berkeley.edu

Mark Rosegrant   International Food Policy Research Institute, Washington, DC, USA, m.rosegrant@cgiar.org

Jürgen Scheffran   University of Illinois, Urbana-Champaign, IL, USA, scheffra@illinois.edu; Hamburg University, Germany, juergen.scheffran@zmaw.de

S. Secchi   Southern Illinois University Carbondale, Carbondale, IL, USA, ssecchi@siu.edu

Teresa Serra   Centre de Recerca en Economia i Desenvolupament Agroalimentari (CREDA-UPC-IRTA), Parc Mediterrani de la Tecnologia, Edifici ESAB, C/ Esteve Terrades 8, 08860 Castelldefels, Spain, teresa.serra-devesa@upc.edu

Steven Sexton Department of Agricultural and Resource Economics, University of California, Berkeley, CA, USA, ssexton@are.berkeley.edu

Guilherme Signorini Escola Superior de Agricultura Luiz de Queiroz, Department of Economics, Management and Sociology, University of Sao Paulo, Sao Paulo, Brazil, signorin@esalq.usp.br

Daniel A. Sumner Department of Agricultural and Resource Economics, University of California, Berkeley, CA, USA, dasumner@ucdavis.edu

Kenneth R. Szulczyk Department of Economics, Suleyman Demirel University, Almaty, Kazakhstan, kszulczyk@hotmail.com

Farzad Taheripour Department of Agricultural Economics, Purdue University, West Lafayette, IN, USA, tfarzad@purdue.edu

Wyatt Thompson Food and Agricultural Policy Research Institute (FAPRI), University of Missouri, Columbia, MO, USA, thompsonw@missouri.edu

Simla Tokgoz International Food Policy Research Institute, Washington DC USA, s.tokgoz@cgiar.org

David Tréguer INRA, UMR Economie Publique, F-75005 Paris, France, treguer@gmail.com

Deniz Ü. Tursun Department of Civil and Environmental Engineering, University of Illinois, Urbana-Champaign, IL USA, utursu2@uiuc.edu

Wallace E. Tyner Department of Agricultural Economics, Purdue University, West Lafayette, IN, USA, wtyner@purdue.edu

Tun-Hsiang Yu Department of Agricultural Economics, University of Tennessee, Knoxville, TN, USA, tyu1@utk.edu

David Zilberman Department of Agricultural and Resource Economics, University of California, Berkeley, CA, USA, zilber11@berkeley.edu

Part I
Introduction

Chapter 1
Bioenergy Economics and Policy: Introduction and Overview

Madhu Khanna, Jürgen Scheffran, and David Zilberman

Concerns about energy security, high oil prices, declining oil reserves, and global climate change are fuelling a shift towards bioenergy as a renewable alternative to fossil fuels. Public policies and private investments around the globe are aiming to increase national capacities to produce biofuels. A key constraint to the expansion of biofuel production is the limited amount of land available to meet the needs for fuel, feed, and food in the coming decades. Large-scale biofuel production raises concerns about food versus fuel trade-offs, demands for natural resources such as water, and its potential impacts on environmental quality. Policies to support biofuel production have distributional implications for consumers and producers, farm and nonfarm sectors, global trade in food and biofuels, and the price of land and other scarce resources. Moreover, the potential to gain significant independence from foreign oil for most countries, including the United States, by relying simply on corn as a feedstock for biofuels is limited. This has increased interest in second-generation, lignocellulosic feedstocks that can increase the energy productivity of the land resource. These feedstocks include crop residues, perennial grasses, and woody biomass. The competitiveness of cellulosic biofuels and their land use requirements have implications for the costs of meeting advanced biofuel mandates in the United States, for the land diverted from food to fuel production, and for food prices. Chapters in this handbook use economic modeling tools to provide insights into these issues.

The introductory part of this handbook provides a context for the emerging economic and policy challenges related to bioenergy and the motivations for biofuels as an energy source. It includes chapters that explain the current state of knowledge about second-generation feedstocks for advanced biofuels and the technology for the deconstruction and conversion of lignocellulosic biomass to fuel. Part II of the handbook includes chapters that examine the implications of expanded production of first-generation biofuels for the allocation of land between food and fuel,

M. Khanna (✉)
Department of Agricultural and Consumer Economics, University of Illinois, Urbana-Champaign, IL, USA
e-mail: khanna1@illinois.edu

M. Khanna et al. (eds.), *Handbook of Bioenergy Economics and Policy*,
Natural Resource Management and Policy 33, DOI 10.1007/978-1-4419-0369-3_1,
© Springer Science+Business Media, LLC 2010

for food/feed prices for and trade in biofuels as well as the potential for technology improvements to mitigate the food versus fuel competition for land. These chapters discuss the implications of a growing biofuel industry for agricultural markets, food prices, and commodity price volatility. Part III examines the infrastructural and logistical challenges posed by large-scale biofuel production and the factors that will influence the location of biorefineries and the mix of feedstocks they use. Part IV includes chapters that examine the environmental implications of biofuels, their implications for the design of policies, and the unintended environmental consequences of existing biofuel policies. These chapters assess the implications of biofuels and related policies for greenhouse gas (GHG) mitigation and the challenges in determining these effects using life cycle analysis. They examine the trade-offs among different environmental impacts, such as water quality and biodiversity, caused by biofuel policies that focus on achieving particular social and environmental goals. The last part discusses the economic and distributional implications of existing biofuel policies. These chapters present economic analysis of the market, social welfare, and distributional effects of biofuel policies, including tax credits, tariffs, and mandates such as the Renewable Fuel Standard (RFS) in the United States. The differences in the distributional effects of the emerging biofuel industry across developed and developing countries are explored.

This handbook is of value for various groups in academia, education, industry, and governmental and nongovernmental organizations. It is also a useful reference book for analysts in developed and developing countries working on the socioeconomic impacts of the emerging bioeconomy as well as its implications for land use, carbon emissions, natural resources, energy, and food prices. This handbook will also help practitioners and managers in industry and agriculture to deepen their understanding about theoretical and practical issues associated with implementation and use of bioenergy and economic and policy dimensions of a growing bioeconomy.

Interest in the topics presented in this handbook is strong among policy makers both in the developed and developing world and in international organizations such as The World Bank and various United Nations agencies. Policy makers will find useful insights on the economic consequences of various policy alternatives to support biofuel production. Another major group that will benefit from this handbook consists of scholars in agriculture, trade, economic development, resource economics, and public policy that are interested in issues of renewable energy policy. They will appreciate the international dimensions presented in this handbook, in particular issues of trade and the interaction between developed and developing nations.

While several chapters rely on economic models, we have sought to maintain a standard of accessibility for a wider audience by de-emphasizing technical content and expert jargon and emphasizing conceptual, applied, and policy issues that are of great interest for society. As a result, this handbook can be used as a textbook for courses and curricula associated with the emerging field of bioenergy economics. As universities develop more specialized curriculum centered around bioenergy, this book will serve as a reading to familiarize students with applications of economic tools to analyze the economic and environmental implications of bioenergy development and policies.

This handbook provides an integrated and comprehensive perspective on economics and policies related to biofuels. It covers a breadth of issues related to economic and policy analysis at local, regional, and global levels including crop and feedstock choices, transportation and infrastructure, processing and production, markets and trade, as well as societal implications (welfare effects, agriculture, food), and environmental impacts (climate change, land use). This handbook specifies these issues for selected regions of the world (United States, Europe, and Brazil), which are likely to be important players in the biofuel arena in the near to medium term.

1.1 Next-Generation Energy Technologies: Options and Possibilities

Nelson examines whether biofuels are the best use of sunlight for generating energy usable by humans. Biofuels are produced from plants that capture energy from the sun through photosynthesis and convert it to starch, sugar, and cellulose that can then be converted to liquid fuels. An alternative way to use solar energy is to convert it to electricity using photovoltaic technology. Plants are typically able to capture less than 5% of the solar radiation intercepted with perennials, such as *Miscanthus*, having higher radiation use efficiency. Even conservative estimates suggest that photovoltaics can generate about twice as many kilowatt-hours per square meter as miscanthus and about three times more than current corn to ethanol technologies. However, the use of photovoltaics for providing energy for transportation requires the development of cost-effective engine and battery technologies that have the power, longevity, and safety needed for automotive applications.

Dohleman, Heaton, and Long discuss the potential of perennial grasses as second-generation feedstocks that increase the productivity of land in producing biofuels without compromising environmental sustainability. As compared to corn ethanol, which requires large nitrogen inputs and has debatable potential to reduce GHG emissions, low-input high-diversity systems, such as restored mixed-prairies, have the benefit of requiring low inputs. The productivity of these systems, however, appears too low to make them economically viable. High-yielding perennial grass species, such as miscanthus and switchgrass, have many features that make them "ideal" feedstocks. They are relatively high-yielding, have a high-energy output to input ratio, high nitrogen and water use efficiency and can be grown on marginal land with conventional farm equipment. This chapter explores the features of dedicated energy crops that, if managed properly, will allow integration of biofuel production into existing agricultural systems with a reduced impact on food production compared to grain-based biofuels. Expected breakthroughs in cellulosic biomass production, deconstruction, and conversion to liquid fuels will help accomplish these goals.

Blaschek, Ezeji, and Price provide an introduction into present and future possibilities for the deconstruction and utilization of lignocellulosic biomass, the most abundant renewable energy resource on the planet. The "Billion Ton Study" published by the U.S. Department of Energy (DOE) in 2005 indicated that 1.3 billion dry tons of biomass (including agricultural residues, municipal paper wastes, dedicated energy crops) is available per year in the United States, enough to produce biofuels to meet more than one-third of the country's current demand for transportation fuels. The status of current deconstruction technologies for lignocellulosic biomass and the role of genomics for producing feedstocks and tailor-made microbes for fermentation-based processes are discussed. This chapter focuses on the use of biomass-based hydrolysates for the production of bio-butanol, a second-generation biofuel that can be directly used as a liquid fuel or blended with fossil fuels. Metabolic engineering and systems biology approaches offer a new toolbox for the development of microbial strains which are able to grow and ferment in the presence of inhibitors produced during the deconstruction process, thereby eliminating a major bottleneck in biomass-based fermentations. These new technologies are expected to play a major role in the successful commercialization of second-generation biofuels which have an improved energy-carbon footprint and do not directly compete with food crops.

1.2 Integration Between Energy and Agricultural Markets

Part II of the handbook examines the interactions between biofuels and agricultural markets and the trade-offs that biofuels pose for food production. As a result of biofuel production and the resulting integration between energy markets and agricultural markets, energy prices now affect food prices in two ways; directly by raising the costs of production of agricultural products and indirectly by diverting land away from food to biofuels. Studies differ in their estimates of the extent to which biofuel production contributed to the recent increase in food prices; growing incomes, population, and urbanization coupled with the sharp increase in the oil price in 2007–2008 and the devaluation of the dollar make it difficult to isolate the impact of increasing the share of cropland used for biofuel production on commodity prices.

Serra et al., Zilberman, Gil et al. use time series data to empirically examine the linkages among the prices of corn, ethanol, and crude oil. They present a framework that is based on demand and supply relationships in the corn and fuel markets to formulate testable hypotheses about the positive relationship between each pair of the above three prices. They estimate a multivariate vector error correction model using daily futures prices for corn, ethanol, and crude oil over the 2005–2007 period and examine the long-run relationships among the variables of the model while allowing for nonlinear adjustment paths toward long-run equilibrium. They find that the price of ethanol is positively related to the price of corn and to the price of crude oil, with changes in corn price having a larger impact on ethanol price than changes in crude oil price. The implications of this finding for the long-run competitiveness of the US ethanol industry are discussed.

Msangi, Ewing, and Rosegrant present a simple framework to show the linkages between energy and agricultural markets, the implications of growth in demand for energy and food for land use, and the need for technology improvements to increase the productivity of agricultural crops and to increase fuel conversion efficiencies which would reduce the need to divert land away from food production to produce biofuels. They examine the impact of the RFS in the United States on global cereal prices in 2015 and show the extent to which policy-driven investments in agricultural yield growth can mitigate the increase in cereal prices and reduce the need for expansion in total cultivated area.

Ferris and Joshi also examine the implications of the RFS for ethanol and biodiesel production in the next decade (2008–2017) and its impacts on production and prices of food, fuel, and feed. Their analysis uses an econometric model (AGMOD) of US agriculture and the international sector encompassing grain and oilseeds to present a "baseline" scenario and three alternative scenarios to embrace a wide range of crude oil prices. The authors find that acreage under coarse grains, wheat, and soybeans will increase in the United States and globally even with projected higher yields, but that overall expansion of land in the United States will be dampened somewhat by a reduction in land under hay, silage, and the Conservation Reserve Program, resulting in a net increase in total crop acreage of only 5%. Global crop acreage is expected to increase by 12% between 2008 and 2017. The authors find that under a high oil price scenario, cropland acreage, land prices, and the consumer price index would be much higher than in the baseline. The authors describe the existing barriers to significant switching of land from conventional crops to energy crops. Cellulosic conversion technologies continue to face major uncertainties including the uncertainty about the best pathway for conversion of biomass to ethanol—is it thermochemical or biochemical or a combination of the two? In addition, problems related to system integration, commercial scale-up, and overall process optimization remain unresolved. Projected capital requirements for cellulosic ethanol plants are much higher than for corn−ethanol dry mills. Significant new investments will also be necessary for establishing appropriate biomass supply chains, including harvesting and storage infrastructure.

Fabiosa et al., Beghin, Dong, et al. use the FAPRI model to estimate the impact of expanding corn ethanol production and consumption in the United States and other countries on changes in acreage of biofuel feedstock crops and crops that compete with these feedstocks for land, as well as on consumption of food and feed globally in 2016–2017. Impact multipliers showing the responsiveness of crop acreage to ethanol production and consumption indicate that an increase in corn ethanol production in the United States has its largest impact on corn acreage in the United States and is accompanied by significant reductions in the acreage under wheat and soybeans. A simulation of a 100% increase in ethanol production in the United States shows that it would increase overall crop acreage in the United States by about 3% and in the world by about 2%, with the magnitude of the effects being largest in Brazil and South Africa. Another simulation of the land use effects of a global expansion in ethanol demand shows that it would largely result in an expansion of land under sugarcane in Brazil with modest effects on other countries, in large part due to the availability of land in Brazil. These results have implications for the magnitude of the indirect land use changes likely to occur with the expansion of corn ethanol production in the United States and ethanol consumption in the world.

While a number of studies have examined the influence of biofuels and biofuel policy on the prices of agricultural commodities, the impact of greater biofuel production on commodity price volatility is less understood. Meyer and Thompson examine how demand behavior and commodity price volatility change under evolving biofuel markets and policies. Rising or falling oil, natural gas, and other energy prices cause corresponding changes in production costs, including fertilizer, transportation, and processing costs. In recent years, changes in market conditions, technology, petroleum prices, and policies for both energy and agriculture brought about an explosion in feedstock demand. While a change in the ethanol quantity relative to the base value may have a very small effect on motor fuel prices, movement in the large motor fuel market will cause large changes in ethanol volumes, changes so large that the gasoline price will drive the ethanol price. On the other hand, use of corn for ethanol drives the corn market and creates a link that transmits volatility in petroleum prices to corn prices. Growing biofuel processing in the United States has brought about greater integration between energy and agricultural markets and thus contributed to volatility in agriculture markets. Their analysis suggests that biofuel use mandates increase corn price volatility if they are binding. Political actors and market participants must gauge the consequences of this market volatility for farm income, food security, and biofuel investments.

1.3 Designing the Infrastructure for Biofuels

Increases in biofuel mandates pose enormous challenges to the infrastructure needed across all stages of the supply chain − from crop production, feedstock harvesting, storage, transportation, and processing to biofuel distribution and use. The chapter by Kang et al., Onal, Ouyang et al. focuses on the biofuel transportation and distribution network infrastructure needed to meet given biofuel targets. Building on an optimal land use allocation model for feedstock production, a mathematical programming model is introduced to determine optimal locations and capacities of biorefineries, delivery of bioenergy crops to biorefineries, and processing and distribution of ethanol and coproducts. The model aims to minimize the total system costs for transportation and processing of feedstock, transportation of ethanol from refineries to blending terminals, shipping ethanol from blending terminals to demand destinations, capital investment in refineries, and transportation of the coproduct DDGS to livestock producing areas in a multiyear planning horizon for the period of 2007–2022. Using Illinois as a case study, it lays the ground for future expansion of the analysis to the Midwest and the United States and the whole supply infrastructure. Their analysis shows that certain locations may be more suitable for corn and corn stover-based ethanol plants, while others may be more suitable for producing ethanol using perennial grasses (miscanthus). The availability of feedstock and location of ethanol demand influence the optimal location and capacity of biorefineries. Cellulosic refineries are expected to be located in central Illinois where much of the corn stover is produced and in Southern Illinois where miscanthus is expected to be profitable to produce.

Bioenergy feedstock production based on extensive farming systems may lead to inefficiency in the use of capital assets when biofuel production systems are spatially dispersed and involve fuels and feedstocks that have relatively low-energy densities. In their analysis of the capital efficiency challenge of bioenergy models, Goldsmith et al. apply the Liquid Fuel Bioenergy Model to ethanol production in Mato Grosso, Brazil, to demonstrate the key concepts of density and capital intensity that are critical for the efficient use of capital. This chapter shows that asset utilization improves by including maize as a complementary feedstock, because it would improve the spatial, volumetric, and/or gravimetric densities of feedstock for a Mato Grosso flex mill.

1.4 Environmental Effects of Biofuels and Biofuel Policies

A key motivation for promoting biofuel production is its potential to reduce the environmental externalities associated with transportation fuel. The first four chapters in this part of the handbook examine the potential for biofuels to mitigate GHG emissions and the regulatory framework that needs to be designed to fully account for the carbon mitigation benefits of biofuels. Two of these chapters also examine the unintended effects of existing biofuel policies on GHG emissions and social welfare. While biofuels have the potential to mitigate GHG emissions, they can worsen other externalities associated with expanding the land under biofuel production. The last three chapters in this part examine the multiple environmental effects associated with biofuel production and the trade-offs that biofuels pose among these effects.

McCarl, Maung, and Szulczyk discuss the potential for biomass- and food-based fuels to reduce GHG emissions by providing biopower, biofuels, and soil carbon sequestration. Their chapter describes the range of options and technologies for bioenergy and uses the FASOMGHG model to simulate the future production levels of various types of bioenergy and GHG mitigation under alternative and fossil fuel and GHG prices. Using life cycle analysis, the model accounts for direct mitigation by displacing fossil fuels with bioenergy and for leakages due to indirect land use changes in other parts of the world. When the prices of fossil fuels and GHG emissions are low, agricultural soil sequestration is the dominant mitigation strategy and as carbon prices increase, production of cellulosic ethanol can be expected to increase. The chapter considers the implications of carbon taxes for crop prices, meat prices, and agricultural exports and recommends that biofuel and GHG mitigation policies take leakage effects into account. Their analysis suggests that these leakage effects can be reduced by relying on crop residues and waste products for cellulosic ethanol and on bio-based electricity.

Rajagopal, Hochman, and Zilberman describe a regulatory framework that could be developed to regulate the direct and indirect GHG emissions from biofuels, while incorporating the heterogeneity in biofuel production sources and the uncertainties that influence production methods and indirect land use changes. While direct GHG emissions can be computed ex-post using life cycle analysis, indirect GHG emissions need to be computed ex-ante using multimarket or general equilibrium models. Given this heterogeneity and uncertainty, regulators may pursue a precautionary approach by designing certification standards that use these estimates to set an upper bound on emissions from biofuels. Biofuels that seek to qualify for government subsidies or account toward mandates would need to be certified.

Ando, Khanna, and Taheripour examine the social welfare and environmental implications of the RFS by developing a framework that considers the demand for gasoline and ethanol to be derived from demand for vehicle miles traveled (VMT) and the imperfect substitutability between ethanol and gasoline in producing VMT. They incorporate the disutilities from VMT (congestion and air pollution) and from fuel (GHG emissions) and show that an optimal policy includes a tax on carbon and on VMT. The extent to which a biofuel mandate leads to displacement of gasoline and reduction in GHG emissions and VMT depends on the elasticity of supply of gasoline and the elasticity of substitution between gasoline and ethanol. They compare the welfare and environmental effects of a mandate with those of a carbon tax and find that the former imposes high welfare costs and results in higher miles and GHG emissions than a carbon tax policy. The magnitude of these effects is sensitive to the gasoline supply elasticity.

Lasco and Khanna examine the effects of current US biofuel policies for corn ethanol, namely tax credits and import tariff, for the imports of biofuels, fuel prices, GHG emissions, and social welfare in the United States. They review the development of the sugarcane ethanol industry in Brazil and compare the costs and life-cycle GHG emissions of corn ethanol in the United States and sugarcane ethanol in Brazil. They show that current biofuel tariffs lead to welfare losses relative to no government intervention even if one accounts for their terms of trade effects for the United States, assuming that the United States has market power in the world ethanol market. Moreover, the current tariff and tax credit policy results in higher GHG emissions as compared to non-intervention because the subsidy induces greater demand for VMT while the tariff causes a substitution away from the less carbon-intensive to a more carbon-intensive biofuel.

The nexus between energy and agricultural markets that has been forged by biofuels has implications for multiple environmental externalities generated by energy and agricultural production. Hochman, Sexton, and Zilberman examine these interactions between energy, food, and the environment by developing a general equilibrium framework in which households obtain utility from food, an environmental amenity, and a convenience good produced using capital and energy. They examine the trade-offs posed by biofuels as they reduce GHG emissions but increase demand to convert non-cropland that provides biodiversity services to energy production. They show that a carbon tax policy would put additional pressure on natural habitat unless biodiversity preservation is valued and carbon tax policy is accompanied by a tax on converting natural land to cropland or payments for environmental services. The chapter also discusses the need to link biofuel policies to food inventories to ease the food versus fuel trade-off and emphasizes the importance of innovations that increase crop productivity and that enable conversion of cellulosic feedstocks into fuel.

Khanna et al., Onal, Chen, et al. examine the competitiveness of second-generation cellulosic feedstocks and the allocation of land among food and energy crops for meeting biofuel targets such as those under the RFS at a regional level. They examine the trade-offs that the RFS poses for reduced GHG emissions but higher nitrogen use as production of corn ethanol and harvesting of corn stover increases. Using a dynamic optimization model coupled with a biophysical model that simulates the yields of dedicated energy crops, switchgrass and miscanthus, and county-specific data on costs of production of conventional and energy crops, they show that biofuel targets are likely to lead to a significant shift in acreage from soybeans and pasture to corn and a shift toward conservation tillage and continuous corn rotation as demand for corn and corn stover for biofuels increases. The economic viability of miscanthus is found to vary spatially depending on its yields per acre and the opportunity cost of land.

Kurkalova, Secchi, and Gassman assess the environmental implications of the removal of corn residue for cellulosic biofuels. Crop residues left on the field after harvest provide several environmental benefits such as soil organic matter, nutrient recycling, control of nutrient runoff, and preventing erosion. The authors use detailed field-level GIS data together with economic and environmental models to examine the spatial distribution of corn production, corn residue availability, and soil and water quality indicators under alternative prices of corn, soybeans, and corn stover with 50% removal of corn residue. The authors find that as corn stover prices increase, residue removal increases and so do the sediment and nitrogen losses. These losses are larger under continuous corn rotations. Soil carbon stocks decrease as corn stover removal increases and levels are higher with continuous corn rotations than with corn–soybean rotations due to the larger corn biomass compared to soybeans.

1.5 Economic Effects of Biofuel Policies

Lee and Sumner review the history of biofuel policies in the United States under various Energy and Farm Bills. The authors discuss conditions under which the US elasticity of demand for imported ethanol ranges from high elasticity (with a low share of ethanol in total US fuel use) to low elasticity (under a binding mandate). Using historical data on ethanol imports and the prices of ethanol, crude oil, and corn, the authors estimate the elasticity of supply of exports of ethanol from Brazil to the United States to lie between 2.5 and 3.0. They examine the implications of these elasticities for removal of the import tariff and argue that it is more likely to lead to a change in the quantity of imports (of about 68%) and not much change in the domestic price of ethanol unless imports have a large share and US ethanol capacity is low.

de Gorter and Just describe the existing biofuel policies in the United States and the objectives they seek to promote. They show the linkage between ethanol and gasoline price and analyze the implications of tax credits, mandates, and ban on MTBE as an additive to gasoline for the relationship between ethanol and gasoline prices. The chapter also examines the factors leading to a link between ethanol and corn prices and the social welfare effects of the ethanol subsidy and the loan rate program for corn. The authors discuss policy reforms that can better achieve the multiple goals of biofuel policies.

Taheripour and Tyner discuss the contribution of US biofuel policies to the biofuels boom since 2005 and the consequences of expansion in the ethanol industry for the agricultural and energy markets under alternative policy options. Their estimates of the break-even combinations of corn and ethanol prices that keep a representative dry mill ethanol plant at zero profit condition show why the ethanol industry's profits declined after 2006 after the ethanol premium fell and corn prices rose. Analysis of the impacts of alternative policies, the tax credit, a variable subsidy, and the corn ethanol mandate under a range of crude oil prices shows the linkage between crude oil prices and corn prices, as well as the contribution of the biofuels subsidy to the price of corn. The costs of these policies to the government (in the case of the subsidy) and the consumers (in the case of the mandate) are examined. The authors then describe the results of their analysis of the global changes in land use due to biofuel policies in the United States and EU using a general equilibrium model, GTAP. They show the importance of accounting for biofuel by-products while analyzing the effects of biofuel policies on agricultural production patterns around the world.

Kahrl and Holst examine the distributional incidence of energy and food price increase across countries that differ in their per capita income levels, using a variety of empirical techniques. Their analysis shows the dichotomy in North–South energy and food dependence. While the share of income spent on food is negatively related to per capita income, the per capita energy use increases with income. The authors use various methods to examine the effects of energy and food on household cost of living and incomes. They estimate the elasticity of the poverty gap with respect to food and energy prices for various income deciles in Thailand and Vietnam and show that energy price vulnerability is relatively low in both countries and much lower than food price dependence. Using the social accounting matrix, they examine how food and energy price changes are transmitted through the economy and affect households. An application of this method to Morocco shows that the total impact of food price increase on the household consumer price index is much higher than its direct effect on the food price index alone and that both are relatively high for the poor. Finally, using a general equilibrium model, the authors explore both the income and expenditure effects of food price increases for households in Senegal. They show that high food prices benefit some rural deciles, while they hurt the urban poor. Their findings have implications for equity effects of biofuel policies across high- and low-income countries and groups within a country.

Bureau et al., Guyomard, Jacquet, et al. assess European biofuel policy and public support for it, which resulted in two biofuel directives in 2003 and led to an unprecedented growth in biofuel production over the past 5 years. The main driving force has been the measures taken at the member state level aiming at increasing the use of biofuels, including tax exemptions, subsidies, and mandatory blending with transport fuel. This, together with significant import barriers, at least for ethanol, has led to a considerable increase in domestic production since 2003. Meanwhile, concerns about the overall environmental and social effects of biofuels, including the effects of indirect land use changes and potential competition for land with food production, have triggered intense debates and an erosion in the public image of biofuels. This has led some member states to review their initial ambitions. The 2008 reform of the Common Agricultural Policy still favors the widespread development of biofuels, but ended the subsidies for the production of energy crops. The new energy policy directives now focus on renewable energy mandates rather than biofuel blending mandates. This together with the current economic crisis adds to uncertainty about the future of biofuels in the EU.

1.6 In Sum

This handbook covers a wide range of issues that have emerged with the advent of biofuels and presents a diverse set of economic models and approaches to analyze their implications for food and fuel prices, consumers, producers, and the environment. It shows that food-based biofuels have led to a competition for land and integration between energy and agricultural markets, while the environmental benefits of current biofuel policies are ambiguous or even negative. New technologies that increase crop productivity and fuel-conversion efficiencies and enable the use of cellulosic feedstocks together with policies targeted at sources of market failures need to be designed to induce a shift toward biofuels that are economically, socially, and environmentally sustainable.

Chapter 2
Are Biofuels the Best Use of Sunlight?

Gerald C. Nelson

Abstract Biofuels are liquid sunlight. In effect, we use plants to convert raw solar energy into a liquid (ethanol or biodiesel) that can be used as an energy source for our transportation systems. The question this chapter asks is whether this conversion process is the best way to make use of solar energy. Photovoltaics clearly dominate plants in terms of technical conversion efficiency, with conversion rates for commercial cells of the mid-2000s that are 2–10 times higher than plants and operate throughout the year rather than just during the growing season. But photovoltaics provide electricity, which is not currently cost-effective for use in transportation. As research into photovoltaics and battery technology is still in its infancy, the potential for commercially viable technology breakthroughs seems high.

2.1 Introduction

Biofuels are liquid sunlight. In effect, we use plants to convert raw solar energy into a liquid (ethanol or biodiesel) that can be used as an energy source for our transportation systems. The question this chapter asks is whether this conversion process is the best way to make use of solar energy. The answer to this question has three parts – what is the technical efficiency of the conversion process from solar power to useful energy relative to other methods of transforming solar power, to what extent are the resulting products substitutable, and what are the fixed and operating costs of the conversion process? This chapter compares biofuels with the most widely used alternate conversion process – photovoltaics.

G.C. Nelson (✉)
International Food Policy Research Institute, Washington, DC, USA
e-mail: g.nelson@cgiar.org

I would like to thank Madhu Khanna for very helpful comments on this chapter. Any errors or omissions remain my responsibility.

M. Khanna et al. (eds.), *Handbook of Bioenergy Economics and Policy*,
Natural Resource Management and Policy 33, DOI 10.1007/978-1-4419-0369-3_2,
© Springer Science+Business Media, LLC 2010

Plants capture the energy of the sun through photosynthesis and transform it to starches, sugars, and cellulose. Liquid biofuels are created with various industrial processes that combine the plant material with additional energy and water and produce a biofuel and byproducts, including polluted water. The output, either ethanol or biodiesel, can relatively easily be used in the world's transportation systems.

Much attention has been paid to whether the energy used in the industrial processes that produce ethanol is greater or less than the energy available in the final product, and whether that energy is derived from fossil fuels or renewable sources such as bagasse. Farrell et al. (2006) found that ethanol produced with today's conversion technology required 0.774 MJ of direct and indirect fossil fuel inputs to produce 1.0 MJ of fuel (Summary table of Farrell et al., at http://rael.berkeley.edu/ebamm) for a net energy yield (NEY) of 1.30. Liska et al. (2009) argue that the Farell et al. results are based on older technology and find NEYs of 1.29–2.23 depending on feedstock for the conversion process and corn yields (see Table 2.1 below). However, little attention has been paid to the efficiency of converting the underlying energy source − solar − to a form directly useful to humans.

Table 2.1 The annual energy output of various biofuels

	Liters per hectare* (1)	MJ per hectare** (2)	KWh per hectare (3)	KWh per sq m (4)
Ethanol from				
Corn[1]	3,730	89,517	24,886	2.489
Corn[2]	3,003	72,063	20,033	2.003
Corn stover[1]	1,544	37,051	10,300	1.030
Miscanthus[1]	6,945	166,676	46,336	4.634
Switchgrass[1]	2,009	48,208	13,402	1.340
Sugar cane[2]	6,744	161,861	44,997	4.500
Biodiesel[3]				
Oil palm	4,752	156,810	43,593	4.359
Coconut	2,151	70,997	19,737	1.974
Rapeseed	954	31,485	8,753	0.875
Peanut	842	27,781	7,723	0.772
Sunflower	767	25,312	7,037	0.704
Soybean	524	17,286	4,806	0.481

*1 US gallon = 3.78541178 l and 1 acre = 0.404685642 ha so 1 gallon per acre = 9.354 l/ha.

**Ethanol contains 23.4 MJ/l and biodiesel contains 35.7 MJ/l. 1 gigajoule (GJ) = 278 kWh. Source: Bioenergy Feedstock Information Network (http://bioenergy.ornl.gov/papers/misc/energy_conv.html).

Sources: Khanna et al. (2008), http://www.choicesmagazine.org/magazine/article. php?article=40, Woodrow Wilson Brazil Inst. Special report − pdf Brazil_SR_e3.pdf http://gristmill.grist.org/story/2006/2/7/12145/81957

[1] Khanna, 2008, http://www.choicesmagazine.org/magazine/article.php?article=40
[2] Woodrow Wilson Brazil Inst. Special report – pdf Brazil_SR_e3.pdf
[3] http://gristmill.grist.org/story/2006/2/7/12145/81957

In the photovoltaic process, photons from the sun energize electrons in a special material that channels them into an electric current. The electricity can be used
immediately, for heat or to power electric devices, or stored in batteries. A different solar technology with some promise, but not discussed further involves using
parabolic reflectors to concentrate solar energy and convert it to heat to drive steam
or gas turbines to generate electricity. Both of these types of solar technologies
produce electricity rather than liquid fuel.

This chapter has three goals. First, I present information on the raw availability
of solar energy and compare that to the effective energy available at the end of the
biological and PV processes. Second, I look at available evidence on the progress
of photovoltaic technological improvements in converting solar energy into useful energy. Finally, I discuss briefly the issue of the technologies needed to use
electricity in transportation instead of liquid fuels, i.e., plug-in cars and batteries.

2.2 From Solar Energy Input to Useful Energy Output

The sun delivers massive amounts of energy to the earth's surface. According to
Wilkins et al. (2004), "Using current solar technology, an area just 100 miles by 100
miles (10,000 square miles) in the southwestern United States could generate as
much energy as the entire nation currently consumes." Figure 2.1 shows the number
of kilowatt hours (kWh) [1] delivered per m^2 per year, adjusted for the extent of cloud

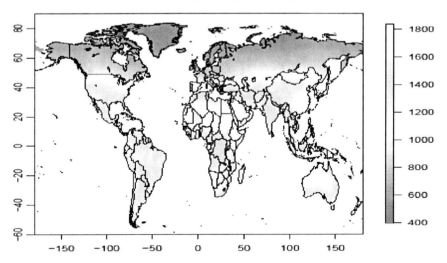

Fig. 2.1   Solor energy flux, kWh per m2 per year
Source: Personal communication from Robert Hijmans, International Rice Research Institute,
based on a data set by Mark New and colleagues (New, Lister et al. 2002).

---

[1] The units used to report energy density of various sources vary. Biofuels energy densities are
often reported in megajoules (MJ) or British thermal units (btu). Electricity is reported in kilowatt
hours (kWh). One kWh is equivalent to 0.278 mj and 3,412 btus.

cover. The values range from 400 to 1,800 kWh per m^2 per year. Further from the equator, the energy delivers diminish somewhat because the tilt of the earth in its orbit around the sun reduces day length for part of the year. However, cloud cover plays a more important role in reducing effective solar energy. The highest values are found in desert regions, even those that are relatively far from the equator. Both plants and photovoltaics capture part of this raw energy and convert it into useful energy output.

2.3 Biofuel Energy Conversion

Table 2.1 reports several estimates of liquid fuel production and the resulting energy output from various biofuels crops. For ethanol, the largest reported volumes are from sugarcane and *Miscanthus*, although the *Miscanthus* value is based only on experimental results. Both crops can produce 6,500–7,000 l/ha under optimal conditions. Corn is a distant third at 3,700–4,000 l/ha. For biodiesel, oil palm performs best, producing almost 5,000 l/ha. These numbers are not typically reported by location but Fig. 2.2, which reproduces Fig. 2.1 in Khanna et al. 2008, provides some indirect evidence of the effects of the north-south gradient of solar energy availability. The cost per ton of dry matter from switchgrass is substantially lower in southern Illinois than northern Illinois principally because the dry matter yield is higher in the south.

The key column in Table 2.1 is column 4, which reports the effective energy availability per square meter of land used to grow the crop. *Miscanthus*, sugar cane, and oil palm have the highest effective energy outputs at 4.4–4.6 kWh per m^2 per year. Other crops produce much less energy. It is important to note that the values in column 4 do not take into account the energy used in processing the plant material into biofuels. Without attempting to provide quantitative estimates, it seems likely that palm oil requires the least energy input in processing, sugar cane somewhat more and *Miscanthus* the most because the cellulose must first be converted to a fermentable material.

Comparing the range of values of solar energy delivered to the earth's surface given in Fig. 2.1 (400 to 1,500 kWh per m^2 per year) with the effective energy delivered by the biofuels process (about 4.5 kWh per m^2 per year), it is clear that the combination of the crop conversion plus the processing conversion results in very little of the solar energy being made available for human use in liquid form. Loomis and Amthor (1996) (as cited in Reynolds et al. (2000)) report that the energy contained in a mature crop typically represents less than 5% of incident radiation received (radiation use efficiency, RUE). This inefficiency is caused by several factors. Chlorophyll has evolved to absorb wavelengths between 400 and 700 nm. It cannot take advantage of longer or shorter wavelengths, which photovoltaics can. In addition, some photosynthetically active radiation falls on nonphotosynthetically active cell components or structures such as dead leaves. Furthermore, plants can capture solar energy only during their growing season while photovoltaics work

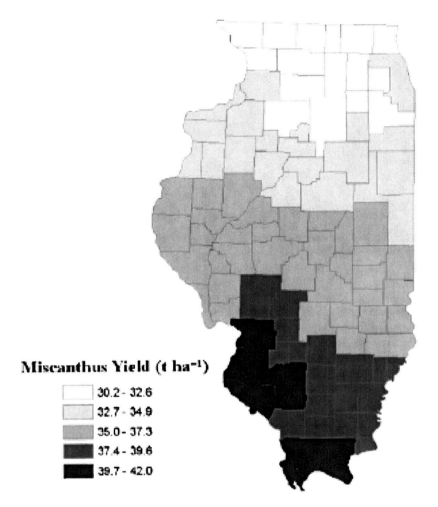

Fig. 2.2 Simulated Miscanthue yield
Source: Figure 3a in Khanna et al. (2008).

year round, and in fact are more efficient in colder temperatures. Finally, not all of the products of photosynthesis can be converted to biofuels.

Improvements in biofuels conversion can come from three sources – increasing productivity of existing fuel crops with improvements in RUE and biomass composition, switching to biofuels crops that extend their solar capture period by having a longer growing season, and improvements in the efficiency in converting the feedstocks to fuel. An example of the first source, increasing productivity of existing fuel crops, can be seen in productivity improvements in the Brazilian sugarcane industry (Fig. 2.3) where the ethanol yield per hectare increased over 3.5% per year between 1975 and 2003, at least some of which is because of increased sucrose production within the plant (although improved conversion efficiencies in the fermentation

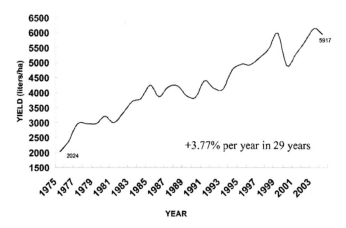

Fig. 2.3 Ethanol form sugarcane productivity increase in Brazil, 1975–2003
Source: Goldemberg (2008).

processes may have also contributed). For maize, Liska et al. (2009) report steady increases in maize grain productivity of over 0.1 mt/ha per year.

An example of the second source of productivity increase – switching to biofuels crops that extend their solar capture period by having a longer growing season – underlies the efforts to develop cellulosic ethanol, which allows any plant-based material to be a potential source of biofuels. In temperate climates, *Miscanthus* has two advantages over annual crops such as maize; as a perennial, it does not need to repeat the process of developing a root system every year, and it is cold tolerant allowing it to take advantage of more of the solar energy available throughout the year. Heaton et al. (2008) report harvestable dry matter yields of 10 to 30 mt/ha. An example of the third source of productivity increase – improvements in the efficiency in converting the feedstocks to fuel – is reported in Liska et al. (2009) who observe that "newer biorefineries have increased energy efficiency and reduced GHG emissions through the use of improved technologies, such as thermocompressors for condensing steam and increasing heat reuse; thermal oxidizers for combustion of volatile organic compounds (VOCs) and waste heat recovery; and raw-starch hydrolysis, which reduces heat requirements during fermentation" (p. 2).

2.4 Photovoltaic Energy Conversion

Photovoltaics convert solar energy directly into electricity. The amount of current depends on a wide range of factors including the design of the cell, the materials used to construct it, ambient temperature, and its orientation toward the sun. The photovoltaic phenomenon was first observed in the early 1800s and the first photovoltaic cell was constructed in the late 1800s.

The efficiency of a solar cell is defined as "the percentage of power converted (from absorbed light to electrical energy) and collected, when a solar cell is connected to an electrical circuit."[2] Figure 2.4 shows a history of solar cell efficiency improvements. Silicon wafer-based solar cell technologies, which make up most commercially available systems today, have efficiencies ranging from 6 to 20% commercially. High-efficiency thin-film technologies, currently available only in laboratories, achieve efficiencies of over 40%.

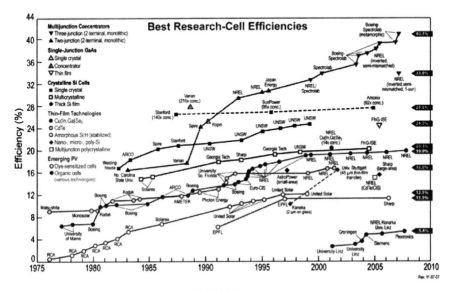

Fig. 2.4 Improvements in Research Sola Cell Efficiency
Source: Figure 1 in Kurtz (2008).

It is difficult to compare directly the efficiency of the biofuels process to photovoltaics in terms of converting solar (and other) energy into effective energy. The biofuels process includes processing energy not reported in Table 2.1. The photovoltaic efficiency values are essentially for systems installed to track the sun so that the optimal angle of incidence is always obtained. But even if we assume that today's commercial photovoltaic systems effectively deliver only one tenth of their official efficiency rating (say 0.6% instead of 6%), they would still produce 2.4 to 10.8 kWh per m^2 per year. A useful comparison is in the US Midwest where maize to ethanol technology is widespread and research on *Miscanthus* to ethanol is being undertaken. The average solar energy incidence is around 1,400 kWh per m^2 per year. With extremely conservative estimates of conversion efficiency, photovoltaics of the mid-2000s will generate about 8.4 kWh per m^2 per year. By contrast, current

---

[2]"Photovoltaic efficiency is calculated using the ratio of the maximum power point, P_m, divided by the input light *irradiance* (E, in W/m^2) under standard test conditions (STC) and the *surface area* of the solar cell (A_c in m^2)." (Source: http://en.wikipedia.org/wiki/Solar_cell)

maize to ethanol technologies yield 2.0 to 2.5 kWh per m^2 per year and experimental results for *Miscanthus* are equivalent to 4.6 kWh per m^2 per year.

2.5 Photovoltaics and the Transportation Sector

Biofuels find their primary use in transportation, as partial or total substitutes for gasoline or diesel. Can photovoltaics play a similar role? The liquid fuel industry has tremendous advantages – widespread distribution networks (filling stations), an enormous pool of labor resources specialized in providing services through processing to distribution and repair, massive infrastructure for storage and transport, and a large installed base of vehicles. But addition of ethanol to this system in large volume presents technical difficulties. Transportation pipelines for oil cannot be readily changed to ethanol. Ethanol is hydrophilic, and the resulting water contamination can cause problems in storage facilities and engines not designed to use ethanol as fuel. Biodiesel presents problems in cold climates where it can congeal and block fuel lines.

There are numerous technical challenges to be met before solar-based electricity can compete commercially with liquid fuels in transportation (Graham 2001; MacLean and Lave 2003; Gaines et al. 2007; Karplus 2008; Bradley and Frank 2009). Commercially available vehicles that make partial use of electricity have been available only since the mid-1990s so both the engine and battery technologies needed to utilize solar-based electricity are still relatively new and evolving rapidly. Large-scale solar (arrays located in deserts) would require access to the grid, which is readily available in some places but not in all places. Distributed PVs (solar panels located on roof tops or back yards) are an option in locations where the grid is less well developed, but would need cost-effective battery technology to be useful for transportation.

The key technology, however, to make solar-based transportation technology possible, is improved batteries. Battery makers must develop battery technology that essentially emulates the energy-storage densities of liquid fuels. The batteries used in hybrid vehicles in the mid-2000s have insufficient capacity to provide the 300-mile range that would make it competitive with a gasoline-powered vehicle. However, substantial research is underway in battery technology, motivated in large part currently by the need for high-energy density storage for portable consumer electronics, in particular cell phones.

Axsen et al. (2008) provide an overview of the state of battery technology for use in transportation. The most popular hybrid vehicle, the Toyota Prius, initially used a lead acid battery technology but switched to an NiMH battery technology for its second-generation product. The consumer electronics industry has almost entirely shifted to lithium-ion (Li-ion) battery technology, which has much better performance characteristics. The much anticipated Chevrolet Volt hybrid (commercial release planned for 2010) is expected to use Li-ion batteries. According to Axsen et al. (p. iv), "Li-Ion battery technologies hold promise for achieving much higher power and energy density goals, due to lightweight material, potential for high voltage, and anticipated lower costs relative to NiMH. NiMH batteries could play an interim role in less demanding blended-mode designs, but it seems likely that falling Li-Ion battery prices may preclude even this role. However, Li-Ion batteries face drawbacks in longevity and safety which still need to be addressed for automotive applications."

2.6 Comparing the Costs of Energy from Biofuels and Photovoltaics

A meaningful comparison of the costs of delivering a unit of useful energy derived from sunlight via plants or photovoltaics is extremely complex. Creating ethanol from plants requires both significant fixed investments in capital equipment (processing facilities, tractors, transport equipment, etc.) and annual recurring costs (fuel for transport, applied nutrients, water, energy for processing). Manufacture of photovoltaics and their installation also have significant upfront costs but recurring expenditures are almost nonexistent. Said another way, the marginal cost of energy from photovoltaics is close to zero. Figure 2.5 taken from Wilkins et al. (2004) provides one estimate of the costs of photovoltaics (and solar water heating). These estimates, made in 2004, show photovoltaic-based electricity under 15 cents per kWh by 2010. This is still roughly 50% higher than the typical consumer rate of around 10 cents per kWh, but if plans to impose carbon emissions caps are implemented a 50% increase in the price of coal-based electricity would not be out of the question.

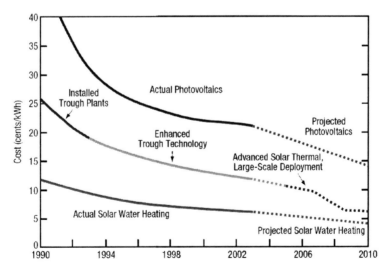

Fig. 2.5 Historical and Projected costs of various solar energy technologies
Source: Wilkins et al., (2004).

2.7 Concluding Remarks

This chapter has focused on one aspect of technical efficiency – the conversion of raw solar energy to a form that is useful to humans. Photovoltaics clearly dominate plants, with conversion rates with the commercial cells of the mid-2000s that are 2–10 times higher and operate throughout the year rather than just during the growing season. But photovoltaics provide electricity, which is not currently cost-effective for use in transportation. As research into photovoltaics and battery technology is still in its infancy, the potential for commercially viable technology breakthroughs seems high. The numerous negative externalities associated with bio-fuels (use of land that could otherwise be devoted to food, high water consumption, potentially noxious by-products) (Pimentel 2000; Gurgel et al. 2007; Rajagopal and Zilberman 2007; Hellegers et al. 2008; Searchinger et al. 2008), the still unproven commercial potential of cellulosic ethanol, and the inherent inefficiencies in capturing solar energy suggest we would be remiss not to continue substantial research efforts into cost-effective photovoltaics and automotive battery technology.

References

Axsen, J, Burke A, and Kurani KS (2008) Batteries for Plug-in Hybrid Electric Vehicles (PHEVs): Goals and the State of Technology circa 2008. Davis, Institute of Transportation Studies, University of California, Davis: 26.

Bradley TH and Frank AA (2009) Design, demonstrations and sustainability impact assessments for plug-in hybrid electric vehicles. Renew Sustain Energy Rev 13(1): 115–128.

Farrell AE, Plevin RJ, Turner BT, Jones AD, OHare M, and Kammen DM (2006) Ethanol Can Contribute to Energy and Environmental Goals, Am Assoc Advance Sci 311: 506–508.

Gaines L, Burnham A, Rousseau A, and Santini D (2007) Sorting through the many total-energy-cycle pathways possible with early plug-in hybrids. Center for Transportation Research, Argonne National Laboratory: 32.

Graham R (2001) Comparing the Benefits and Impacts of Hybrid Electric Vehicle Options. Electric Power Research Institute (EPRI), Palo Alto, CA, Report 1000349.

Gurgel A, Reilly JM, and Paltsev S (2007) Potential land use implications of a global biofuels industry. J Agri Food Ind Org 5(2):1–34.

Heaton EA, Dohleman FG, and Long SP (2008) Meeting US Biofuel goals with less land: the potential of Miscanthus. GCB 14(9): 2000–2014.

Hellegers P, Schoengold K, and Zilberman D (2008) Water resource management and the poor. Economics of Poverty, Environment and Natural-Resource Use. R. B. Dellink and A. Ruijs. Wageningen. 25: 41.

Karplus VJ (2008) Prospects for plug-in hybrid electric vehicles in the United States: a general equilibrium analysis, Massachusetts Institute of Technology.

Khanna M, Dhungana B, Clifton-Brown J (2008) Costs of producing Miscanthus and switchgrass for bioenergy in Illinois. Biomass Bioenerg 32: 482–493.

Liska AJ, Yang HS, Bremer VR, et al. (2009) Improvements in life cycle energy efficiency and greenhouse gas emissions of corn-ethanol. J Ind Eco 5:423–437.

Loomis RS and Amthor JS (1996) Limits to yield revisited. Increasing yield potential in wheat: breaking the barriers. M. P. Reynolds, S. Rajaram and A. McNab. Mexico, CIMMYT: 76–89.

MacLean HL and Lave LB (2003) Evaluating automobile fuel/propulsion system technologies. Prog Energy Combus Sci 29(1): 1–69.

Pimentel D (2000) Soil erosion and the threat to food security and the environment. Ecosystem Health 6: 221–226.

Rajagopal D and Zilberman D (2007) Review of Environmental, Economic and Policy Aspects of Biofuels. Policy Research Working Paper. Washington, DC, World Bank: 107.

Reynolds MP, van Ginkel M and Ribaut J-M (2000) Avenues for genetic modification of radiation use efficiency in wheat. J Exp Bot 51(suppl_1): 459–473.

Searchinger T, Heimlich R, Houghton RA, Dong F, et al. (2008) Use of U.S. Croplands for Biofuels Increases Greenhouse Gases Through Emissions from Land Use Change. Science: 1151861.

Wilkins F, Klimas P, McConnell, R et al. (2004) Solar Energy Technologies Program: Multi-Year Technical Plan 2003–2007 and beyond. U. S. DOE.

Chapter 3
Perennial Grasses as Second-Generation Sustainable Feedstocks Without Conflict with Food Production

Frank G. Dohleman, Emily A. Heaton, and Stephen P. Long

Abstract Biofuel production from maize grain has been touted by some as a renewable and sustainable alternative to fossil fuels, while being criticized by others for removing land from food production, exacerbating greenhouse gas emissions, and requiring more fossil energy than they produce. The use of second-generation feedstocks for cellulosic biofuel production is widely believed to have a smaller greenhouse gas footprint than first-generation feedstocks. In particular, perennial grasses may provide a balance between the high productivity necessary to minimize the amount of land area necessary for feedstock production and the sustainability of the perennial growth habit.

3.1 Introduction

In 2005, the U.S. Department of Energy (DOE) and U.S. Department of Agriculture (USDA) released the "Billion Ton Study," which investigated the feasibility of producing 1 billion tons of biomass annually in the United States by 2030 for conversion into liquid fuels such as ethanol (Perlack et al. 2005). This amount of biomass is expected to produce enough fuel to displace 30% of US petroleum usage. The "Billion Ton Study" breaks down the billion tons of biomass into a wide variety of different feedstocks: crop residues, in particular maize stover, would provide more than 30% of the total, dedicated perennial feedstocks more than 25%, and maize grain contributing over 6% of the biomass. The remainder of the biomass is expected to come from forestry resources and process residues. In 2007, 6.5 billion gallons of ethanol was produced, almost exclusively from maize grain (RFA 2008). It is projected that by 2022, 36 billion gallons of ethanol will be produced, but only 15 billion gallons will be produced from grain as a feedstock. The remaining 21

F.G. Dohleman (✉)
University of Illinois, Urbana, IL, USA; Monsanto Company, St. Louis, MO, USA
e-mail: frank.g.dohleman@monsanto.com

M. Khanna et al. (eds.), *Handbook of Bioenergy Economics and Policy,*
Natural Resource Management and Policy 33, DOI 10.1007/978-1-4419-0369-3_3,
© Springer Science+Business Media, LLC 2010

billion gallons are expected to come from second-generation processes which convert cellulose and hemi-celluloses into ethanol. One assumption which is included in many projections of future ethanol production is an increase in the productivity of feedstocks per unit land area (Perlack et al. 2005). As land use decisions are made, productivity should not come at the expense of environmental sustainability. Conversely, sustainability cannot come at the expense of productivity. Maize is productive, but has been criticized widely for its intensive agronomic inputs that result in a small ratio of energy in to energy out, and debatable greenhouse gas mitigation (Lal 2006; Lynd et al. 2006; Wilhelm et al. 2007). At the other end of the spectrum, low-input high-diversity systems, such as restored mixed-prairies, have been touted for their low inputs, but the productivity of these systems appears too low to make them economically viable (Hill et al. 2006; Tilman et al. 2006). Is there a middle ground? Adding cover crops to food production rotations can provide biomass and environmental services at times of the year when main crops are not in the field (Anex et al. 2007; Snapp et al. 2005). One compromise between sustainability and productivity, however, may be found within the highest yielding monocultures of perennial grass species such as *Miscanthus* x *giganteus* (Miscanthus) and *Panicum virgatum* (switchgrass) (Heaton et al. 2008).

3.2 Ideal Feedstock Characteristics

Critics have suggested that biofuels from crops have a low energy ratio (energy in:energy out) and a large greenhouse gas (GHG) footprint and remove grain from the food system, thus driving up commodity prices (Searchinger et al. 2008; Patzek and Pimentel 2005; Hill et al. 2006). These analyses have focused largely on food and feed crops used for biofuels, so-called first-generation feedstocks, such as maize grain for ethanol production and soybean-based biodiesel production. Second-generation feedstocks developed specifically for bioenergy production, such as the perennial grasses Miscanthus and switchgrass, could provide many advantages over maize-based ethanol production. Perennial grasses have an improved energy ratio from reduced energy inputs and in some cases increased outputs (Farrell et al. 2006). They may also be grown on land that is marginal for food crop production or not used in grain production (Lemus and Lal 2005). This section will explore the features of dedicated energy crops that, if managed properly, will allow integration of biofuel production into existing agricultural systems without negatively impacting food production.

3.3 Perennial Growth Habit

Diesel fuel used for field operations like annual tillage and planting together with annual additions of nitrogen fertilizer produced by the energy-intensive Haber process constitute major energy inputs into annual row crops such as *Zea mays* (maize).

The use of perennial grasses not only eliminates the need for annual tillage and planting, but also reduces soil erosion and allows for carbon capture within the soils (Hansen et al. 2004; Schneckenberger and Kuzyakov 2007; Lal 2006). Because nutrients are translocated from the annual crop of stems to the perennial root system in the fall, nutrients are in effect recycled, minimizing fertilizer requirements (Christian et al. 2006). Many of the advantageous characteristics of cellulosic biofuel feedstocks that are detailed below result from the perennial growth habit.

3.4 C_4 Photosynthetic Pathway

Photosynthesis is the pathway by which plants are able to capture sunlight energy and convert it into stored chemical energy in biomass. Plants are broadly divided into two classes of photosynthetic pathway, C_3 and C_4. All plants use the enzyme ribulose-1,5-bisphosphate carboxylase oxygenase (Rubisco) to fix carbon dioxide (CO_2) into two three-carbon compounds which are the building blocks for plant biomass. Under current atmospheric conditions, oxygen competes with CO_2 for the active site of Rubisco, causing inefficiency in the photosynthetic process. C_4 plants have evolved a mechanism by which they are able to pump CO_2 to the active site of Rubisco, causing the oxygenation reaction to be suppressed and, therefore, allowing more efficient uptake of CO_2. Plants with the C_4 pathway have been shown to be up to 40% more efficient at fixing carbon than those with the C_3 pathway, causing them to produce more biomass per unit land area (Long et al. 2006).

Not surprisingly, C_4 plants are some of the most productive species known, including sugarcane, sorghum, napier grass, maize, and Miscanthus. Typically, C_4 plants are found in tropical and subtropical areas of the world, and many are unable to tolerate cold temperatures (Sage 2004). However, Miscanthus and switchgrass include genotypes which are exceptional in their ability to tolerate the cool temperatures prevalent in the upper latitudes of the United States in the spring and fall (Heaton et al. 2008).

3.5 Long Canopy Duration

While C_4 photosynthesis increases the efficiency with which intercepted solar energy is converted into biomass, yield will also depend on the plant producing a canopy of leaves that covers the ground for as much as the year as possible. Annual crops such as maize and *Glycine max L.* (soybean) are planted once soil temperatures are high enough for effective seed germination and field conditions are dry enough for equipment to enter the field. However, initial canopy formation is limited by seed reserves and as a result it may take 4−8 weeks after planting until leaves cover the ground and intercept the majority of incident solar radiation. For maize in the Midwest, this often does not occur until after the summer solstice, i.e., the peak of solar energy receipt. Herbaceous perennial grasses like Phalaris, Arundo, switchgrass, and Miscanthus have large root reserves of carbohydrates that are used for rapid leaf growth as soon as the growing season begins. This allows them to cover the ground rapidly and function as more efficient solar collectors. Current maize varieties have been bred to senesce and dry down prior to the date of first frost in the fall to avoid drying costs if frost occurs earlier than expected, but at the cost of failing to collect the solar energy of late summer and early fall. In the Maize Belt, Miscanthus and switchgrass emerge and begin capturing sunlight weeks before maize and continue to produce green leaves until the first frost in the fall, up to 8 weeks after grain filling in maize is complete. Measurements in central Illinois show that canopy duration for Miscanthus can be as much as 45% longer than maize (Dohleman and Long, 2009).

3.6 Limited Pest and Disease Incidence

Miscanthus has been present throughout the United States as a garden plant for over 100 years. Trials as a bioenergy crop in the United States began only in 2002 and in the European Union in the early 1980s. As of 2008, there have been no reports of pests or diseases which have caused yield loss in mature Miscanthus field stands. While there is evidence of *Miscanthus sinensis* and *Miscanthus sacchariflorus* providing a host for Barley yellow dwarf virus and certain types of aphids, these have not been shown to affect yield (Christian et al. 1994). The longest running Miscanthus yield trials in the United States have been conducted without the need for pest management and have produced large amounts of biomass without pesticide inputs, improving the energy ratio of the system (Heaton et al. 2008). Maize, soybean, and switchgrass each have a number of pests and diseases that have been reported on extensively in the literature (Parrish and Fike 2005).

3.7 Nutrient Recycling

Fertilizer use in maize grain ethanol production has been shown to be the major agricultural input in life cycle analysis. Maize has been bred over the past half century to be able to respond to fertilization. Because maize is traditionally used as a food and feed crop, much of this fertilizer necessarily ends up as nutrition when it is eaten, although fertilizer that is not taken up by the immature root system early in the growing season can be lost in surface water runoff or leached through the soil profile into ground water. Mature perennial grasses have root systems that spread below the entire surface, effectively scavenging available fertilizer. In the case of Miscanthus, negligible amounts of nitrate applied could be detected using resin lysimeters, indicating minimal leaching (Beale and Long 1997). An ideal feedstock, however, does not need high levels of nutrients. Due to their perennial nature, grasses such as Miscanthus and switchgrass are able to recycle nutrients, translocating them from root storage to the growing shoot in the spring and then back to the root system in

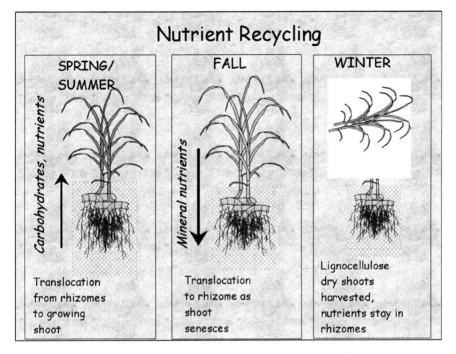

Fig. 3.1 Schematic diagram of nutrient cycling through a perennial grass life cycle

the fall as the annual crop of shoots senesce (Fig. 3.1). This allows for low input of energy-intense fertilizer and also makes for a cleaner burning fuel. Miscanthus seems to be exceptional at this process; recent trials at Rothamsted in England showed no response to added nitrogen over 15 years of cultivation (Christian et al. 2008). This low nitrogen requirement will greatly improve the net GHG emissions relative to grain crops. This process also means that plants of this life-form will be well suited for more marginal soils than annual row crops, which can help to produce energy without taking land out of food production.

3.8 High Water Use Efficiency

Irrigation represents yet another energy input; therefore, an ideal feedstock should have high water use efficiency (WUE), i.e., ratio of dry mass formed per unit water of evapotranspiration. Plants undergo a constant struggle between assimilating as much CO_2 as possible without losing too much water to cause drought stress. Due to their CO_2 concentrating mechanism, C_4 plants have an inherently higher WUE than C_3 plants when grown at the same location. Miscanthus has been shown to have a similar WUE to pearl millet (Beale et al. 1999). However, the amount of water used will be the biomass yield divided by the WUE. So although it has a high WUE, its high yields still mean that the crop has a high annual water requirement. So, while a Miscanthus field uses more water than maize, soybean, or switchgrass fields, the amount of biomass return on that water investment is much higher in Miscanthus, giving it the highest WUE of these crops (Clifton-Brown and Lewandowski 2000).

3.9 Low Herbicide Requirement

Perennial grasses can form a tall closed canopy rapidly, with the result that weeds are quickly overtopped and shaded out. In the case of Miscanthus, this means no weed control is required after establishment, as will be the case with other tall perennial grasses (Heaton et al. 2008). However, this does not mean that no herbicide will be needed at all. Evidence is accumulating to suggest that weed control is critical in successful establishment of both Miscanthus and switchgrass. A grower's guide available from the University of Illinois explains procedures for weed control when establishing Miscanthus (Pyter et al. 2010). Standard weed-control practice in an establishing switchgrass field includes herbicide application and frequent mowing of the plots prior to seed set of annual weeds in order to reduce the seed bank of these weeds (Parrish and Fike 2005).

3.10 Noninvasive, Easily Eradicated from Existing Land

There have been reports suggesting that the low-input high-productivity charac-teristics of an ideal biomass crop would also give it a high potential for invading the surrounding landscape (Raghu et al. 2006; Barney and Ditomaso 2008). This is a legitimate concern as invasive species can outcompete native species in nat-ural systems, with high environmental and economic costs. Current lead varieties of switchgrass are simply native populations that have undergone seed increase for commercialization, and as such are considered lower risk than nonnative species. In response to the demand for increased biomass production in switchgrass, however, development of hybrid switchgrass varieties that may be more competitive than the existing varieties is now underway (Vogel and Mitchell 2008). Miscanthus is not native to the United States and the genus is large and diverse. It is the naturally occurring interspecific hybrid $M.$ x $giganteus$ discussed here that is of most interest for bioenergy production. This triploid hybrid is unable to produce a viable seed (Scally et al. 2001), eliminating one key avenue of plant reproduction. While spread through the rhizome system is possible, nearly 30 years of research trials in the European Union, 7 years of trials in the United States, and over 100 years of use as a garden plant have yet to show invasiveness as an issue for concern (Greenlee 1992). This empirical evidence is supported by risk-assessment models which sug-gest that $M.$ x $giganteus$ is less likely to escape than other types of second-generation biofuel feedstocks (Barney and Ditomaso 2008). It is important to bear in mind, however, that related species, in particular *M. sinensis*, does produce seed, and some genotypes have spread in the southeast of the United States

Eradication of existing Miscanthus is of concern to farmers if they want to remove it from their fields and revert back to standard row-crop systems. There have been no reports of eradication trials in the literature, but preliminary data from the University of Illinois suggest that Miscanthus can be eradicated by glyphosate followed by conventional tillage, followed by planting of glyphosate-resistant soybeans.

3.11 Uses Existing Farm Equipment

Farmers would be less likely to adapt new cropping systems if they are required to purchase special equipment for harvest. Perennial grasses can be harvested using conventional hay or maize silage equipment (Pyter et al. 2010; Jones and Walsh 2001). Because perennial grasses are harvested during the late fall and winter months, this equipment can be used at a time that it generally sits idle. Storage of biomass is a potential problem with large-scale bioenergy production (Fales et al. 2007). However, perennial grasses are able to stand over winter and be harvested as needed from full senescence until the emergence of the following year's crop, which allows for a more consistent stream of biomass from the field to the processing station with a limited need for storage costs. Leaving biomass to stand in the field may be limited in temperate growing regions of the United States that experience severe winters, as heavy snowfall can lead to lodging and reduce harvestable yield (Adler et al. 2006) (Table 3.1).

Table 3.1 List of ideal biofuel feedstock characteristics with first- and second-generation feedstocks. Adapted from (Heaton et al. 2004)

The "Ideal" biomass crop?	Maize	Soybean	Miscanthus	Switchgrass	Forestry
High yielding	★		★★	★	
Perennial			★	★	★
C$_4$ photosynthesis	★		★	★	
Long canopy duration			★	★	★
Low fertilization requirements		★	★		
Low herbicide requirements			★	★	
Low pesticide requirements			★		
Sterile − noninvasive	★	★	★		
Winter standing			★		★
Easily removed	★★	★★	★	★	
High water use efficiency	★		★	★	
Uses existing farm equipment	★	★	★	★	
Ease of establishment	★★	★★		★	

3.12 Feedstock Yield, Greenhouse Gas Mitigation, and the World Food Supply

First-generation biofuels have been criticized for causing increases in food prices worldwide (Hill et al. 2006; Runge and Senauer 2007). Recent work shows that Miscanthus is more than 250% more productive than switchgrass and maize grain in the Midwestern United States (Heaton et al. 2008), allowing for the production of more ethanol per unit land area (Fig. 3.2). Maize stover is another potential feedstock which is able to be harvested as well, but even with grain and stover, Miscanthus still produces nearly double the amount of ethanol in a more sustainable manner. Switchgrass and Miscanthus have the added advantage that they are able to be grown on marginal cropland and would not take away from food production.

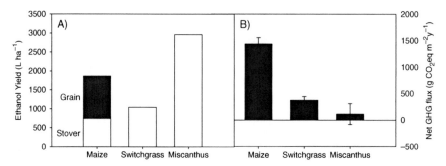

Fig. 3.2 A) Potential ethanol yield of three biofuel feedstocks. Maize grain represents first-generation feedstock technology. Maize stover, switchgrass cv. Cave-in-Rock, and Miscanthus represent second-generation cellulosic feedstocks. Adapted from Heaton et al. (2008). **B)** Modeled GHG flux of three biofuel cropping systems. Adapted from Davis et al., *in press*

Highly productive monocultures of Miscanthus and switchgrass represent two potential biofuel feedstock sources which provide a compromise between high yield and environmental sustainability necessary to reach US biofuel goals without exacerbating global climate change. Federal mandates will require 36 billion gallons of ethanol in the next few decades, production of which could be feasible on far less land if yields shown here for Miscanthus are reached on a commercial scale (Table 3.2). High yield that equates to high biofuel production would also provide economic incentive for large-scale adaptation of these crops by farmers (Khanna et al. 2008).

Recent analysis has shown that over a 10-year time period, the cultivation of perennial crops for energy production in the Midwestern United States is expected to have substantially lower greenhouse gas (GHG) emissions than cultivation of maize (Fig. 3.2b; Davis et al., *in press*). This modeled analysis predicts that Miscanthus GHG flux will be nearly 0, while switchgrass will have a slightly positive flux and due to its intensive management and inputs, maize will have threefold higher fluxes of GHGs than switchgrass.

Table 3.2 Percentage of US-harvested cropland necessary to reach the US advanced energy initiative goal of 36 billion gallons of ethanol production by 2022 for first- and second-generation biofuel feedstocks. Adapted from Heaton et al. (2008)

Feedstock	% of harvested cropland needed to reach US biofuel goals
Maize grain	25.1
Maize stover	38.3
Maize total	15.2
Switchgrass	27.3
Miscanthus	9.6

3.13 Conclusion

Renewable energy which balances high productivity, environmental sustainability, and the potential to be deployed at a scale which can make a noticeable reduction in US oil consumption is not feasible with first-generation grain ethanol technologies. As further research leads to breakthroughs in cellulosic biomass production, deconstruction, and conversion to liquid fuels, these goals become increasingly feasible. Second-generation high-yielding perennial feedstocks such as Miscanthus represent a more environmentally and economically sustainable means by which energy can be produced on a large scale in the near term.

References

Adler PR, Sanderson MA, Boateng AA, Weimer PI, Jung HJG (2006) Biomass yield and biofuel quality of switchgrass harvested in fall or spring. Agron J **98**, 1518–1525.

Anex RP, Lynd LR, Laser MS, Heggenstaller AH, Liebman M (2007) Potential for enhanced nutrient cycling through coupling of agricultural and bioenergy systems. Crop Sci **47**, 1327–1335.

Barney JN, Ditomaso JM (2008) Nonnative species and bioenergy: Are we cultivating the next invader? Bioscience **58**, 64–70.

Beale CV, Morison JIL, Long SP (1999) Water use efficiency of C4 perennial grasses in a temperate climate. Agric For Meteor **96**, 103–115.

Beale CV, Long SP (1997) Seasonal dynamics of nutrient accumulation and partitioning in the perennial C4-grasses Miscanthus x giganteus and Spartina cynosuroides. Biomass Bioenerg **12**, 419–428.

Christian DG, Riche AB, Yates NE (2008) Growth, yield and mineral content of Miscanthus x giganteus grown as a biofuel for 14 successive harvests. Ind Crop Prod **28**, 320–327.

Christian DG, Lamptey JNL, Forde SMD, Plumb RT (1994) First report of barley yellow dwarf luteovirus on Miscanthus in the United Kingdom. Eur J Plant Pathol **100**, 167–170.

Christian DG, Poulton PR, Riche AB, Yates NE, Todd AD (2006) The recovery over several seasons of 15 N-labelled fertilizer applied to Miscanthus x giganteus ranging from 1 to 3 years old. Biomass Bioenerg **30**, 125–133.

Clifton-Brown JC, Lewandowski I (2000) Water use efficiency and biomass partitioning of three different Miscanthus genotypes with limited and unlimited water supply. Annals of Bot **86**, 191–200.

Davis SC, Parton WJ, Dohleman FG, et al. (*in press*) Sustainability of nutrient budgets in bioenergy agroecosystems. Ecosystems.

Dohleman FG, Long SP (2009) More productive than maize in the Midwest. How does Miscanthus do it? Plant Physiol **50**, 2104–2115.

Fales SL, Hess JR, Wilhelm WW (2007) Convergence of Agriculture and Energy: II: Producing Cellulosic Biomass for Biofuels. CAST Commentary QTA2007-2,

Farrell AE, Plevin RJ, Turner BT, Jones AD, O'Hare M, Kammen DM (2006) Ethanol can contribute to energy and environmental goals. Science **311**, 506–508.

Greenlee J (1992) *The encyclopedia of ornamental grasses: how to grow and use over 250 beautiful and versatile plants*. Rodale Press, Emmaus, PA, 186 pp.

Hansen EM, Christensen BT, Jensen LS, Kristensen K (2004) Carbon sequestration in soil beneath long-term Miscanthus plantations as determined by 13C abundance. Biomass Bioenerg **26**, 97–105.

Heaton EA, Dohleman FG, Long SP (2008) Meeting US biofuel goals with less land: The potential of miscanthus. GCB **14**, 2000–2014.

Heaton EA, Clifton-Brown J, Voigt TB, Jones MB, Long SP (2004) Miscanthus for renewable energy generation: European union experience and projections for Illinois. Mitig Adapt Strat Glob Change **9**, 433–451.

Hill J, Nelson E, Tilman D, Polasky S, Tiffany D (2006) Environmental, economic, and energetic costs and benefits of biodiesel and ethanol biofuels. Proc Natl Acad Sci **103**, 11206–11210.

Jones MB, Walsh M (2001) *Miscanthus for Energy and Fibre*. James and James Ltd., London, 271 pp.

Khanna M, Dhungana B, Clifton-Brown J (2008) Costs of producing miscanthus and switchgrass for bioenergy in Illinois. Biomass Bioenerg **32**, 482–493.

Lal R (2006) Soil and environmental implications of using crop residues as biofuel feedstock. Int Sugar J **108**, 161–167.

Lemus R, Lal R (2005) Bioenergy crops and carbon sequestration. Crit Rev Plant Sci **24**, 1–21.

Long SP, Zhu XG, Naidu SL, Ort DR (2006) Can improvement in photosynthesis increase crop yields? PCE **29**, 315–330.

Lynd L, Greene N, Dale B, Laser M, Lashof D, Wang M, Wyman C (2006) Energy returns on ethanol production. Science **312**, 1746–1747.

Parrish DJ, Fike JH (2005) The Biology and Agronomy of Switchgrass for Biofuels. Crit Rev Plant Sci **24**, 423–459.

Patzek TW, Pimentel D (2005) Thermodynamics of energy production from biomass. Crit Rev Plant Sci **24**, 327–364.

Perlack RD, Wright LL, Turhollow A, Graham RL, Stokes B, Erbach DC (2005) Biomass as feedstock for a bioenergy and bioproducts industry: the technical feasibility of a billion-ton annual supply. USDA/DOE report.

Pyter RJ, Heaton EA, Dohleman FG, Voigt TB, Long SP (2010) Agronomic Experiences with Miscanthus x giganteus in Illinois, USA. In: *Biofuels: Methods and Protocols* (ed. J. Mielenz). Springer, New York.

Raghu S, Anderson RC, Daehler CC, Davis AS, Wiedenmann RN, Simberloff D, Mack RN (2006) Adding biofuels to the invasive species fire? Science **313**, 1742.

RFA (2008) U.S. Ethanol Statistics, Renewable Fuel Association.

Runge CF, Senauer B (2007) How biofuels could starve the poor. Foriegn Affairs, **86**, 41.

Sage RF (2004) The evolution of C-4 photosynthesis. New Phyto, **161**, 341–370.

Scally L, Hodkinson TR & Jones MB (2001) Origins and Taxonomy of Miscanthus. In: *Miscanthus for Energy and Fibre* (eds M.B. Jones & M. Walsh). James & James, London.

Schneckenberger K, Kuzyakov Y (2007) Carbon sequestration under Miscanthus in sandy and loamy soils estimated by natural C-13 abundance. J Plant Nutr Soil Sci Zeitschrift Fur Pflanzenernahrung Und Bodenkunde **170**, 538–542.

Searchinger T, Heimlich R, Houghton RA, et al. (2008) Use of U.S. croplands for biofuels increases greenhouse gases through emissions from land use change. Science **319**, 1238–1240.

Snapp SS, Swinton SM, Labarta R, Mutch D, Black JR, Leep R, Nyiraneza J, O'Neil K (2005) Evaluating cover crops for benefits, costs and performance within cropping system niches. Agron J **97**, 322–332.

Tilman D, Hill J, Lehman C (2006) Carbon-negative biofuels from low-input high-diversity grassland biomass. Science **314**, 1598–1600.

Vogel K.P., Mitchell K.B. (2008) Heterosis in switchgrass: Biomass yield in swards. Crop Sci **48**, 2159–2164.

Wilhelm WW, Johnson JMF, Karlen DL, Lightle DT (2007) Corn stover to sustain soil organic carbon further constrains biomass supply. Agron J **99**, 1665–1667.

Chapter 4
Present and Future Possibilities for the Deconstruction and Utilization of Lignocellulosic Biomass

Hans P. Blaschek, Thaddeus Ezeji, and Nathan D. Price

Abstract Current technologies for the deconstruction of lignocellulosic biomass rely on physical–chemical pretreatment processes followed by enzymatic hydrolysis. These technologies, while able to efficiently produce sugars, also allow the formation of degradation products that are inhibitory to microbes such as yeast or bacteria which are used for fermentation. The status of current deconstruction technologies and the role of genomics and the "New Biology" for producing feedstocks and tailor-made microbes with characteristics that make them more amenable to fermentation-based processes are discussed.

4.1 Introduction: Current State of Technology

In order to discuss what the future may hold when it comes to the production of biofuels and the bioeconomy, it is important to examine where we are today with respect to the current bioenergy platforms. Both dry and wet mill ethanol production from corn starch (US) and ethanol production from sugarcane (Brazil) are regarded as essentially mature technologies for producing bioethanol. Currently, dry-grind ethanol plants produce the majority of fuel ethanol in the United States. Annual ethanol production capacity reached 9.0 billion gallons in 2008 (Renewable Fuels Association; www.ethanolrfa.org). Approximately, 20% of the US corn crop was used for ethanol production in 2006, while ethanol production consumed 51% of Brazil's sugarcane crop in 2004. Given the concerns regarding the net energy balance and the sustainability of corn ethanol in the context of the food vs. fuel debate, ethanol production from corn is expected to level off (von Braun, 2007). However, some incremental increases in energy efficiency of these processes can be expected

H.P. Blaschek (✉)
Center for Advanced BioEnergy Research, University of Illinois Urbana-Champaign, IL, USA
e-mail: blaschek@illinois.edu

M. Khanna et al. (eds.), *Handbook of Bioenergy Economics and Policy*,
Natural Resource Management and Policy 33, DOI 10.1007/978-1-4419-0369-3_4,
© Springer Science+Business Media, LLC 2010

as coproduct utilization (e.g., distillers' grains and bagasse) is incorporated into next-generation plants. Currently, distiller's grains from corn ethanol production is used as animal feed, while most of the bagasse from sugarcane production is burned for power generation.

While lignocellulosic biomass from agricultural residues, municipal paper wastes, dedicated energy crops, and other sources is expected to be a major renewable feedstock for sustainable production of biofuels, *Saccharomyces cerevisiae*, the best known ethanologenic microorganism, does not naturally have the metabolic capacity to utilize five carbon sugars such as xylose and arabinose. A team led by Professor Nancy Ho, a leader of the molecular genetics group in Purdue University's Laboratory of Renewable Resources Engineering (LORRE), developed *Saccharomyces* yeast in the 1990s that simultaneously converts glucose and xylose to ethanol. In 1993, Professor Ho's group became the first to produce genetically engineered *Saccharomyces* yeast that can ferment both glucose and xylose to ethanol (Ho et al., 1999). Iogen Corporation Canada obtained a nonexclusive license from the Purdue Research Foundation for the yeast and related patents. The first commercial fuel ethanol from lignocellulosic feedstock was produced by Iogen corporation using the Purdue yeast to produce ethanol from wheat straw hydrolysates. It should be noted that xylose makes up about 30% of sugars in typical agricultural residues, and thus the ability of *S. cerevisiae* to ferment xylose to ethanol should increase ethanol yield by a corresponding margin. Since that time, many laboratories have explored numerous metabolic engineering strategies to develop recombinant *S. cerevisiae* strains that can efficiently utilize both xylose and arabinose for ethanol production (Hahn-Hagerdal et al., 2001; Jin et al., 2003; Sedlack and Ho, 2004;). Recently, arabinose-fermenting *S. cerevisiae* strains were generated by overexpression of a bacterial L-arabinose utilization pathway consisting of *Bacillus subtilis* AraA, *Escherichia coli* AraB and AraD, and simultaneous overexpression of the L-arabinose-transporting yeast galactose permease (Becker and Boles, 2003). This was followed by selection of an L-arabinose-utilizing yeast strain by sequential transfer in L-arabinose media. Other challenges, such as tolerance of the recombinant *S. cerevisiae* strains to fermentation inhibitors generated during pretreatment and hydrolysis of the lignocellulosic feedstock still exist. Many laboratories, including those of the authors, are currently involved in research directed toward development of osmo- and inhibitor-tolerant fermenting microorganisms.

4.2 Advantages of Lignocellulosic-Based Biofuels

Cellulosic biomass represents the most abundant renewable energy resource on the planet. The "Billion Ton Study" published by the US Department of Energy (DOE) in 2005 indicated that there were 1.3 billion dry tons of biomass available per year — enough to produce biofuels to meet more than one-third of the current demand for transportation fuels. The "Roadmap for Biomass Technologies in the United States" of 2002 indicated that bio-based transportation fuels are anticipated to increase from

0.5% of US consumption in 2001 to 4% in 2010, 10% in 2020, and further to 20–30% in 2030 or approximately 60 billion gallons of gasoline equivalent per year. The development of ethanol as a biofuel, beyond its role as a fuel oxygenate, will require developing lignocellulose as a feedstock. The production of biofuels from low-cost lignocellulosic biomass has the added advantage of not directly competing with food crops. This second-generation approach is key to meeting the DOE target and would be more consistent with ethanol as an economically viable and sustainable energy source.

Interest in the cultivation of lignocellulosic crops such as switchgrass and *Miscanthus* for subsequent conversion into fermentable sugars is receiving considerable attention. In addition, industrial and agricultural coproducts such as corn fiber, corn stover, distillers' dried grains with solubles (DDGS), wheat straw, rice straw, soybean residues, as well as various types of agricultural and industrial wastes are presently considered as potential feedstocks for the production of fermentable sugars (Ezeji et al., 2005). The depolymerization of these renewable and abundant resources to fermentable sugars represents a challenge for microbiologists and chemical engineers due to the recalcitrance of these materials. Various strategies have been developed to utilize alternative substrates for conventional and unconventional ethanol fermentations.

Seven million metric tons of DDGS are expected to be produced from corn ethanol processing in the United States by the end of 2008. Some experts are predicting that DDGS production in the United States will reach up to 15 million metric tons in a few years. In addition to starch, distiller's grains contain fiber, which is composed of cellulose, xylan, and arabinan. If these coproducts were further hydrolyzed and converted into liquid fuels or other bioproducts, the efficiency and profitability of corn ethanol plants would be expected to improve even further. In order to accomplish this goal, technologies have to be developed for deconstruction and enzyme treatment of the fiber component present in DDGS. Members of The Midwest Consortium for Biobased Products recently completed a comprehensive study on the utilization of DDGS and cellulose conversion that has been published in a special issue of Bioresource Technology (Ladisch et al., 2008; Ezeji and Blaschek, 2008). As part of this study, the fermentation of DDGS hydrolysates to biobutanol, a 4-carbon liquid fuel, by the solvent-producing clostridia was also examined (Ezeji and Blaschek, 2008). In this study, fermentation of AFEX- and hot water-pretreated DDGS by solventogenic clostridia was unimpeded until all the sugars were fermented into acetone-butanol. The sugars were utilized concurrently throughout the fermentation, although the rate of sugar utilization was sugar specific (Ezeji and Blaschek, 2008).

4.3 Status of Current Conversion Technologies: Pretreatment

Various pretreatment technologies are available to fractionate and separate cellulose, hemicellulose, and lignin components (Wyman et al., 2005). These include concentrated acid, dilute acid, alkali, alkaline peroxide, steam explosion, ammonia fiber explosion, liquid hot water and organic solvent treatments (Saha, 2003). These different approaches all work to open up the physical structure of the biomass to allow for depolymerization by enzymes.

One of the key steps in the lignocellulosic biomass-to-fermentable sugars conversion is pretreatment. The goal of pretreatment is to alter the biomass macroscopic and microscopic size and structure, as well as its submicroscopic chemical composition so that enzymatic hydrolysis of the carbohydrate fraction to monomeric sugars can be achieved more rapidly and with greater yield (Mosier et al., 2005a). A number of pretreatment technologies as given above can be applied to bring about hydrolysis of agricultural feedstock. Unfortunately, during acid hydrolysis, lignins are oxidized or degraded to form phenolic compounds and some of the sugars that are released during hydrolysis are degraded into products that inhibit cell growth and fermentation (Ezeji et al., 2007a,b). Examples of such inhibitory compounds include furfural and hydroxymethyl furfural (HMF) and ferulic, glucuronic, p-coumaric acids, etc. Organic acids such as acetic acid are produced in significant amounts during acid hydrolysis of DDGS, corn fiber, and stover (Ezeji et al., 2007b). The acetic acid in agricultural residues hydrolysate is formed when O-acetyl groups are released from hemicellulose during acid hydrolysis. We have investigated the effect of some of these inhibitors on *Clostridium beijerinckii* growth and butanol fermentation. Degradation products such as furfural and HMF have stimulatory effects up to 3.0 g L^{-1} and do not have a negative effect on acetone-butanol production by *C. beijerinckii* BA101, while syringaldehyde, ferulic and p-coumaric acids were potent inhibitors of butanol production by *C. beijerinckii* BA101 (Fig. 4.1). Glucuronic acid had only a slight effect on acetone-butanol production at the highest concentration tested. Acetate (up to 9.0 gL^{-1}) was found to be stimulatory for

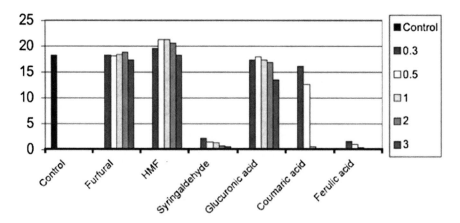

Fig. 4.1 Effect of degradation and hydrolysis products on the ABE fermentation by *Clostridium beijerinckii* BA101 (Y-axis in units of g/l ABE; Legend contains values in g/l of inhibitor)

both cell growth and butanol production (Ezeji et al., 2005, 2007b and Ezeji and Blaschek, 2008) by *C. beijerinckii* BA101.

A coordinated study supported through funding by the USDA Initiative for Future Agricultural and Food Systems (IFAFS) compared the performance of five promising biomass pretreatment methods. Using a single feedstock (corn stover), common analytical protocols, and consistent data interpretation, five research teams documented the technical and economical feasibility of selected pretreatment techniques (Wyman et al., 2005; Eggeman and Elander, 2005). They found that among the dilute acid (Lloyd and Wyman, 2005), hot water (controlled pH) (Mosier et al., 2005b), ammonia fiber/freeze explosion (AFEX) (Teymouri et al., 2005), ammonia recycle percolation (ARP; Kim and Lee, 2005), and lime (Kim and Holtzapple, 2005) pretreatments, low-cost pretreatment reactors are often counterbalanced by the higher costs associated with either pretreatment catalyst recovery or higher costs for ethanol product recovery (Eggeman and Elander, 2005). Pretreatment adds approximately 30% to the cost of processing of the biomass. A summary of the pretreatment methods examined in the IFAFS project is presented in Table 4.1.

Table 4.1 Summary of promising biomass pretreatment methods for biofuel production

Pretreatment	Increases surface area	Decrystalizes cellulose	Removes hemicellulose	Removes lignin	Limitations[6]
Dilute acid[1]	Yes		Yes		Corrosion, neutralization, formation of inhibitors, and cost of neutralization salts disposal
Hot water[2]	Yes		Yes		High temperature, need to add alkali to control pH, and relatively high cost in product recovery
AFEX[3]	Yes	Yes		Yes	Cost of ammonia
ARP[4]	Yes	Yes		Yes	Cost of ammonia and relatively low conversion of xylose
Lime[5]	Yes	ND		Yes	Long pretreatment time and relatively low conversion of xylose

[1] Lloyd and Wyman (2005).
[2] Mosier et al. (2005b).
[3] Teymouri et al. (2005).
[4] Kim and Lee (2005).
[5] Kim and Holtzapple (2005).
[6] Hao Feng (Personal communication, 2005).

In recent years, studies have been conducted using corn fiber as a low-cost feed-stock for the production of bioethanol and the effectiveness of different pretreatment methods were examined (Saha, 2003). Due to the relatively low lignin content of corn fiber, a hot water pretreatment carried out at 121°C for 1 h released about 87% of the glucose (Saha and Bothast, 1999). A pilot scale test employing pH-controlled hot water pretreatment was conducted by Mosier et al. (2005c) for ethanol production from corn fiber. Another pilot scale operation was performed by Schell et al. (2004) at the National Renewable Energy Laboratory (NREL) employing a dilute sulfuric-acid pretreatment to achieve simultaneous saccharification and fermentation to produce bioethanol. Efforts were also made to improve the enzymatic hydrolysis efficiency by utilizing new enzyme solutions (Li et al., 2005; Koukiekolo et al., 2003). There are relatively few investigations on the use of corn fiber for the production of chemicals and biofuels other than ethanol.

A closer look at Table 4.1 reveals that the acidic pretreatment methods (including hot water) are effective in removing hemicellulose, while the alkali methods function to remove lignin and decrystallize cellulose; both approaches are known to improve enzymatic digestibility. In addition, an economic analysis of the pretreatment methods has shown that the relatively high costs of ethanol production from lignocellulosic biomass arise mainly from costs associated with pretreatment conditions, enzymes, and recovery of end products from dilute streams (Eggeman and Elander, 2005). Technologies that lead to improvement in any of these areas will help to improve the profitability of a biofuel production operation.

Recently, we collaborated with Dr. Hao Feng at the University of Illinois to study the use of acidic- and alkaline-electrolyzed water to treat corn fiber. Electrolyzed water is a technique developed in Japan to decontaminate food and agricultural products. A major advantage of using electrolyzed water is that it is produced using pure water with no added chemicals other than dilute sodium chloride. Because one can neutralize acidic water-treated samples with alkaline water, no salt is generated following such a pretreatment. As the treatment with acidic-electrolyzed water is conducted at relatively low temperatures, the production of potential microbial inhibitors is minimized, although the sugar yields are comparable to samples treated with sulfuric acid.

In order to examine the feasibility of acidic- and alkaline-electrolyzed water pretreatments, we conducted preliminary experiments using corncobs, which consist of a combination of lignin (14%), cellulose (35%), and hemicellulose (37%). In the first test, the thermal stability of both water components was confirmed. After autoclaving at 121°C for 2 h, no change in pH was observed for both the acidic- and alkaline-electrolyzed water. The pH values of corncob slurry (solid to water ratio of 1:8 w/w) before and after a 6-h steaming in different reaction agents at 125°C were recorded. The total sugar, reducing sugar, and the degree of polymerization (DP) were also determined and are shown in Table 4.2 (Wang et al., 2006). The total sugar yield from acidic water-treated corncobs was 34.3%, which is comparable to a yield of 35% for samples treated with sulfuric acid at the same pH.

There is considerable hope and hype focused on using enzymes to convert lignocellulosic biomass to fermentable sugars. The complete enzymatic hydrolysis

Table 4.2 Pretreatment of corncobs using electrolyzed alkaline/acidic water and comparison with acid, base, or hot water pretreatments (125°C, for 6 h)

One-step method	Initial pH	Final pH	Total sugar (%)	Reducing sugar (%)	Degree of polymerization (%)
Hot water	6.1	4.4	27.5 ± 0.7	14.5 ± 0.7	1.9 ± 0.0
H_2SO_4	2.3	3.9	35.0 ± 1.4	17.5 ± 1.1	2.0 ± 0.0
NaOH	11.7	4.0	27.5 ± 0.8	13.2 ± 0.6	2.0 ± 0.0
Acidic water	2.3	4.5	34.3 ± 0.6	16.3 ± 0.8	2.1 ± 0.1
Alkaline water	11.7	4.7	25.9 ± 0.6	12.6 ± 0.6	2.1 ± 0.1
Two-step method					
Steep in alkaline water at 60°C, then autoclave in acid water	Alkaline water, pH 11.7 Acidic water, pH 2.6		39.0 ± 0.6	15.5 ± 1.0	2.5 ± 0.1

of hemicellulose requires xylanase, β-xylosidase, and several other complimentary enzymes such as acetylxylan esterase, α-arabinofuranosidase, α-glucuronidase, α-galactosidase, ferulic and/or p-coumaric acid esterase (Ezeji et al., 2007b). The activities of these enzymes, in addition to the activities of cellulases on the cellulose component of the biomass, result in the generation of complex mixtures of acids (ferulic, p-coumaric, acetic, glucuronic) in addition to monomeric sugars such as glucose, galactose, xylose, and arabinose in the biomass hydrolysates. Therefore, for complete depolymerization of lignocellulosic biomass, it is difficult to totally avoid the generation of inhibitory compounds irrespective of the pretreatment and hydrolysis method utilized. Fermentability studies using hot water- or AFEX-pretreated DDGS hydrolysates suggest that strains of solventogenic clostridia can be "adapted" to produce butanol at levels that are consistent with solvent production on mixed sugar solutions not containing inhibitors.

While the focus has been on technologies that produce few inhibitors, it may not be possible to completely prevent the formation of inhibitors during biomass deconstruction. As an alternative, strategies are being developed to selectively remove inhibitors after they are formed. The down side to this approach is that it would add an additional processing step.

4.4 Genomics for Producing New Microbes with Enhanced Characteristics for Fermentation: Synthetic Biology and Production of Advanced Biofuels

The acetone−butanol−ethanol (ABE) fermentation employing the solventogenic clostridia was carried out industrially throughout the United States during the first half of the last century. However, this fermentation was discontinued in the early 1960s due to unfavorable economic conditions brought about by competition with the petrochemical industry (Ezeji et al., 2004a). *Clostridium acetobutylicum* and

C. beijerinckii are among the prominent solvent-producing strains. The production of butanol by fermentation processes is of interest for chemical/fuel production from renewable resources. Butanol can be directly used as a liquid fuel or may be blended with fossil fuels. This 4-carbon alcohol has higher energy content than ethanol, is more miscible with gasoline and diesel fuel, and has a lower vapor pressure. Butanol is currently used as a feedstock in the chemical industry.

The Blaschek laboratory has focused on seeking ways to make the ABE fermentation more economically competitive with petrochemical processes. Some of the problems associated with butanol production via fermentation of lignocellulose-based hydrolysates have been the low concentration of butanol in the fermentation broth due to the toxicity of inhibitors produced during pretreatment (Ezeji et al., 2007b). These factors negatively impact the economics of biomass-derived butanol relative to petrochemical-derived butanol. Recently, we developed a highly efficient laboratory-scale integrated ABE fermentation and recovery system employing gas stripping technology (Provisional U.S. Patent #60/504,280). Gases (CO_2 and H_2) produced during the fermentation are used for the stripping process and recycled. The ABE vapors are condensed and removed. Elimination of butanol inhibition resulted in an elevated cell concentration, a high substrate-utilization rate, together with the complete utilization of substrate and acids. The productivity of the integrated fed-batch process employing gas stripping was improved up to fourfold when compared to the batch process. The productivity of the traditional ABE batch bioreactors ranges from 0.1 to 0.3 g/Lh^{-1} and requires larger bioreactors, which produce larger effluent volumes. The integrated system offers the advantages of high productivity, use of concentrated feed streams, reduction in reactor size, efficient recovery of products and is relatively easy and inexpensive to operate (Ezeji et al., 2004b). The potential of inexpensive agricultural raw materials can be realized when this new technology is applied to the bioconversion processes for ABE production.

4.5 Genomics

Studies on the solvent-producing clostridia have been hindered largely because of difficulties with the genetic manipulation of these microorganisms. Because of the availability of genomic information for *C. acetobutylicum* ATCC 824, initial efforts have focused on this strain. However, the functional characterization of genes and the metabolic pathways associated with solvent formation in the clostridia remain to be understood (Nolling et al., 2001; Alsaker and Papoutsakis, 2005). Recently, genome sequencing of *C. beijerinckii* NCIMB 8052 was completed by the Joint Genome Institute of the Department of Energy. This opened up the exciting possibility of using *C. beijerinckii* as an alternative model organism to investigate the molecular mechanisms of solventogenesis.

The recent availability of genomic sequence information for *C. beijerinckii* NCIMB 8052 has allowed for an examination of gene expression using DNA microarray analysis of the wild-type strain and the *C. beijerinckii* BA101 hyper-butanol-producing mutant strain over the time course of a batch fermentation (Shi and Blaschek, 2008). The results revealed marked differences between the *C. beijerinckii* wild-type and BA101 with respect to transcription of a diverse set of genes representing multiple functional classes necessary for solventogenesis, sporulation, cell motility, and sugar transport. Taken together, the variations in expression patterns appear to collectively contribute to enhanced butanol formation in *C. beijerinckii* BA101. This study provides a road map for further genetic modification of the solventogenic clostridia in order to achieve higher efficiency for the production of biobutanol. The long-term goal is to improve butanol yield, titer, and productivity by developing genetically modified clostridia strains for economical biobutanol production from renewable resources.

An examination of gene expression in the recently sequenced *C. beijerinckii* NCIMB 8052 indicates that transcriptional activities are coordinated with various physiological changes over the course of the butanol-acetone fermentation. Temporal patterns of gene expression were clustered into two major classes that correspond to the metabolic states of acidogenesis and solventogenesis. *C. beijerinckii* 8052 and *C. beijerinckii* BA101 exhibited differential expression of genes representing various functional gene categories. Interestingly, in contrast to the transient mode of transcriptional activation observed in *C. beijerinckii* 8052, several key genes associated with butanol formation were constitutively expressed at high levels in *C. beijerinckii* BA101 during the solventogenic phase.

4.6 Systems Biology and Metabolic Engineering

With the completion of the genome sequence of *C. beijerinckii* and *C. acetobutylicum*, we can now employ the tools of systems biology to gain increased insight into the metabolic and regulatory networks relevant to solvent production. A systems understanding of these processes will provide the basis for rational design of this organism in future studies, both in terms of modification beneficial to decreasing the effects of the inhibitors produced during deconstruction and in terms of maximizing solvent production.

A systems approach of particular interest is constraint-based modeling of genome-scale metabolic networks that can be utilized as a basis for interpreting the outcome of experiments and for metabolic engineering. Modeling genome-scale metabolic networks of this type is crucial for interpreting data in the context of known metabolism, for identifying gaps in current knowledge, and for providing predictive capacity for the outcome of genetic modifications to the organism. The basis for the constraint-based approach is that microbial cells operate under governing constraints that limit their range of possible functions. With the availability of annotated genome sequences, it has become possible to reconstruct genome-scale biochemical reaction networks for microorganisms (Reed and Palsson, 2003; Duarte et al., 2004; Thiele et al., 2005; Feist et al., 2007). The imposition of governing constraints on a reconstructed biochemical network leads to the definition of achievable cellular functions. In recent years, a substantial and growing toolbox of computational analysis methods has been developed to study the characteristics and capabilities of microorganisms using a constraint-based reconstruction and analysis approach (Price et al., 2004a). This approach provides a biochemically and genetically consistent framework for the generation of hypotheses and the testing of functions of microbial cells.

When the necessary biochemical detail is known, as it now is for *C. beijerinckii* and *C. acetobutylicum*, we can reconstruct genome-scale reaction networks. These networks represent the underlying chemistry of the system, and thus at a minimum represent stoichiometric relationships between interconverted biomolecules. Biochemical reaction networks are thus directly based on chemistry, rather than only a reflection of statistical associations (Price and Shmulevich, 2007). These stoichiometric reconstructions have been most commonly applied to small-molecule interconversions in metabolic networks, but this formalism can easily incorporate biological transformations of all types, including for metabolic, signaling, protein−protein interactions, and gene regulatory networks (Palsson, 2004). Consistent with our goal of increasing metabolic flux of butanol, we will focus primarily on the metabolic network. For metabolism, the reconstruction of the biochemical reaction network is a well-established procedure (Reed and Palsson, 2003; Reed et al., 2003; Duarte et al., 2004; Francke et al., 2005; Heinemann et al., 2005; Thiele et al., 2005; Duarte et al., 2007). These biochemical reaction networks represent many years of accumulated experimental data and can be simulated in silico to determine their functional states. Genome-scale models based on biochemical networks provide a comprehensive, yet concise, description of cellular functions and have proven very valuable for the study to the research communities focused on many different organisms. We anticipate that this approach will be crucial in the development of osmo- and lignocellulosic inhibitor-tolerant solventogenic clostridium strains with the ability to effectively convert heterogeneous lignocellulosic hydrolysates to acetone-butanol-ethanol.

4.7 Conclusion

The utilization of plant-based lignocellulosic biomass as a substrate for biochemical conversion to biofuels has the potential for reducing our energy carbon footprint. Current chemical−physical technologies for cell wall deconstruction result in the formation of degradation products that inhibit yeast, as well as other fermentation microbes. This chapter specifically focused on the use of biomass-based hydrolysates for the production of biobutanol, a second-generation bioproduct produced by the solventogenic clostridia. While chemical−physical technologies for cell wall deconstruction have been developed which have demonstrated limited success, metabolic engineering and systems biology approaches offer a new toolbox for the development of microbial strains which are able to grow and ferment in the presence of inhibitors produced during the deconstruction process, thereby eliminating a major bottleneck in biomass-based fermentations. These new technologies are expected to play a major role in the successful commercialization of second-generation biofuels.

References

Alsaker KV, Papoutsakis ET (2005) Transcriptional program of early sporulation and stationary-phase events in *Clostridium acetobutylicum*. J Bacteriol 187:7103–7118.

Becker J, Boles E (2003) A modified *Saccharomyces cerevisiae* strain that consumes L-arabinose and produces ethanol. Appl Environ Microbiol 69 (7): 4144–4150.

Duarte NC, Palsson Bo, Fu P, et al. (2004) Integrated analysis of metabolic phenotypes in Saccharomyces cerevisiae. BMC Genomics 5(1): 63.

Duarte NC, Herrgard MJ, Palsson BO, et al. (2004) Reconstruction and validation of Saccharomyces cerevisiae iND750, a fully compartmentalized genome-scale metabolic model. Genome Res 14: 1298–1309.

Duarte NC, Becker SA, Jamshidi N et al. (2007) Global reconstruction of the human metabolic network based on genomic and bibliomic data. Proc Natl Acad Sci USA 104(6): 1777–1782.

Eggeman T, Elander RT (2005) Process and economic analysis of pretreatment technologies. Bioresource Technol 96(18): 2019–2025.

Ezeji TC, Qureshi N, Blaschek HP (2004a) Butanol fermentation research: Upstream and downstream manipulations. Chem Rec 4:305–314.

Ezeji TC, Qureshi N, Blaschek HP (2004b) Acetone-Butanol-Ethanol (ABE) Production from concentrated substrate: Reduction in substrate inhibition by fed-batch technique and product inhibition by gas stripping. Appl Microbiol Biotechnol 63:653–658.

Ezeji TC, Qureshi N, Blaschek HP (2005) Butanol Production from Agricultural Residues: Impact of Degradation Products on *Clostridium beijerinckii* Growth and Butanol Fermentation. The 2nd annual world congress on industrial biotechnology and bio-processing, held on April 20–22, 2005, in Orlando, Florida, Poster No. 36.

Ezeji TC, Qureshi N, Blaschek HP (2007a) Bioproduction of butanol from biomass: from genes to bioreactors. Curr Opin Biotechnol 18:220–227.

Ezeji TC, Qureshi N, Blaschek HP (2007b) Butanol production from agricultural residues: impact of degradation products on *Clostridium beijerinckii* growth and butanol formation. Biotechnol Bioeng 97:1460–1469.

Ezeji TC, Blaschek HP (2008) Fermentation of dried distillers grains and solubles (DDGS) hydrolysates to solvents and value-added products by solventogenic clostridia. Bioresour J 99:5232–5242.

Feist AM, Henry CS, Reed JL, et al. (2007) A genome-scale metabolic reconstruction for Escherichia coli K-12 MG1655 that accounts for 1260 ORFs and thermodynamic information. Mol Syst Biol 3:121.

Francke C, Siezen RJ, Teusink B et al. (2005) Reconstructing the metabolic network of a bacterium from its genome. Trends Microbiol 13(11): 550–558.

Hahn-Hagerdal B, Wahlbom CF, Gardonyi M, van Zyl WH, Cordero Otero RR, Jonsson LF (2001) Metabolic engineering of Saccharomyces cerevisiae for xylose utilization. Adv Biochem Eng Biotechnol 73:53–84.

Heinemann M, Kummel A, Ruinatscha R, Panke S (2005) In silico genome-scale reconstruction and validation of the Staphylococcus aureus metabolic network. Biotechnol Bioeng 92(7): 850–864.

Ho NWY, Chen Z, Brainard AP, Sedlak M (1999) Successful design and development of genetically engineered Saccharomyces yeasts for effective cofermentation of glucose and xylose from cellulosic biomass to fuel ethanol. Adv Biochem Eng 65:163–192.

Jin Y-S, Ni Haiying Ni, Laplaza JM, Jeffries TW (2003) Optimal growth and ethanol produc-
    tion from xylose by recombinant *Saccharomyces cerevisiae* require moderate d-xylulokinase
    activity. Appl Environ Microbiol 69:495–503.
Kim S, Holtzapple MT (2005) Lime pretreatment and enzymatic hydrolysis of corn stover.
    Bioresource Technol 96:1993–2006.
Kim TH, Lee YY (2005) Pretreatment and fractionation of corn stover by ammonia recycle
    percolation (ARP) process. Bioresource Technol 96:2007–2013.
Koukiekolo R, Cho H-Y, Kosugi A, Inui M, Yukawa H, Doi RH (2003) Degradation of corn fiber by
    Clostridium cellulovorans cellulases and himicellulases and contribution of scaffolding protein
    CpbA. Appl Environ Microbiol 71:3504–3511.
Ladisch M, Dale B, Tyner W, et al. (2008) Cellulose conversion in dry grind ethanol plants.
    Bioresour Technol 99:5157–5159.
Li X-L, Dien BS, Cotta MA, Wu YV, Saha BC (2005) Profile of enzyme production by
    Trichoderma reesei grown on corn fiber fractions. Appl Biochem Biotechnol 121:321–334.
Lloyd TA, Wyman CE (2005) Combined sugar yields for dilute sulfuric acid pretreatment of
    corn stover followed by enzymatick hydrolysis of the remaining solids, Bioresour Technol 96:
    1967–1977.
Mosier N, Wyman C, Dale BE, Elander R, Lee YY, Holtzapple M, Ladisch M (2005a) Features
    of promising technologies for pretreatment of lignocellulosic biomass. Bioresour Technol 96:
    673–686.
Mosier N, Hendrickson R, Ho N, Dedlak M, Ladisch MR (2005b) Optimization of pH controlled
    liquid hot water pretreatment of corn stover. Bioresour Technol 96:1986–1993.
Mosier NS, Hendrickson R, Brewer M, Ho N, Sedlak M, Dreshel R, Welch G, Dien BS, Aden A,
    Ladisch MR (2005c) Industrial scale-up of pH-controlled liquid hot water pretreatment of corn
    fiber for fuel ethanol production. Appl Biochem Biotechnol 125:77–98.
Nolling J, Breton G, Omelchenko MV, et al. (2001) Genome sequence and comparative analysis of
    the solvent-producing bacterium *Clostridium acetobutylicum*. J Bacteriol 183(16): 4823–4838.
Palsson B (2004) Two-dimensional annotation of genomes. Nat Biotechnol 22(10): 1218–1219.
Price ND, Reed JL, Palsson BO, et al. (2004a) Genome-scale models of microbial cells: evaluating
    the consequences of constraints. Nat Rev Microbiol 2(11): 886–897.
Price ND, Shmulevich I (2007) Biochemical and statistical network models for systems biology.
    Curr Opin Biotechnol 18(4): 365–370.
Reed JL, Palsson BO (2003) Thirteen years of building constraint-based in silico models of
    Escherichia coli. J Bacteriol 185(9): 2692–2699.
Reed JL, Vo TD, Schilling CH, Palsson BO (2003) An expanded genome-scale model of
    Escherichia coli K-12 (iJR904 GSM/GPR). Genome Biol 4(9): R54.
Saha BC (2003) Hemicellulose bioconversion. J Indust Microbiol Biotechnol 30: 279–291.
Saha BC, Bothast RJ (1999) Pretreatment and enzymatic saccharification of corn fiber. Appl
    Biochem Biotechnol 76:65–77.
Schell DJ, Riley CJ, Dowe N, et al. (2004) Bioethanol process development unit: Initial operating
    experiences and results with a corn fiber feedstock. Bioresour Technol 91:179–188.
Sedlack M, Ho NWY (2004) Production of ethanol from cellulosic biomass hydrolysates using
    genetically engineered saccharomyces yeast capable of cofermenting glucose and xylose. Appl
    Biochem Biotechnol 114:403–416.
Shi Z, Blaschek HP (2008) Transcriptional analysis of *Clostridium beijerinckii* NCIMB 8052 and
    the hyper-butanol producing BA101 mutant during the shift from acidogenesis to solventogen-
    esis. Appl Environ Microbiol 74:7709–7714.
Teymouri F, Laureano-Perez L, Alizadeh H, Dale BE (2005) Optimization of the ammonia fiber
    explosion (AFEX) treatment parameters for enzymatic hydrolysis of corn stover. Bioresour
    Technol 96:2014–2018.
Thiele I, Vo TD, Price ND, Palsson BO (2005) Expanded metabolic reconstruction of Helicobacter
    pylori (iIT341 GSM/GPR): an in silico genome-scale characterization of single- and double-
    deletion mutants. J Bacteriol 187(16): 5818–5830.

von Braun J (2007) *The world food situation: New driving forces and required actions*. Food Policy Report. Washington, DC: IFPRI.

Wang B, Feng H, Ezeji T, Blaschek HP (2006) An environmentally friendly pretreatment for enhancing sugar yield and enzymatic digestiblity of lignocellulosic biomass. Corn Utilization & Technology Conference, Dallas, Texas. USA.

Wyman CE, Dale BE, Elander RT, Holtzapple M, Ladisch MR, Lee YY (2005) Coordinated development of leading biomass pretreatment technologies. Bioresour Technol 96:1959–1966.

Part II
Interactions Between Biofuels, Agricultural Markets and Trade

Chapter 5
Price Transmission in the US Ethanol Market

Teresa Serra, David Zilberman, José M. Gil, and Barry K. Goodwin

Abstract We use nonlinear time series models to assess price relationships within the US ethanol industry. Daily ethanol, corn, and crude oil futures prices observed from mid-2005 to mid-2007 are used in the analysis. Our results suggest the existence of an equilibrium relationship between the three prices studied. Only ethanol prices are found to adjust to deviations from this relationship. The evolution of ethanol prices in relation to corn and crude oil prices may have important implications for the long-run competitiveness of the US ethanol industry.

5.1 Introduction

Worldwide biofuels market is dominated by ethanol. World ethanol output has grown very quickly in recent years, doubling its amount during the first half of the 2000s and reaching 13,000 million gallons in 2007. The two main ethanol-producing countries in 2007 were the United States, with almost 50% of total output, and Brazil, with a share of 38% (Renewable Fuels Association, 2009). While there is a wide variety of feedstocks from which ethanol can be produced, sugarcane in Brazil and corn in the United States are the most relevant ones.

While sugar and starch-based crops (and their associated conversion technologies) are the most mature feedstocks today, their low yields per hectare, intensive use of agricultural inputs, and the fact that these crops are also used for food have increased concerns about their suitability for large-scale ethanol production. Though not yet technically and commercially mature, cellulose-based fuels are considered more promising for this latter purpose and are expected to displace sugar and corn as the major ethanol feedstock in the future.

T. Serra (✉)
Centre de Recerca en Economia i Desenvolupament Agroalimentari (CREDA-UPC-IRTA), Parc Mediterrani de la Tecnologia, Edifici ESAB, C/ Esteve Terrades 8, 08860, Castelldefels, Spain
e-mail: teresa.serra-devesa@upc.edu

M. Khanna et al. (eds.), *Handbook of Bioenergy Economics and Policy*,
Natural Resource Management and Policy 33, DOI 10.1007/978-1-4419-0369-3_5,
© Springer Science+Business Media, LLC 2010

Recent increases in ethanol production have been coupled with increases in demand as a response to different market and regulatory changes. Late spikes registered by crude oil prices and the volatility of these prices have reduced crude oil competitiveness and increased the demand for alternative energy sources. Policy measures that aim to reduce greenhouse gas emissions and increase the quantity of domestically produced fuel have also contributed to heightened interest in ethanol.

The momentum gained by the ethanol industry has motivated a series of economic studies on biofuels (Rajagopal and Zilberman, 2007). The objective of this research is to assess price linkages and price transmission patterns in the US ethanol industry during the second half of the 2000s. Specifically, by using time series modeling techniques, we assess the relationships between ethanol, corn, and crude oil prices. These relationships have not been addressed by previous research, which represents the major contribution of our work. The extent to which price shocks are transmitted along the marketing chain may have significant implications for pricing practices and permit a better understanding of the overall operation of the market.

The analysis is also relevant given the important structural changes affecting the US ethanol and related markets during the period studied, which constitute an important change from previous market conditions, and that may have altered price relationships. A formal analysis of the issue is thus warranted.

5.2 The US Bioenergy Market

The US ethanol industry is responsible for producing half of worldwide ethanol. This industry uses maize as the primary feedstock and has an important role in the reformulated gasoline (RFG) market, where ethanol is used as an oxygenate additive to improve fuel combustion and bring pollution down. The use of ethanol as a conventional gasoline extender is currently one of the most attractive expansion opportunities for the US ethanol industry (Energy Information Administration, 2007).

The development of ethanol production in the United States is closely related to federal government programs. The recent history of these programs starts with the 1978 Energy Tax Act that was passed by the US Congress as a reaction to the petroleum scarcity caused by the 1973 oil crisis. The Act provided excise tax exemptions on gasoline blended with 10% ethanol. Tax exemptions were later extended to other blend ratios and to biodiesel. Since its introduction, the ethanol subsidy has ranged between 40 and 60 cents per gallon of ethanol and has been complemented with other federal and state subsidies and import tariffs. The total subsidy for ethanol in 2006 has been estimated to range between USD 1.42 and USD 1.87 per gallon of gasoline equivalent (Koplow, 2006).

As noted above, both policy and market changes have recently encouraged an increase in the demand for ethanol. The 1990 Clean Air Act established a 2% oxygen content requirement for RFG specifications. Despite tax exemptions and the oxygen constraint, only a small percentage of the RFG contained ethanol before the second half of the 2000s. Refiners instead were mainly using methyl-tertiary-butyl ether (MTBE). As a response to rising concerns about groundwater pollution with MTBE and the potential risks for both public health and the environment, several states passed legislation to ban or restrict its use.

The demise of MTBE was further accelerated by the revocation of the 2% oxygen requirement by the 2005 Energy Policy Act and the elimination of the liability protection that gasoline marketers had petitioned. Refiners continuing to use MTBE after the oxygen requirement had been eliminated may face liability claims. This forced the gasoline industry to quickly eliminate MTBE from gasoline in spring 2006. MTBE was mainly replaced by ethanol, leading to an unprecedented increase in its demand.

An additional incentive to ethanol demand and production in the United States came from the Renewable Fuel Standard (RFS) signed into law in 2005 (US Energy Policy Act of 2005) and its amendment by the Energy Independence Security Act of 2007. The RFS program is expected to substantially contribute to the ethanol industry expansion (United States Department of Agriculture, 2008).

Several analyses have evaluated ethanol production costs (see, for example, Shapouri and Gallagher, 2005; or OECD, 2006). Though results vary across studies, countries, and feedstocks, a common finding is the relevance of feedstock that represents the biggest share of total cost and the major determinant of the ethanol industry's economic sustainability and expansion capacity. Also, available data suggest that Brazil has the lowest ethanol production costs, allowing its production to become competitive at world crude oil prices at and above USD 39 per barrel. Brazil is followed by the United States, whose production costs make ethanol profitable when crude oil prices exceed USD 44 (OECD, 2006).

Since the introduction of the US federal ethanol subsidy and until recently, crude oil prices had ranged between USD 20 and 30 per barrel. Historically high prices have been reached in the second half of the 2000s: prices were around USD 60–70 per barrel from mid-2005 to mid-2007 and went well above these levels afterward, leading to an increase in ethanol competitiveness relative to gasoline and strengthening ethanol demand. Ethanol investments have been very profitable at these prices, which has attracted new capital into the industry. The market grew from around 2.8 billion gallons in 2003 to 9 billion gallons in 2008 (Renewable Fuels Association, 2009).

There are several requirements that need to be met for ethanol to realistically compete with crude oil: supplies must be cost-competitive and readily accessible and transition costs from crude oil to ethanol must be bearable. The period of record growth and expansion recently undergone by the ethanol industry has evidenced a number of shortcomings that challenge ethanol's immediate and widespread use. These weaknesses include a deficient infrastructure for transporting and blending ethanol with gasoline. Other downsides are bottlenecks in the distribution infrastructure, such as a lack of ethanol fueling stations. Further, the small relevance of flex-fuel vehicles (that can run on gasoline–ethanol blends up to 85% ethanol) within the automotive fleet considerably limits the expansion capacity of ethanol demand and constitutes another obstacle for the industry (Schmidhuber, 2006). In the next section, we elaborate a framework to explain price relationships within the US ethanol industry.

5.3 Price Relationships in the US Ethanol Industry

There has been a growing body of literature on the economic impact of biofuel, both on the fuel and food markets. Rajagopal et al. (2007) suggest that the introduction of biofuels increases the demand for corn and the supply of fuels. Their analysis aimed to assess the welfare implications of introducing biofuels, and they found that the outcomes were dependent on demand and supply elasticities in the relevant crop and fuel markets. They found that the introduction of corn ethanol in 2006 reduced the price of fuel by 3% and increased the price of corn by about 21%, benefiting farmers and drivers and hurting food consumers. Tyner and Taheripour (2007) and de Gorter and Just (2009) recognized that the interaction between ethanol, crop, and fuel prices should incorporate subsidies, tariffs, and mandates, as well the use of corn residues as animal feed. They incorporated these factors in analyzing the welfare effect of introducing biofuels and identifying the conditions under which farmers will allocate their corn to fuel or food production. Biofuels were not treated in most of the literature as a distinct product with a market and price of its own but, rather, as a quantity that affects fuel and food prices. However, ethanol and other biofuels are not perfect substitutes to gasoline and in reality have their own markets, which are affected by the prices of corn and fuel as well as regulations, such as biofuel mandates, and constraints on the capacity to utilize biofuels.

Not much attention has been given to the behavior of the price of biofuel and its interaction with the prices of fossil fuels and food. Figure 5.1 analyzes these relationships. Let B denote the quantity of biofuel and P_B the price of biofuel. The unconstrained demand for biofuel is $D(P_B, P_G)$, where P_G is the price of fossil fuel (gasoline). The unconstrained demand is negatively sloped and increases with the price of fossil fuel. Let P_G^0 be the initial price of fossil fuel and let P_G^H be a higher price of fossil fuel, $P_G^0 < P_G^H$; $D(P_B, P_G^H)$ is above $D(P_B, P_G^0)$, denoting that demand increases with the price of fossil fuel. Let $S(P_B, P_C)$ be the unconstrained supply of biofuel, which is a declining function of the price of corn P_C. P_C^0 denotes the initial price of corn, and P_C^H is a higher price. $S(P_B, P_C^0)$ is below $S(P_B, P_C^H)$ as supply contracts with the higher corn price. Now consider two constraints, a mandate and a use capacity. Suppose there is a biofuel mandate, denoted by B_M, requiring that the quantity of biofuel produced is $B \geq B_M$. Similarly, with a use capacity constraint, B_U, the total biofuel supplied will be $B \leq B_U$ because the capacity to process and use biofuel is limited. These constraints are presented in Fig. 5.1.

The figure considers three types of equilibria. The first is when both constraints are not binding, as is the case in point A in Fig. 5.1. In these situations, the price of the biofuel is determined by the interaction of demand and supply for biofuel. Small changes in the price of fossil fuel and food that will not cause one of the constraints to bind will shift either demand or supply and affect the price of biofuel. The figure suggests that an increase in the price of fossil fuel, increasing demand, or an increase in the price of corn that will reduce supply, will increase the price of biofuel under this scenario.

Another equilibrium occurs when the mandate is binding (point C in Fig. 5.1). When this happens, the price is determined by the intersection of the supply and the mandate level (P_B^C in Fig. 5.1). The price of biofuel in this case is affected by the price of food (corn) that affects supply. When this equilibrium occurs, a higher price of corn will increase the price of ethanol. The third equilibrium is when the capacity constraint is binding (point B in Fig. 5.1) and the price of ethanol is determined in the intersection of the demand for ethanol and the constraint (P_B^B). In this case, the price of fossil fuel (gasoline) and not the price of corn affects the price of ethanol, and an increase in the price of gasoline leads to higher ethanol prices.

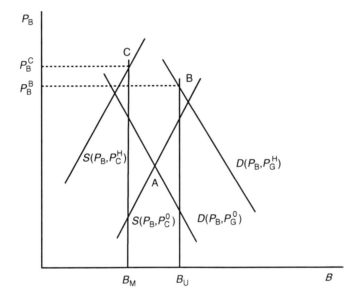

Fig. 5.1 Determination of the price of ethanol

Our analysis thus suggests that given the biofuel mandate and capacity constraints, there is a range of low fossil fuel prices and high corn prices, where the mandate constraint holds and changes in the price of ethanol are mainly affected by the changes in the price of food. Similarly, there is a range with low prices of food and high prices of fuel where the capacity constraint holds and the price of ethanol is mainly affected by the price of fossil fuel. In a wide range of prices between the two above-mentioned regions, the constraints do not hold and the price of ethanol is changing as a response to changes in both food and fossil fuel prices. Our analysis suggests that an increase in the mandates will increase the range where the price of ethanol depends only (or to a large extent) on the price of corn, while an increase in the capacity to process ethanol will reduce the capacity constraint and reduce the range of prices where the price of ethanol mostly depends on the price of fossil fuel.

This chapter assesses the relationships between ethanol, corn, and crude oil prices. In this regard, we assume that gasoline is the generic fuel. Our analysis and the description of the US ethanol industry presented above suggest several hypotheses on price relationships between ethanol, oil, and corn prices. First, we hypothesize that feedstock, ethanol, and crude oil prices move together in the long run.

Second, we anticipate that movements of corn and ethanol prices are likely to be positively correlated. Third, we hypothesize the existence of a positive relationship between ethanol and crude oil prices. Fourth, regulations limiting the use of MTBE and setting de facto binding mandates on the use of ethanol during the period studied lead to a stronger influence of corn prices on ethanol prices (point C in Fig. 5.1). Fifth, we hypothesize that the adjustment toward the equilibrium may be of a nonlinear nature. Mandates, as well as transaction costs, adjustment costs, or risk may create a band of price differentials within which arbitrage activities are inactive.

The economic literature has acknowledged several causes for nonlinearities in price transmission. Apart from public regulation (such as mandates, etc.), two main causes dominate the literature: adjustment costs and market power, which may both be relevant to explain price links within the US ethanol industry.

It is well known that adjustment costs are generated as a result of firms changing the quantities and/or prices of inputs and/or outputs. Azzam (1999) shows that adjustment costs can have implications for price transmission and may result in a less-than-full pass-through of prices. The aforementioned rigidities affecting the ethanol industry are likely to increase adjustment costs and cause nonlinearities. Market power has been pointed as another cause for nonlinear price behavior (McCorriston et al., 1998, 2001; Lloyd et al., 2006). The market power particularly exercised by the refining industry may further contribute to nonlinear price transmission in the US corn–ethanol–crude oil nexus.

While price transmission within the US ethanol marketing chain has not been formally analyzed, the possibility of nonlinear price behavior in the Brazilian ethanol industry has been studied by Rapsomanikis and Hallam (2006) and Balcombe and Rapsomanikis (2008). Based on the use of a bivariate threshold vector error correction model (TVECM), Rapsomanikis and Hallam (2006) show that while sugar–crude oil and ethanol–crude oil price linkages display nonlinear threshold-type behavior, sugar and ethanol prices are linearly cointegrated. Further, crude oil prices are found to Granger-cause (Granger, 1969) sugar and ethanol prices, while sugar prices Granger-cause ethanol prices and are exogenous to ethanol. Balcombe and Rapsomanikis (2008) analyze price transmission within the Brazilian ethanol industry by developing generalized bivariate error correction models and obtain results in accord with those of Rapsomanikis and Hallam (2006).

5.4 Methodology

Recent developments in time series econometrics allow for changing behavior of economic variables as economic conditions change and have been increasingly used to explain different phenomena. The economics literature has focused a great deal of attention on threshold-type models and especially on TVECMs (Goodwin and Piggott, 2001; Chavas and Metha, 2004; Rapsomanikis and Hallam, 2006; Balcombe et al., 2007).

A TVECM is built upon the assumption that the shift from one regime to another is sudden and discontinuous (Chan and Tong, 1986). This assumption may be too restrictive if the objective of the analysis is to capture aggregate economic behavior, and commodity points represent transaction costs. The more general Smooth Transition Autoregressive (STAR) types of models allow for a smooth transition between regimes by replacing the discrete regime-changing function by a smooth function. STAR type of models nest discontinuous regime-switching models as a special case of very fast adjustment.

Our empirical analysis is based on a multivariate smooth transition vector error correction model (STVECM). This multivariate model allows for both long-run relationships among the variables of the model and for nonlinear adjustment paths toward long-run equilibrium. An STVECM has not been previously used to assess price relationships within biofuel markets, which represents a contribution of our analysis.

Assume that there exists a single long-run (cointegration) relationship between ethanol, corn, and crude oil prices that, by using the ethanol price as the normalization variable, can be represented as $\beta + \beta_C P_C + \beta_G P_G + P_B = 0$ (P_C is the price of corn, P_G the price of crude oil, and P_B the price of ethanol). Our STVECM can be regarded as a regime-switching model that allows for two regimes, one that corresponds to the industry being close to this long-run equilibrium, and another that is characterized by large deviations from the equilibrium. The transition from one regime to the other is allowed to occur smoothly. A three-dimensional STVECM can be expressed as (van Dijk et al., 2002):

$$\Delta \mathbf{P}_t = \left(\boldsymbol{\phi}_{1,0} + \boldsymbol{\alpha}_1 z_{t-1} + \sum_{j=1}^{p-1} \boldsymbol{\phi}_{1,j} \Delta \mathbf{P}_{t-j} \right) (1 - G(z_{t-1};\gamma,c)) +$$
$$\left(\boldsymbol{\phi}_{2,0} + \boldsymbol{\alpha}_2 z_{t-1} + \sum_{j=1}^{p-1} \boldsymbol{\phi}_{2,j} \Delta \mathbf{P}_{t-j} \right) (G(z_{t-1};\gamma,c)) + \boldsymbol{\varepsilon}_t \tag{1}$$

where $\mathbf{P}_t = (P_{Bt}, P_{Ct}, P_{Gt})$ is the (3×1) vector of nonstationary prices, subindex t indicates time, $\boldsymbol{\alpha}_i = (\alpha_{iB}, \alpha_{iC}, \alpha_{iG})$ $i = 1,2$ are (3×1) matrices that represent the speed of adjustment (under each regime i) of each price to disequilibrium from the long-run relationship, $z_{t-1} = \boldsymbol{\beta}' \mathbf{P}_{t-1}$ contains deviations from the long-run equilibrium (error-correction term), and $\boldsymbol{\beta}$ (3×1) contains the parameters of the long-run (cointegration) relationship. The short-run price dynamics are represented by $\boldsymbol{\phi}_{i,0}$ (3×1) and $\boldsymbol{\phi}_{i,j} j = 1,...,p-1$ (3×3), and $\boldsymbol{\varepsilon}_t$ is a k-dimensional vector white noise process.

$G(z_{t-1};\gamma,c)$ is a continuous function that takes values between zero and one (if $i = 1$ then $G(.) = 0$, while if $i = 2$ then $G(.) = 1$). It depends on the transition variable z_{t-1} and the associated parameters γ and c that reflect the speed of transition from one regime to another and the threshold between regimes. Our analysis assumes that the transition variable represents the lagged deviations from the long-run equilibrium relationship between prices.

Based on the hypotheses of price relationships formulated in the previous section, we can now make predictions about some of the parameters of the model. We anticipate ethanol prices to be positively related with corn and crude oil prices, involving negative parameter estimates for β_C and β_G. The existence of transaction costs, fixed costs of adjustment, or other rigidities may cause mean-reverting mechanisms to be active only for large deviations from the long-run relationship. As a result, we do not expect vector $\alpha_1 = (\alpha_{1B}, \alpha_{1C}, \alpha_{1G})$ to be statistically significant. Conversely, vector $\alpha_2 = (\alpha_{2B}, \alpha_{2C}, \alpha_G)$ should contain statistically significant parameters. Specifically, and following previous research (Balcombe and Rapsomanikis, 2008), we expect crude oil prices to be exogenous, which entails that α_{2G} cannot be statistically significant. Conversely, ethanol is expected to be endogenous, which requires α_{2B} to be statistically significant and negative.

It is not easy to anticipate whether corn prices will be exogenous or not. During the period of analysis, ethanol-derived demand for maize represented still a relatively small portion of total maize demand: 14% (19%) in the 2005–2006 (2006–2007) marketing year. On the contrary, 50% of corn was used for livestock feed (Westcott, 2007; United States Department of Agriculture, 2008). It is important to note that we focus on modeling the US ethanol market and do not allow for other variables that can explain corn prices such as feed demand. Consequently, we may find that α_{2C} is not statistically significant.

As noted, parameter γ determines the speed of transition from one price behavior regime to another. As the value of γ increases, the transition speed becomes quicker. The above-mentioned rigidities characterizing the US ethanol industry are likely to cause a slow shift from one regime to another. The threshold parameter c determines the value of the transition variable that activates arbitrage activities. Since the transition variable is assumed to be a lag of the error correction term, c is expected to be near zero.

Our model estimation strategy consists of three main steps. First, the existence of a long-run relationship between prices is tested using the Johansen (1988) cointegration test. Second, the STVECM is specified and estimated following Teräsvirta (1994), Saikkonen and Luukkonen (1988), and Eitrheim and Teräsvirta (1996). Finally, once the model is estimated, we examine dynamic price relationships by computing Generalized Impulse Response Functions (GI) following Koop et al. (1996) and Weise (1999).

5.5 Results

Our empirical analysis uses daily futures prices for corn, ethanol, and crude oil observed from July 21, 2005 to May 15, 2007.[1] Prior to the second half of the 2000s,

---

[1] Corn, ethanol, and oil futures prices are quoted in cents per bushel, dollars per gallon, and dollars per barrel, respectively.

ethanol played only a minor role in both the US energy and agricultural markets. Only after recent market and regulatory changes has the relevance of ethanol as well as the interest in economic studies on biofuels increased.

Information on corn and ethanol futures was obtained from the Chicago Board of Trade (CBOT) and crude oil (light-sweet, Oklahoma) futures prices were obtained from the New York Mercantile Exchange (NYMEX). While contract months for crude oil and ethanol futures are the 12 consecutive calendar months (from January to December), contract months for corn are limited to December, March, May, July, and September. To make the three series comparable, we use corn contract months as the reference to define maturity months and nearby contracts when constructing our futures price series.[2]

Following previous research (Jin and Frechette, 2004; Booth and Tse, 1995), when a futures contract comes to its maturity month, the nearby contract is used to compile the data. A jump is made to the new nearby contract when the old contract is close to expiration. Specifically, on the 15th day of the month previous to the expiration of the contract (day t), the price is computed as the average settle price for the old contract for days t and $t-1$. On day $t + 1$, the price corresponds to the average settle price for the new contract for days t and $t + 1$.

The price series used in the empirical implementation are presented in Fig. 5.2. To assess the time series properties of the prices, we tested for nonstationarity by using standard Augmented Dickey-Fuller (ADF) (Dickey and Fuller, 1979), Perron (1997), and Kwiatkowski et al. (1992) statistics. Test results confirmed the presence of unit roots in all price series.

Johansen (1988) tests for cointegration suggest the existence of a single cointegration relationship between the three prices. Since several important market and

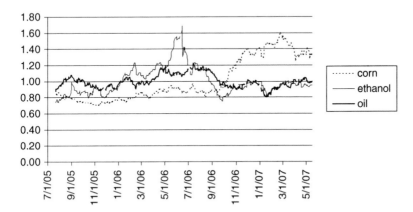

Fig. 5.2 Price series
Note: The series have been divided by their respective means.

policy changes characterize the period of analysis, we test for the constancy of the cointegration relationship by following the method proposed by Hansen and Johansen (1999). The method consists of applying the fluctuation test of Ploberger et al. (1989) to test for constancy of the estimated eigenvalues and the Lagrange Multiplier (LM) test to test for constancy of the cointegration parameters (Hansen and Johansen, 1999).

Results suggest the existence of a structural break at the end of April 2006, when the refining industry shifted from MTBE to ethanol as a gasoline oxygenate, which caused ethanol demand to expand nationally and the ethanol price to reach unprecedented high levels.[3] Based on these results, the cointegration vector is estimated with a break in the intercept at April 27, 2006. The resulting cointegration relationship as well as cointegration tests are presented in Table 5.1. After introducing the break, the Hansen and Johansen (1999) method suggests constancy of the cointegration relationship.

Table 5.1 Johansen λ_{trace} test for cointegration and cointegration relationship

H_o	H_a	λ_{trace}	P-value
$r = 0$	$r < 0$	50.697	0.008
$r \leq 1$	$r > 1$	21.352	0.184
$r \leq 2$	$r > 2$	8.401	0.250
Cointegration relationship			
$P_B - 0.012P_C - 0.036P_G + 2.579 + 1.946D_{\text{break}} = 0$			

As expected, the cointegration relationship shows that the price of ethanol is positively related to the prices of corn and crude oil. The positive relationship between ethanol and crude oil confirms that ethanol is a crude oil substitute in the fuels market. A simple exercise was carried out to determine the long-run sensitivity of ethanol prices to changes in corn and crude oil prices. First, while holding the rest of the variables constant, a 10% increase in the price of corn was considered. The corresponding increase in the price of ethanol was computed using the cointegrating vector at each data point. Results show an average increase in the price of ethanol of 15%. The same exercise was then repeated for a 10% increase in crude oil prices, which was found to cause an average increase in ethanol prices of the same magnitude (10%). Results are compatible with the theoretical model presented above that anticipates corn price changes to have a stronger influence on ethanol prices than oil price changes when mandates are binding.

These results have important economic implications. Since US ethanol prices are expected to increase along the same path as crude oil prices, ethanol competitiveness within the US fuel market is not expected to be compromised by crude oil

---

of 0.987 (CME Group, 2007). Hence, the findings of our paper are not expected to differ from the ones that would be obtained if using cash prices.

[3]Results are compatible with unit-root Perron test results for the ethanol series pointing towards a break in the price series by the end of April 2006.

price increases, all other things remaining constant. It should be noted, however, that findings from Balcombe and Rapsomanikis (2008) suggest that Brazilian ethanol prices will grow at about 60% of the growth rate of crude oil prices. Hence, in a scenario of increasing crude oil prices, US ethanol will become less and less competitive in international markets. This is consistent with previous research (Szklo et al., 2007).

Another issue that could compromise US ethanol competitiveness is the quick path at which US ethanol prices grow in response to corn price increases. Because the area dedicated to corn production cannot be easily expanded, at least in the short run, corn is a quasi-fixed input for which price is expected to rise with an increase in (ethanol-derived) corn demand. Hence, future corn price increases represent a major limitation for the competitiveness and expansion of the US ethanol industry.

Initial tests for STVECM specification support the use of an exponential function as the transition function, which signals a symmetric adjustment around the threshold parameter. The most relevant parameters in STVECM are the speeds of adjustment to disequilibrium from the long-run parity (vectors α_i, $i = 1,2$), the speed of transition from one price regime to another (γ), and the threshold parameter (c). We present these estimates in Table 5.2 together with the Eitrheim and Teräsvirta (1996) test for autocorrelation that suggests nonautocorrelated residuals.[4]

The speeds of adjustment parameters (α_i) imply that ethanol is the only endogenous variable in the system and that corn and crude oil prices are exogenous. As expected, the adjustment only takes place when deviations from the long-run relationship are big, i.e., whereas α_{2B} is negative and statistically significant, α_{1B} is not statistically different from zero.

In accord with the initial hypotheses, the threshold parameter estimate (c) is close to zero, which indicates that both positive and negative deviations from the equilibrium generate smooth mean-correcting mechanisms. The speed of transition (γ) is statistically significant and equal to 3.15. Given that this parameter can take any positive value (from zero to infinity), an estimate of 3.15 can be considered as a relatively slow transition speed (see Fig. 5.3).

The evolution of the transition function over time, together with the transition variable and the estimated threshold, is presented in Fig. 5.4. As can be observed, regimes associated with relatively high values of the transition function are prevalent. The changes that have affected the US ethanol market during the period of analysis have shocked the system, thereby increasing the error correction term. The highest G values are registered during the spring-summer of 2006 coinciding with the replacement of MTBE by ethanol, and during the relevant corn price increases registered by the end of our sample. These periods coincide with high positive and negative deviations from the long-run parity relationships.

---

[4]To preserve space, parameters showing the short-run dynamics of the series are not presented.

Table 5.2 ESTVECM parameter estimates

Speed of adjustment parameters (α)

Equation	Parameter	Regime $G = 1$ ($i = 2$)		Regime $G = 0$ ($i = 1$)		Autocorrelation test	p-value
		Parameter estimate	Standard error	Parameter estimate	Standard error		
Ethanol	α_{iB}	−0.0081**	0.0038	−0.0004	0.0375	1.6200	0.1547
Corn	α_{iC}	0.4144	0.3364	0.3140	4.0630	1.0800	0.3708
Crude oil	α_{iG}	−0.0259	0.0729	−1.1771	0.7650	0.7400	0.5961

Speed of transition and threshold variable

		Parameter estimate		Standard error
Transition function	γ	3.1520**		0.7210
	c	0.5593**		0.0497

** (*) denotes statistical significance at the 5 (10) percent significance level.
LM(10) is the F variant of the LM test for no remaining autocorrelation described above.

Our STVECM suggests that the only endogenous price in the system is ethanol. As a result, when assessing the dynamic properties of the model by means of GI Functions, we focus on studying how ethanol price responds to positive shocks to crude oil and corn prices.[5] The forecast horizon is set to $H = 30$, i.e., a period of approximately 1.5 months. We select two starting points, one where the transition function is close to 1 ($t = 220$) and another

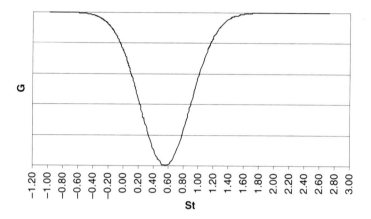

Fig. 5.3 Estimated exponential transition function

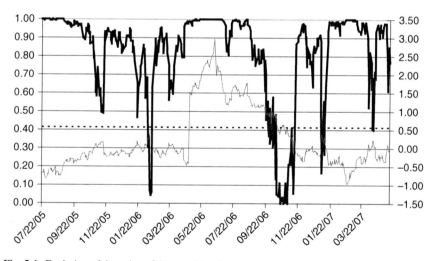

Fig. 5.4 Evolution of the value of the transition function over time
Notes: The transition function (G) is represented by the solid thick line and is plotted on the left-hand side axis. The transition variable (z_{t-1}), represented by the solid thin line, and the threshold value (*dotted line*) are plotted on the right-hand side of the axis.

---

[5]Since the exponential function implies symmetric adjustments around the threshold parameter, responses to both negative and positive shocks should be identical.

where the transition function is close to 0 ($t = 316$). Ethanol prices should only adjust to disequilibrium when the system is in the regime corresponding to large deviations from equilibrium (see Table 5.2). Results are presented in Figs. 5.5, 5.6, 5.7, 5.8.

If the corn price is shocked when the system is distant from the equilibrium (G close to 1), the ethanol price responds moving in the same direction (Fig. 5.5). The response magnitude increases with time and reaches a peak at about 25 days (1.25 months) after the shock. Afterward the response ceases to be statistically significant indicating that the shock dies out. If corn prices are shocked when the system is

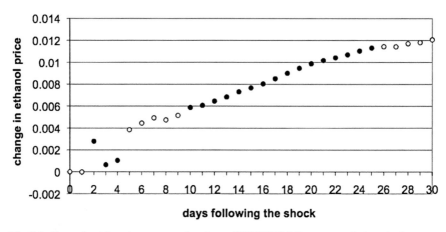

Fig. 5.5 Generalized impulse response functions of ESTVECM: Responses of ethanol prices to a positive shock to corn prices. Regime corresponding to G close to 1
Note: ● (○) indicates the response is (is not) statistically significant at the 5% level.

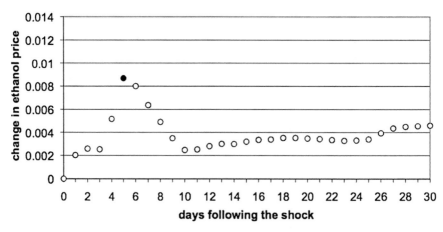

Fig. 5.6 Generalized impulse response functions of ESTVECM: Responses of ethanol prices to a positive shock to corn prices. Regime corresponding to G close to 0
Note: ● (○) indicates the response is (is not) statistically significant at the 5% level.

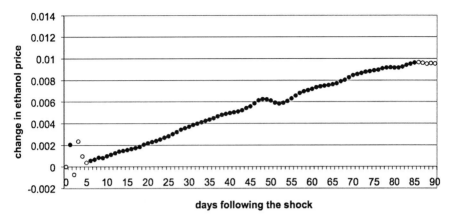

Fig. 5.7 Generalized impulse response functions of ESTVECM: Responses of ethanol prices to a positive shock to crude oil prices. Regime corresponding to *G* close to 1
Note: • (o) indicates the response is (is not) statistically significant at the 5% level.

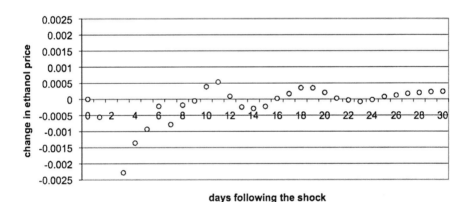

Fig. 5.8 Generalized impulse response functions of ESTVECM: Responses of ethanol prices to a positive shock to crude oil prices. Regime corresponding to *G* close to 0
Note: • (o) indicates the response is (is not) statistically significant at the 5% level.

close to the equilibrium (*G* close to 0) and since mean-reverting mechanisms are only active for relatively large deviations (Table 5.2), we find very small responses from ethanol prices (Fig. 5.6).

Figure 5.7 shows that an increase in crude oil price when *G* is close to 1 originates a response in the ethanol price in the same direction. Responses to crude oil price changes are considerably slow relative to responses to corn price shocks. Since at day 30 the responses still persisted, we expanded the forecast horizon to 100 days. We found that the responses reached a peak after approximately 85 days (4.25 months) and became statistically nonsignificant thereafter. Note, however, that although responses are slower to occur, they are not larger than responses to corn, which is consistent with the US ethanol market being characterized by binding mandates during the period of analysis, as a result of limitations on the use of MTBE. If crude oil prices are shocked when the system is close to equilibrium (Fig. 5.8), arbitrage activities are not activated and ethanol prices do not experience statistically significant price changes.

5.6 Concluding Remarks

This research studies price transmission patterns in the US ethanol industry during the second half of the 2000s by making use of nonlinear time series modeling. These relationships have not been addressed by previous research, which represents the major contribution of our work. Under the hypothesis that corn, ethanol, and crude oil prices have a long-run relationship and that adjustment toward this equilibrium is likely to be of a nonlinear nature, we estimate a STVECM. We use daily futures prices observed from mid-2005 to mid-2007.

Results suggest that the three prices have a long-run (cointegration) relationship. This cointegration relationship experiences a structural break by the end of April 2006 and shows ethanol prices to be positively related to corn and crude oil prices. Cointegration results also suggest that, in a scenario of crude oil price increases, US ethanol will see its competitiveness progressively reduced in international markets. Further, results indicate that future corn price increases represent a major limitation for the competitiveness and expansion of the corn-based US ethanol industry.

We find ethanol prices to be the only endogenous variable in the price system, which involves that only ethanol adjusts to correct disequilibria from the long-run parity. Mean-reverting mechanisms are only activated when disequilibria are relatively large and the adjustment path is relatively slow.

Generalized impulse response functions (GI) suggest that a shock to both crude oil and corn prices when the system is far from the equilibrium causes changes in ethanol prices in the same direction. Ethanol responses to crude oil price shocks are considerably slower and smaller than reactions to corn price changes.

To conclude, we would like to emphasize that the US ethanol industry continues to undergo an expansion process that has not yet ended. Further research will thus be needed to assess to what extent price patterns found in our analysis are maintained over time.

Acknowledgments The authors gratefully acknowledge financial support from Instituto Nacional de Investigaciones Agrícolas (INIA) and the European Regional Development Fund (ERDF), Plan Nacional de Investigación Científica, Desarrollo e Innovación Tecnológica (I+D+i), Project Reference Number RTA2009-00013-00-00.

References

Azzam A M (1999) Asymmetry in rigidity in farm-retail price transmission. Am J Agri Econ, 813): 525–533.
Balcombe K, Bailey A, Brooks J (2007) Threshold effects in price transmission: the case of Brazilian wheat, maize and soya prices. Am J Agri Econ 89(2): 308–323.

Balcombe K, Rapsomanikis G (2008) Bayesian estimation of nonlinear vector error correction models: the case of sugar-ethanol-oil nexus in Brazil. Am J Agri Econ 90(3): 658–668.

Booth G G, Tse Y (1995) Long memory in interest rate futures markets: A fractional cointegration analysis. J Futures Markets 15(5): 573–584.

Chan K S, Tong H (1986) On estimating thresholds in autoregressive models. J Time Ser Anal 7(3): 179–190.

Chavas J P, Metha A (2004) Price dynamics in a vertical sector: The case of butter.Am J Agri Econ 86(4): 1078–1093.

CME Group (2007) Ethanol Derivatives, Key Charts and Data. Available at http://www.cbot.com/cbot/docs/87237.pdf

de Gorter H, Just D (2009) The welfare economics of a biofuel tax credit and the interaction effects with price contingent farm subsidies. Am J Agri Econ 91(2): 477–488

Dickey D A, Fuller W A (1979) Distribution of the estimators for autoregressive time series with a unit root. J Am Stat Assoc 74(336): 427–431.

Eitrheim Ø, Teräsvirta T (1996) Testing the adequacy of smooth transition autoregressive models. J Econometrics 74(1): 59–75.

Energy Information Administration (EIA) (2007) US Energy Information Administration Before the Committee on Agriculture. Testimony of Dr. Howard Gruenspecth Deputy Administrator. Available at http://www.eia.doe.gov/neic/speeches/howard101807.pdf

Goodwin B K, Piggott N E (2001) Spatial marketing integration in the presence of threshold effects. Am J Agri Econ 83(1): 302–317.

Granger C W J (1969) Investigating causal relations by econometric methods and cross-spectral methods. Econometrica 37(3): 424–438.

Hansen H, Johansen S (1999) Some tests for parameter constancy in cointegrated VAR-models. Econometrics J 2(2): 306–333.

Jin H J, Frechette D L (2004) Fractional integration in agricultural futures price volatilities. Am J Agri Econ 86(2): 432–443.

Johansen S (1988) Statistical analysis of cointegrating vectors. J Econ Dynamics Control 12(2–3): 231–254.

Koop G, Pesaran M H, Potter S M (1996) Impulse response analysis in nonlinear multivariate models. J Econometrics 74(1): 119–147.

Koplow D (2006) Biofuels – at what cost? Government support for ethanol and biodiesel in the United States. International Institute for Sustainable Development, Geneva, Switzerland. Available at http://www.globalsubsidies.org/article.php3?id_article=6.

Kwiatkowski D, Phillips P C B, Schmidt P, Shin Y (1992) Testing the null hypothesis of stationarity against the alternative of a unit root. J Econometrics 54(1–3): 159–178.

Lloyd T A, McCorriston S, Morgan C W, Rayner, A J (2006) Food scares, market power and price transmission: the UK BSE crisis. European Rev Agri Econ 33(2): 119–147.

McCorriston S, Morgan C W, Rayner A J (1998) Processing technology, market power and price transmission. J Agri Econ 49(2): 185–201.

McCorriston S, Morgan C W, Rayner A J (2001) Price transmission: The interaction between market power and returns to scale. Eur Rev Agri Econ 28(2): 143–159.

OECD, Directorate for Food, Agriculture and Fisheries, Committee for Agriculture (2006) Agricultural Market Impacts of Future Growth in the Production of Biofuels. Paris. Available at http://www.oecd.org/LongAbstract/0,2546,en_2649_33727_36074136_119666_1_1_1,00.html.

Perron P (1997) Further evidence on breaking trend functions in macroeconomic variables. J Econometrics 80(2): 355–385.

Ploberger W, Krämer W, Kontrus K (1989) A new test for structural stability in the linear regression model. J Econometrics 40(2): 307–318.

Rajagopal D, Zilberman D (2007) Review of environmental, economic and policy aspects of biofuel production and use. Policy Research Working paper 4341, The World Bank, Washington DC.

Rajagopal D, Sexton S E, Roland-Holst D, Zilberman D (2007) Challenge of biofuel: filling the tank without emptying the stomach? Environ Res Lett 2(4): 1–9.

Rapsomanikis G, Hallam D (2006) Threshold cointegration in the sugar-ethanol-oil price system in Brazil: Evidence from nonlinear vector error correction models. FAO Commodity and Trade Policy Research Papers 22, FAO, Rome. Available at http://www.fao.org/es/esc/en/41470/41522/highlight_110345en.html.

Renewable Fuels Association (2009) Industry Statistics. Available at http://www.ethanolrfa. org/objects/pdf/outlook/outlook_2006.pdf.

Saikkonen P, Luukkonen R (1988) Lagrange multiplier tests for testing non-linearities in time series models. Scandinavian J Stat 15: 55–68.

Schmidhuber J (2006) Impact of an increased biomass use on agricultural markets, prices and food security: A longer-term perspective. Paper presented at the International Symposium of Notre Europe, Paris, 27–29 November.

Shapouri H, Gallagher P (2005) USDA's 2002 Ethanol Cost-of-Production Survey. Agricultural Economic Report 841, U.S. Department of Agriculture, Washington D.C. Available at http://www.ethanolrfa.org/objects/documents/126/usdacostofproductionsurvey.pdf.

Szklo A, Schaeffer R, Delgado F (2007) Can one say ethanol is a real threat to gasoline? Energy Policy 35(11): 5411–5421.5

Teräsvirta T (1994) Specification, estimation, and evaluation of smooth transition autoregressive models. J Amer Stat Assoc 89(425): 208–218.

Tyner W E, Taheripour F (2007) Renewable energy policy alternatives for the future Am J Agri Econ 89(5): 1303–1310.

United States Department of Agriculture (2008) USDA Agricultural Projections to 2017. Projections Report OCE-2008-1, Washington DC. Available at http://www.ers.usda.gov/Publications/OCE081/OCE20081fm.pdf

van Dijk D, Teräsvirta T, Hans Franses P (2002) Smooth transition autoregressive models – A survey of recent developments. Econom Rev 21(1): 1–47.

Weise C L (1999) The asymmetric effects of monetary policy: A nonlinear vector autoregression approach. J Money, Credit Banking 31(1): 85–108.

Westcott P C (2007) Ethanol expansion in the United States. How will the agricultural sector adjust? U.S. Department of Agriculture, Economic Research Service, Washington DC. Available at http://www.ers.usda.gov/Publications/FDS/2007/05May/FDS07D01/

Chapter 6
Biofuels and Agricultural Growth: Challenges for Developing Agricultural Economies and Opportunities for Investment

Siwa Msangi, Mandy Ewing, and Mark Rosegrant

Abstract Global projections for increasing food demand combined with increasing demand for energy from all sources – including crop-based biofuels – point toward greater stress on food systems and their supporting ecosystems. In many parts of the world, increasing household incomes has translated into increasing demands for energy, of which transportation fuel comprises a fast-growing share. Accompanying the world's steady population growth is an increasing demand for food and the necessary feedstuffs to fuel the requisite increases in livestock production. The combination of these two trends will inevitably lead to greater stresses and demands on the natural resource base and eco-systems that underlie the world's food and energy production systems – such as land and water. In this chapter, we examine the increasing demands on agricultural production systems, within the context of both biofuels and demographically driven demand for food and feed products, and the implied stresses that these drivers represent. By looking at the implied crop productivity improvements that are necessary to maintain adequate supplies of food and feed for a growing global population, we are able to infer the magnitude of investments in agricultural research, among other policy interventions (such as irrigation investments), that are needed to avoid worsening food security outcomes in the face of growing biofuel demands. From our analysis, clear policy implications will be drawn as to how to best avoid the deterioration in human well-being, and recommendations for strengthening food systems and their ability to deliver needed services will also be made. By illustrating the policy problem in this way, we hope to better clarify the key issues that connect biofuels growth to agricultural growth, human welfare, and policy-focused interventions and investments.

Keywords Biofuels · Agricultural productivity · Nutrition · Welfare

S. Msangi (✉)
International Food Policy Research Institute, Washington, DC, USA
e-mail: s.msangi@cgiar.org

M. Khanna et al. (eds.), *Handbook of Bioenergy Economics and Policy*,
Natural Resource Management and Policy 33, DOI 10.1007/978-1-4419-0369-3_6,
© Springer Science+Business Media, LLC 2010

6.1 Introduction

As global energy resources become increasingly scarce in the face of growing energy demand for transport fuel and other productive uses, many countries have begun to turn to the possibilities that biofuels from renewable resources could offer in supplementing their domestic energy portfolio. While much of the recent literature has focused on the growth of biofuels in the developed world, there has been growing interest in biofuel production expressed by developing nations as well (Worldwatch Institute, 2006; FAO, 2008a). North America has taken a leading role in biofuels consumption, worldwide, and is followed by Latin America and the European Union (IEA, 2008). While both Brazil and the United States represent over 90% of world's ethanol production, the United States has overtaken Brazil as the world's leading producer of ethanol, since 2004, while the majority of the world's biodiesel production is concentrated within the EU (IEA, 2008). Fast-emerging economies such as China and India are expected to have a growing share of production in these biofuels categories in the coming decades (Fulton et al., 2004), given the increasing concerns about energy security and the need to diversify away from imported fossil fuels. In China, for example, the government's support for biofuels is expected to reach a level of at least US $1.2 billion by 2020, starting from the 2006 level of US $115 million in subsidies to the sector (GSI, 2008).

While a number of other developing countries find the prospect of biofuels attractive, the degree to which they will be able to invest (or attract investment) in building up capacity for their own domestic production remains uncertain, given the fluctuating price of fossil-based energy and the need to have some kind of long-term commitment, on the part of national governments, to support fledgling biofuel-producing industries through subsidies, tax credits, and other producer (and consumer) incentives. There are a number of countries in Sub-Saharan Africa, South and Southeast Asia, and Latin America which have suitable climates and agro-ecological conditions for growing the required feedstock crops needed for biofuel production, as well as having the needed land area and water resource base (Kojima and Johnson, 2005; Fulton et al., 2004). A number of countries within Sub-Saharan Africa, such as within the West African Monetary Union, are considering the prospect of national biofuels programs and trying to balance it with other important development objectives (UEMOA, 2008). The degree of infrastructure development in these countries, however, varies widely, which may facilitate the large-scale production of biofuels in some countries, while leaving other countries noncompetitive – as might be the case with a country like Malawi (Peskett et al., 2007).

One principal concern that is expressed by a number of authors, in directing agricultural resources away from food and feed production is the long-term impact on prices (Oxfam International, 2008; Runge and Senauer, 2008; Eide, 2008). In the last 2 years, the medium-term outlooks of the OECD and FAO have attributed biofuel production as one of the key drivers in long-term trends in commodity prices (OECD/FAO, 2007, 2008). To be sure – there is a wide range of estimates that exist in the literature, with respect to the impact of biofuels on food prices, and so the precise attribution of food price increases to the domestic biofuels programs in the United States or European Union region is not possible. The recent overview that the FAO provided on the linkage between biofuels and food security (FAO, 2008a) described the possible factors leading to this wide divergence in estimates – some coming from the type of prices impacts being measured (e.g., retail-level vs. world market-level) and the types of modeling approaches used.

In the last 6 years, the international market prices of basic grain commodities have more than doubled, whereas the prices of wheat and rice have tripled. While this might represent a different impact upon the consumer price index in various countries, due to the share of these commodities in total consumption – this represents a significant and sharp change in market conditions, nonetheless. While many see the reversal of historically declining real prices of agricultural commodities as an opportunity for the agricultural producers in both developed and developing countries – others remain concerned about the implications of high food prices and increased volatility in food markets on the welfare and well-being of vulnerable populations who consist of mostly net consumers of these products, and who largely reside in the poorest regions of the developing world (Evans, 2008; FAO, 2008b). This increase in prices may be strong enough to shift consumption patterns for the world's poor, rendering more people food insecure. Several comprehensive discussions of the food price issue have appeared in recent literature, and try to assess the relative merit of each of the possible factors contributing to the recent rise in world and domestic food prices. Some have included an overview of the global macroeconomic picture, and the role that the relative decline of the dollar, in relation to other currencies, had to play in the dynamics of international agricultural markets (Abbot et al., 2008). The steady decline in the level of cereal stocks, globally, as a result of the private sector taking over the operation of cereals stocks from government, and adopting a more "just-in-time" management orientation (Trostle, 2008), has also been cited as a factor that has reduced the ability of national governments to stabilize consumer and producer prices (OECD, 2008).

In this chapter, we address the body of literature that looks at the rapidly growing biofuels production and demand within both the developed and developing world and the potential for adverse impacts on global food economies. We discuss both the micro- and macro-level linkages that connect energy to agricultural markets and the implications that exist for the ability of food systems to deliver needed services, such as human energy and nutrition. We use a simple illustration of energy-driven growth, and examine the linkages between agricultural production, food and feed consumption, human welfare and the implications of trying to meet the energy needs of the economy with crop-based biofuels. Through this we can also see the trade-offs in land use, and the implications that arise for food growing capacity. We then follow this illustration with a quantitative assessment of the potential impact of biofuels growth on world food markets.

By taking this approach, we can observe how the dynamics of "food-vs.-fuel" could play out under alternative growth paths for biofuels and the resulting policy implications. Through this exercise we lay out a framework which can allow both policy makers and researchers to better understand how programs which expand biofuel production can synergize with investment and development strategies aimed at strengthening the function of food systems.

6.2 Overview of Current Literature

In this section, we provide a brief overview of the current literature that addresses the food-vs.-fuel issue for biofuels, and summarize the key insights that have been gained into the nature of the underlying trade-offs. This overview will help to orient the policy experiment that will be carried out in the subsequent section, and illustrates some of the key components that drive the outcomes related to this policy question.

Much of the current literature that has focused on the impacts of increased biofuel production on crop prices and land use has tried to separate out the effect of crop-based biofuel growth from other factors affecting food markets. Many studies acknowledge the close connection between energy and agricultural markets – both in terms of the direct effects of energy prices on agricultural production costs and prices, as well as with respect to the effect of added demand for agricultural products to produce energy products, such as crop-based biofuels. The close coincidence of energy price increases, biofuel production increases, and the increases in food prices within the 2005–2007 time period, therefore, makes it somewhat more complicated to attribute the effects of one phenomenon upon the other. Elliot (2008) points out that the higher oil prices in the 2004–2006 period made the production of biofuels more attractive than it would have otherwise been and combined with the generous support policies, encouraged the production of corn-based ethanol within the United States. He cites the congressional testimony of both IFPRI and CARD scientists, who make attributions to the effect of biofuels policies on grain markets which, in the case of IFPRI, constitutes a 30% increase in world cereal prices, and which, in the case of CARD, amounts to a difference of 12% in maize prices, if the support policies of the United States were taken away from the biofuels sector. A UK-based study (DEFRA, 2008) notes that the impact of biofuels on wheat prices is unlikely, given the relatively small amount that is used as feedstock in the European Union, whereas the sizable increase in maize utilization within the United States in the 2005–2007 period makes a price impact inevitable. In their case, they calculate their own biofuels impact, using the OECD Aglink model, to be 5% for wheat and 7% for coarse grains, when comparing the 2010 prices to the baseline case. They note, however, that the weather-related causes of wheat price increase are likely to have an upward-tending price effect on maize, given the substitution in feed consumption that is likely to occur, and that the increase in speculation activity in soybean markets is also likely to be tied to maize prices. Therefore, they urge caution in interpreting model-based biofuel impact estimates, due to the complexity of the interactions and the relatively young state of the empirical literature on this topic (Defra, 2008).

For major food exporters, Ludena et al. (2007) determine that producers in Latin America have enough excess land to meet food requirements and displace 5% of liquid transport fuel demand – although they may not have considered the full environmental implications of land-use change, that is of concern to a number of authors (e.g. Kammen et al., 2003; Searchinger et al., 2008; Fargione et al., 2008). For the key countries within the OECD regions, namely the United States, Canada, and the EU, between 30 and 70% of the current cropland would have to be dedicated to biofuel production to offset 10% of domestic transport fuel demand[1] (OECD, 2006). Concerning prices, an early OECD study predicts that the additional demand for ethanol could increase the world price of sugar by 60% in 2014, when compared to a constant biofuels scenario, while the later outlooks (OECD, 2008; OECD-FAO, 2008) show a 12% price impact for coarse grains and a 15% impact for vegetable oils in 2017, due to an expected doubling of biofuels production over the 10-year projection horizon. Msangi et al. (2007), however, demonstrate that the upward pressure on feedstock prices is lessened as second-generation technologies come on-line, and Rosegrant et al. (2006) show that technology improvements in agriculture can also have a significant effect on further lowering the price impacts. The analyses of OECD (2006) and Taheripour et al. (2008) have also shown that some of the by-products of biofuel production can mitigate the effects on livestock feed costs, by providing additional protein.

As a result of the predicted price increases of key staple crops, there is strong interest in determining how biofuels may affect food security (FAO, 2008a). Ludena et al. (2007), Schmidhuber (2006), and Runge and Senauer (2007, 2008) warn that low-income, food-deficit countries that import both food and fuel are at the largest disadvantage because price increases in both the food and energy sector will strain current accounts, ultimately decreasing the amount of food supply domestically available. Yet, for countries that are able to decrease energy imports by producing bioenergy domestically, there can be an added benefit for consumers in terms of lower energy costs, especially to urban households that spend a greater share of their total income on transportation fuel, and women who spend time procuring scattered biomass. Therefore, the net impact of price changes in energy and agricultural markets on household welfare and food security depends on the overall effect on household expenditure relative to new income earned (FAO, 2008a).

6.3 Interactions Between Energy and Food Markets

In this section, we present a simplified analytical framework to show the linkages between energy and agricultural markets, the implications that growth in energy and food demand have on land, and the need for technology improvements. These linkages show the role that technology plays in improving both the conversion efficiency of crop-based biomass into energy, as well as the productivity of agricultural crops

---

[1] It should be noted that this estimate assumes that conversion technologies, feedstock shares and feedstock crop yields remain unchanged, that no marginal or fallow land is used, and that international trade is neglected.

themselves. Both of these factors underlie the relationship between the growth in biofuel production and its impact on agricultural economies, and how it leads to trade-offs with food availability and welfare.

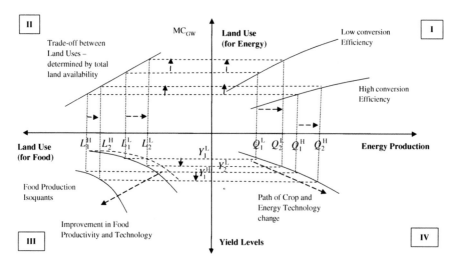

Fig. 6.1 Schematic of derived bioenergy yield curve from agriculture

Figure 6.1 shows a 4-panel schematic, in which the effect of energy production (from biofuels) is linked to land use and ultimately to crop production and productivity. In this illustration, we consider the impacts of two simultaneous technological shifts – both the technology improvement in fuel-conversion technologies, as well as the improvement in agricultural productivity. The shift in fuel-conversion efficiency allows a greater amount of energy to be produced for the same amount of land – or, conversely, the same amount of energy with a lower amount of land. The shifts in energy-conversion efficiency occur in quadrant I. For a given level of efficiency – take the "low" case, for example – we see that an increase in energy production (from Q_1^L to Q_2^L) increases the land-use requirements (for a fixed yield level) and decreases the amount of land available for food production (L_1^L to L_2^L). The curves in quadrant III represent production isoquants – which are curves along which production of food remains constant. So, in order to stay on the isoquant, yield levels have to increase in response to decreases in land use for food production. So, we see that a shift in land use from L_1^L to L_2^L requires that yield levels increase from Y_1^L to Y_2^L in order to maintain food production levels at the same level. Otherwise, production would have to decline, if yields were to remain at Y_1^L, which is shown by the movement to the dashed isoquant in quadrant III.

If there was to be a shift in quadrant I to a biofuels-conversion technology that was more efficient, then we see that an even greater amount of energy could be made available for even lower levels of land use. So a shift from Q_1^H to Q_2^H entails smaller reduction in land for food than before (L_1^H to L_2^H) and from a higher base level (L_1^H, compared to L_1^L). If this shift to higher conversion efficiencies were to coincide with an increase in food demand – as is often seen, over time, as income and population levels increase – then an increase in yield (from Y_1^H to Y_2^H) would also be required to maintain food production at the higher level (whose isoquant has been shifted away from the origin, as shown). This yield shift would also start from a higher base level, compared to that required under lower food demand (i.e., Y_1^H compared to Y_1^L).

From this illustration, we see that the concurrence of increasing food production (to meet the needs of growing and wealthier populations) and increasing energy demand places increasing requirements on the improvement of both energy and crop technologies, in order to keep up. Otherwise, a constant or decreasing food supply, due to less land available for food production and static yields, would cause the "food-vs.-fuel" trade-off that is of such concern to policymakers and analysts. The simultaneous improvements in both fuel conversion and crop productivity trace out an expansion path of crop and energy technology that is shown in quadrant IV of Fig. 6.1, and shows how energy production and yield levels might evolve over time, from a fixed quantity of total agricultural land for food or energy uses.

6.4 Drivers of Change in Food Systems

There are a number of underlining factors contributing to long-term trends in food supply and demand that have contributed to a tightening of global food markets during the past decade. These trends are driven by both environmental and socio-economic changes, as well as by agricultural and energy policy, including those encouraging biofuel production. Figure 6.2 illustrates the interaction between a number of these key drivers of change and the outcomes that they create – in terms

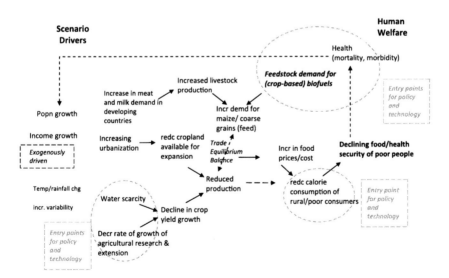

Fig. 6.2 The interrelationships between climatic conditions, agricultural productivity, food prices, and human welfare

of food prices, availability of food (and the energy embodied within it), and human health and welfare.

While we are only able to address a subset of these, in our chapter, we bear in mind the complex interactions that exist between them, and the various points of entry that policy can have in affecting the ultimate outcomes. In this section, the key drivers are examined and discussed, in more detail, in order to better characterize the dynamics within the global food system, and the interactions between energy and agriculture, in particular.

6.4.1 Socio-economic Factors

The main socio-economic factors that drive increasing food demand are population increases, rising incomes, and increasing urbanization. According to the UN medium fertility scenario, global population is set to increase from approximately 6.9 billion in 2010 to 8.3 billion in 2030, with over 90% of this increase in developing countries (UN, 2007). Additionally, a large share of the population increase occurring in developing countries is expected to localize in urban areas (ibid).

The combination of rising income and urbanization is also changing the nature of diets. Rapidly rising incomes in the developing world has led to the increase in the demand for livestock products. In addition, it has been shown that urbanized populations consume less basic staples and more processed foods and livestock products (Rosegrant et al., 2001). Diets with a higher meat content put additional pressure on land resources for pasture and coarse grain markets for feed, including maize. A sizable jump in meat consumption has happened in the last decades, growing from around 27 kg per capita in 1974/1976 to nearly 36 kg per capita in the 1997/1999 period (FAO, 2003) – and now constitutes 8% of the world's calorie intake (Nellemann et al., 2009). As a result of these trends, it is predicted that by 2020 over 60% of meat and milk consumption will take place in the developing world, and the production of beef, meat, poultry, pork, and milk will at least double from 1993 levels (Delgado et al., 1999), which in 2030 will reflect an average per capita consumption of 37 kg per person in developing countries, and 100 kg per person in developed countries (FAO, 2003). These trends constitute a significant challenge for food systems and the environment (Steinfeld et al., 2006) and will place tremendous demands on natural resources, such as water (Molden, 2007).

6.4.2 Policy Drivers

Forward-looking agricultural land requirement projections assume that 70% of food needs will be met through yield enhancements (FAO, 2003). Yet, agricultural research dedicated to productivity enhancement of staple crops has declined over the years (Table 6.1). As the United States and other developed regions have shifted their research focus to reflect consumer preferences for processed, organic,

Table 6.1 Public expenditures in agriculture-related research, 1981–2000

Region/country	Expenditures as a % of Agricultural GDP			Public agricultural R & D spending (2000 int'l dollars, millions)		
	1981	1991	2000	1981	1991	2000
Developing countries	0.52	0.50	0.53	7,223	9,721	13,311
Sub-Saharan Africa	0.84	0.79	0.72	1,239	1,443	1,564
China	0.41	0.35	0.40	1,012	1,745	3,175
Asia and Pacific	0.36	0.38	0.41	3,365	5,281	8,505
Latin America and Caribbean	0.88	0.96	1.16	2,048	2,984	3,086
Middle East and North Africa	0.61	0.54	0.66	751	1,097	1,357
Developed countries	1.41	2.38	2.36	8,644	10,967	10,583
Total	0.79	0.86	0.80	16,084	21,295	23,859

Source: Alston and Pardey (2006); authors calculations based on Alston and Pardey (2006)
Note: Data are provisional estimates and exclude Eastern Europe and countries of the former Soviet Union.

and humane products, the diffusion of more relevant yield enhancing technology in developing countries has slowed (Pardey et al., 2006). Only one-third of global, public agricultural research in the 1990s was in developing countries, over 50% was concentrated in Brazil, China, India, and South Africa (ibid). Therefore, better technology diffusion and more public money dedicated to developing country research programs are critical to meet growing food needs.

Parry et al. (2005) have shown that the regional variation in the number of food insecure is better explained by population increases than climate impacts on food availability. As a result, economic and other development policies – especially policies pertaining to agricultural research and technology – will be critical in influencing future human well-being.

6.5 Quantitative Illustration of Biofuels Impacts on Food

In this section, we present a numerical example to illustrate the impact of rapid biofuels growth on global food prices and to highlight the kinds of interventions which can help to mitigate these effects. We base our simulations on IFPRI's IMPACT model (Rosegrant et al., 2001), which has been applied to a number of biofuels-related studies (Msangi et al., 2007; Rosegrant et al., 2008), and use it to explore some alternative policy-based scenarios.

6.5.1 Model Specification

The International Model for Policy Analysis of Agricultural Commodities and Trade (IMPACT) was developed by the International Food Policy Research Institute

(IFPRI) for projecting global food supply, food demand, and food security to year 2020 and beyond (Rosegrant et al., 2001). The IMPACT model is a partial equilibrium agricultural model for crop and livestock commodities, including cereals, soybeans, roots and tubers, meats, milk, eggs, oilseeds, oilcakes/meals, sugar/sweeteners, and fruits and vegetables. It is specified as a set of 115 country and regional submodels, within each of which supply, demand, and prices for agricultural commodities are determined. IMPACT contains four categories of commodity demand – food, feed, biofuels feedstock, and other uses. The model, therefore, takes into account the growth in demand for the feedstock commodities for biofuel production and determines impact on prices and demand for food and feed for those same agricultural crops. The utilization level of feedstock commodities for biofuel depends on the projected level of biofuel production for the particular commodity, including maize, wheat, cassava, sugarcane, and oilseeds, as well as commodities such as rice, whose demand and supply are influenced by the price of biofuel feedstock crops.

To summarize the essential modeling components in our study, we present below, a simplified system of equations which captures the basic interactions within the model, and highlight some illustrative parameters with which we perform policy experiments. Figure 6.2 showed a number of important interactions between food and energy markets that we try and capture in our numerical simulations. The key relationships within the model can be summarized in the following set of equations, where the yield $(\text{yld}_{\text{cerl}})^2$

$$\text{yld}_{\text{cerl}} = f(p_{\text{cerl}}, a_{\text{irr}}, \tau) \qquad (6.1)$$

and area $(\text{area}_{\text{cerl}})$

$$\text{area}_{\text{cerl}} = f(p_{\text{cerl}}, \text{yld}_{\text{cerl}}, \text{popn}) \qquad (6.2)$$

relationships both contain the price of cereal commodities (p_{cerl}), and contain key drivers which are exogenous to the model – namely population growth (popn) and technological change[3] (τ). Another key driver of yield growth, for cereals, is the share of total area under irrigation a_{irr}, which can change according to defined scenarios for irrigation investment. The per capita levels of food and feed consumption are, respectively, $\text{pcfood}_{\text{cerl}}$ and $\text{pcfeed}_{\text{cerl}}$, and take the form

$$\text{pcfood}_{\text{cerl}} = f(p_{\text{cerl}}, \text{pcInc}) \qquad (6.3)$$

$$\text{pcfeed}_{\text{cerl}} = f(p_{\text{cerl}}, \text{prod}_{\text{mean}}) \qquad (6.4)$$

---

[2] In this specification, there is no modeling of input markets like for fertilizer or labor, so the price is not explicitly included here.

[3] This is an important driver of yield growth, which captures time-dependent technological progress. This will be varied under policy experiments, later in the paper.

where per capita income (pcInc) is one of the key socio-economic "drivers" of change, along with population (popn) and where $\text{prod}_{\text{meat}}$ represents the production of meat products.

$$\text{prod}_{\text{meat}} = f(p_{\text{meat}}, p_{\text{cerl}}) \tag{6.5}$$

The global levels of demand for meat products is given by

$$\text{demd}_{\text{meat}} = f(p_{\text{meat}}, \text{urban}_{\text{share}}, \text{pcInc}) \tag{6.6}$$

and is a function of the price p_{meat}, as well as the share of the total population living in urban areas ($\text{urban}_{\text{share}}$). Urbanization is a key demographic driver of meat consumption and embodies the changes in diet and consumption that are implicit in lifestyle differences between rural and urban populations. A necessary condition for market closure is given by the following two equations:

$$\text{prod}_{\text{meat}} = \text{demd}_{\text{meat}} + \text{trade}_{\text{meat}} \tag{6.7}$$

$$\text{area}_{\text{cerl}} \cdot \text{yld}_{\text{cerl}} = \text{popn} \cdot (\text{pcfood}_{\text{cerl}} + \text{pcfeed}_{\text{cerl}}) + \text{fsdemd}_{\text{bioF}} + \text{trade}_{\text{cerl}} \tag{6.8}$$

which require that supply of meat is equal to its demand, globally – and that the sum of trade also sums to zero at a global level. Likewise, cereal production at the country-level is balanced with food, feed, and biofuel demand for cereal feedstocks ($\text{fsdemd}_{\text{bioF}}$) and trade – and that trade also balances at the world level. It should also be noted that feedstock demand from biofuels is also a function of the feedstock price – namely, that of cereals – such that we can write

$$\text{fsdemd}_{\text{bioF}} = f(p_{\text{cerl}}, \theta) \tag{6.9}$$

where the parameter θ embodies exogenous factors – such as the price of oil. The prices of cereal and meat products ($p_{\text{cerl}}, p_{\text{meat}}$) are the key endogenous variables that allow the global system to remain in balance over time. Figure 6.3, below, illustrates the feedback between these modeling components, within the IMPACT modeling framework.

6.5.2 Baseline Results with Biofuels

Table 6.2 presents the results of the baseline run from year 2000 to 2020, which shows the change in the basic indicators of global agricultural supply and demand. The steady demand for meat products is seen across all regions (and grows at an annual rate of 2.1%, in per capita terms, for the East Asia region) and is driven by steady changes in per capita income (growing globally at an annual average rate of 2%). The fast rate of growth in East Asia stands out, compared to other regions like Sub-Saharan Africa, which grows at less than half a percent over the same

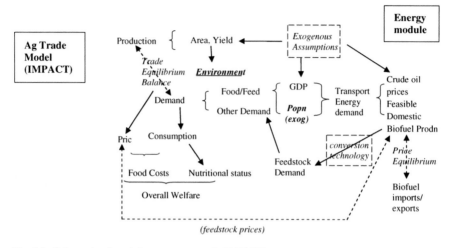

(feedstock prices)

Fig. 6.3 Schematic of modeling components in IMPACT

period. The fast growth of meat demand would be accompanied by an increase in feed demand for cereals, but is sharply contrasted by the small decrease in per capita levels of food demand for cereals in East Asia (while it grows slowly in Sub-Saharan Africa). Nonetheless, the steady 1% growth in global population implies that the total consumption of cereals in all uses grows and causes an increase in the global price of cereals, over time, which rises at an average rate of 1.2%. The growth in cereal supply is realized mostly through yield growth, which increases at a much faster rate (1.2% in East Asia and 1.7% in Africa, compared to that of cereal area (0.3%, globally)). The steady increase in meat demand also causes a similar 1.2% rate of annual growth in the global meat price, over the same period. Therefore, maintaining constant yield growth to meet the needs of the global food system will be critical as key socio-economic drivers of population and income evolve.

The baseline results shown in Table 6.2, above, already include the growth of biofuels – which we consider to be the "business-as-usual" case, against which we will consider other scenario-based alternatives. Embedded in these baseline results – aside from the food and feed demand for grains that we already discussed – is the additional demand from cereals coming from feedstock demand for crop-based biofuels. If we consider alternative paths along which both biofuels-related and agriculture-focused policies could evolve, then we can see what the effect would be on the indicators shown in Table 6.2.

In our baseline case, we assume that the biofuels targets implied by the US Renewable Fuel Standards (RFS) [4] are met, with respect to conventional "1st generation" crop-based biofuels – which peak at a target of 15 billion gal (or 56.8 billion

---

[4]The RFS aims for 36 billion gallons of ethanol by 2022 from all sources, of which the production of "conventional" reaches a peak of 15 billion gal by 2015, which can be met through both production and imports. This is the particular aspect of the RFS target that has been modeled here.

Table 6.2 Model results for baseline and alternative scenarios

	2000	2010	2020	Avg growth (2000–2020) (%)	Effect of mandate reduction (% diff in 2020)	Effect of Enhanced yield growth (% diff in 2020)
World Cereal Price ($/mt)	905.8	1055.4	1153.8	1.2	−3.42	−2.86
World Meat Price ($/mt)	1341.5	1575.9	1695.1	1.2	−0.67	−0.25
Global Cereal Area (millions ha)	700.4	727.5	741.5	0.3	−0.35	−0.21
Average Cereal Yield (mt/ha)						
North America, Europe	4.8	5.2	5.7	0.8	0.3	1.0
East Asia	4.0	4.5	5.0	1.2	0.3	4.9
SS Africa	1.0	1.2	1.4	1.7	0.3	−1.2
Per Capita Food Demand for Cereals (kg/cap)						
North America, Europe	124.4	126.8	130.4	0.2	−0.09	2.76
East Asia	184.5	185.4	178.0	−0.2	0.09	0.99
SS Africa	117.7	119.9	125.2	0.3	−0.51	2.50
Per Capita Food Demand for Meat (kg/cap)						
North America, Europe	93.2	99.8	103.6	0.5	−0.07	0.38
East Asia	47.4	59.8	71.6	2.1	−0.13	0.66
SS Africa	10.3	11.2	12.1	0.8	−0.13	0.73
Global Population (millions)	6084.2	6841.3	7574.5	1.1	–	–
Global Average Per Capita Income (2000 US$/cap)	5587.7	6502.2	8096.2	1.9	–	–

l) by 2015. We assume that this production comes entirely from grains at an average feedstock conversion rate of 400 l of ethanol per metric ton of cereal feedstock. As an alternative to this case, we consider what would happen if a lower target (of 10 billion gal) was set, instead of the 15 billion gal baseline. As would be expected, the most dramatic change of lowering the demand for grain in biofuels production, under this scenario, occurs for global cereals prices, which decrease by 3.4% compared to the baseline 2020 level. Due to the decreased cost of cereals, and its implications for animal feed costs, the prices of meat products also go down slightly, although not nearly as much. This effect is caused by the effect of lower grain prices on animal feed costs, which gets passed on to consumers in the form of meat prices.[5]

---

[5]This effect does not take into account, however, the mitigating effect of by-products from ethanol production, such as dried distiller's grains (DDGs), which can also be used for animal feed, up to a certain proportion. The availability of DDGs as feed would lessen the price impacts on meat.

6.5.3 Impacts of Yield Improvements

By undertaking an additional policy experiment, we can illustrate the impact that policy-driven investments in yield growth can have on the outcomes that we have described, under rapid crop-based biofuels expansion. By returning to the relationship that governs yield, in our simple model, $yld_{cerl} = f(p_{cerl}, a_{irr}, \tau)$, we see that there are several avenues through which yield improvements can be brought about. Cereal yields can be boosted by either increasing the share of cultivated area under irrigation (a_{irr}) or increasing the rate at which yield improvements are made over time, which is captured by the parameter τ.

Table 6.2 shows the impact on the key indicators of supply and demand, when the annual rate of increase in crop yields is augmented over the time horizon of the simulation. We see decrease in cereal prices of similar magnitude to that which occurred when the lower biofuels target is imposed as an alternative to the original baseline case. There is a small decrease in total cereal area, given the land-saving effect that yield enhancements tend to have on agriculture, creating less of a need for expansion in total cultivated area. The regional distribution of yield increases seen in the last column of Table 6.2 also reflects the general trend observed in many countries in Asia, North America, and Europe that have increased supply mostly through yield growth driven largely by increases in irrigation and other technological improvements in seed and production technologies and practices. In response to these price changes, we notice a sizable increase in the per capita consumption levels of cereals – especially in the North America/Europe and Sub-Saharan Africa regions.

Both the decrease in per capita consumption levels of cereals and the increase in price of cereals, for both food and feed, imply that food security is likely to be compromised by the pursuit of the stated targets for first-generation, conventional, crop-based biofuels – such as ethanol from maize – especially for those who rely on cereals for a large share of their calorie intakes, and for whom food represents a large share of their household expenditure. Many of the world's poor fall into this category, as do others who are near the poverty line or within the lower and lower-middle income strata of the world's economy. While there are some increases in agricultural wages and income that result from the expansion of crop-based biofuels, those gains may not permeate widely or deeply enough to offset the loss in human welfare that occurs from sharp increases in the price of basic staple foods. An exhaustive evaluation of these trade-offs would require an economy-wide modeling framework that focuses on the implied shocks for all relevant input and output markets, at the country-level, and which accounts for household-level consumption and expenditure patterns, through the use of micro-level data. A recent study by Arndt et al. (2008) examines the case of biofuels in Mozambique at this level of detail and is able to draw out implications for poverty that are beyond the scope of our analysis to address.

6.6 Implications for Food Security and Policy

The alternative scenarios that we have simulated, in the previous section, have implications for energy policy and agricultural investments that merit some discussion, in
light of their role in affecting agricultural economies, trade, and country-level food
security. Many of these country-level investments coincide with those that we might
consider necessary, in general, for the improvement of food production, distribution,
and delivery systems in developing agricultural economies. In the "yield enhancement" scenario that we considered, there are implications for both the improvement
of crop seed technology as well as the expansion of irrigated area (which also has
an enhancing effect on crop yields). We used a similar scenario design to the "optimistic scenario" described in Rosegrant et al. (2001) and draw the implications
for additional investment from that analysis. The additional expansion of net irrigated area[6] that helps to offset the negative food production impacts of expanded
biofuels production, under this scenario, amounts to nearly a doubling in investments (relative to the baseline) and has the majority of the increases in spending to
2020 accruing to South Asia. The level of investments toward national agricultural
research that also leads to additional technological progress[7] and crop productivity,
under this scenario, amounts to a 7.4% increase over the baseline and has half of the
total global investment levels remaining in Latin America and the West Asia and
North African region. Even under the more optimistic scenario, Sub-Saharan Africa
still lags the levels of investment seen in other regions (only 6.3% of the developing
world total) – but realizes a 3% increase over the baseline, nonetheless, which is
significant.

In light of the analysis that we have done, it might be argued that the "food-
for-fuel" trade-off that some policy analysts use to characterize the prospects for
large-scale expansion of biofuel production need not always occur, if the appropriate investments and efficiency improvements are made in advance – or if moderation
of biofuels targets, in the light of their implications on agricultural markets, is taken
into account at the policy decision-making level. To be sure, there is certainly a
tension that exists between the provision of food and fuel from agricultural production systems – especially when the manufacture of one is supported by policies that
might distort the true cost structure or put up barriers to free trade in either the feedstock or the finished product. Many of the arguments that support liberalization of
trade in agricultural base and finished products might hold as well for fuel products,
as well as food products, themselves.

---

[6]We consider net irrigated area, to take into account the fact that there is multiple cropping in some
regions, which might overstate the actual surface area under irrigation, if we were to simply add
up the statistics of harvested area under irrigation.

[7]The irrigation area increase is decreased when additional technological progress is introduced, due
to the 'land-saving' effects of added yield growth and productivity, as well as the effect induced
by generating additional supply, and thereby dampening the price increases that would otherwise
stimulate area expansion over time.

Without doubt, there will be market-level price effects when there is large-scale expansion of production from feedstock commodities that also have sizeable food and feed use value – and, to be sure, those who are most vulnerable to price increases could be adversely affected. But this should also be considered within the context of specific countries where agricultural (and nonagricultural) wages might improve in the face of such changes, and where there might be possible beneficial spillovers into food production, as was illustrated in the study of Arndt et al. (2008). Therefore, we feel that the need for continued micro- and macro-level economic and policy analysis in this area is clearly evident, and should remain a priority for researchers.

References

Abbot PC, Hurt C, Tyner WE (2008) What's Driving Food Prices? Issue Report. Farm Foundation.

Alston JM, Pardey PG (2006) Developing-Country Perspectives on Agricultural R&D: New Pressures for Self-Reliance? In *Agricultural R&D in the developing world: Too little, too late?*, ed. P.G. Pardey, J.M. Alston, and R.R. Piggott. Washington DC: International Food Policy Research Institute.

Arndt A, Benfica R, Tarp F, Thurlow J, Uaiene R (2008) Biofuels, poverty and growth: A computable general equilibrium analysis of Mozambique. IFPRI Discussion Paper #803. Washington, DC: International Food Policy Research Institute.

Department for Environment, Food and Rural Affairs (Defra) (2008) The impact of biofuels on commodity prices. The Economics Group, Defra, London, United Kingdom.

Delgado C, Rosegrant M, Steinfeld H, Ehui S, Courbois C (1999) Livestock to 2020: The next food revolution. Food, Agriculture, and Environment Discussion Paper #28. Washington, DC: International Food Policy Research Institute.

Eide A (2008) The right to food and the impact of liquid biofuels (agrofuels). Food and Agricultural Organization of the United Nations (FAO), Rome.

Elliot K (2008) Biofuels and the Food Price Crisis: A survey of the issues. Working Paper no. 151, Center for Global Development, Washington, DC.

Evans A (2008) Rising Food Prices: Drivers and Implications for Development. Briefing Paper 08/02, Chatham House, United Kingdom.

FAO (Food and Agricultural Organization of the United Nations) (2003) World agriculture: towards 2015/30. Rome: Food and Agricultural Organization of the United Nations.

FAO (2008a) Biofuels: prospects, risks and opportunities. The State of Food and Agriculture. Rome: Food and Agricultural Organization of the United Nations.

FAO (2008b) Soaring Food Prices: Facts, Perspectives, Impacts and Actions Required. Paper prepared for the High-level Conference on World Food Security "The Challenges of Climate change and Bioenergy". HLC/08/INF/1. FAO, Rome.

Fargione J, Hill J, Tilman JD, Polasky S, Hawthorne P (2008) Land Clearing and the Biofuel Carbon Debt. Science 319 No. 5867, 1235–1238. DOI: 10.1126/science.1152747.

Fulton L, Howes T, Hardy J (2004) 'Bio-fuels for Transport: An International Perspective', International Energy Agency, Paris.

Global Studies Initiative (GSI) (2008) Biofuels – at what cost? Government support for ethanol and biodiesel in China. Report of the Global Studies Initiative of the International Institute for Sustainable Development, Geneva, Switzerland.

International Energy Agency (IEA) (2008) World Energy Outlook 2008. International Energy Agency, Paris.

Kammen DM, Bailis R, Herzog AV (2003) Clean Energy for Development and Economic Growth: Biomass and Other Renewable Energy Options to Meet Energy and Development Needs in Poor Nations (UNDP: New York).

Kojima M, Johnson T (2005) Potential for Biofuels for Transport in Developing Countries. Energy Sector Management Assistance Program (ESMAP). The World Bank, Washington, DC.

Ludena CE, Razo C, Saucedo A (2007) 'Biofuels in Latin America and the Caribbean: Quantitative Considerations and Policy Implications for the Agricultural Sector', paper prepared for the American Association of Agricultural Economics Annual Meeting, July 29–August 1, Portland.

Msangi S, Sulser T, Rosegrant M, Valmonte-Santos R (2007) 'Global Scenarios for Biofuels: Impacts and Implications,' Farm Pol J 4 (2):1–18.

Molden D (ed.) (2007) Water for food, water for life: A comprehensive assessment of water management in agriculture. International Water Management Institute. Earthscan.

Nellemann C, MacDevette M, Manders T, Eickhout B, Svihus B, Prins AG, Kaltenborn BP (eds.) (2009) The environmental food crisis: The environment's role in averting future food crises. A UNEP Rapid Response Assessment. United Nations Environment Program, GRID-Arendal.

OECD (2006) Agricultural Market Impacts of Future Growth in the Production of Biofuels, Directorate for Food, Agriculture and Fisheries, Committee for Agriculture, OECD, Paris. viewed 10 August 2007, http://www.oecd.org/dataoecd/58/62/36074135.pdf

OECD (Organization for Economic Cooperation and Development) (2008) Rising Food Prices: Causes and Consequences.,OECD, Paris.

OECD-FAO (Organization for Economic Cooperation and Development and Food and Agricultural Organization of the United Nations) (2007) Agricultural Outlook 2007–2016. OECD, Paris.

OECD-FAO (Organization for Economic Cooperation and Development and Food and Agricultural Organization of the United Nations) (2008) Agricultural Outlook 2008–2017. OECD, Paris.

Oxfam International (2008) Another Inconvenient Truth: How Biofuels Policies are Deepening Poverty and Accelerating Climate Change. Briefing Paper 114, Oxfam International.

Pardey PG, Alston JM, Piggott RR, eds. (2006) Agricultural R&D in the developing world: Too little, too late? Washington DC: International Food Policy Research Institute.

Parry M, Rosenzweig C, Livermore M (2005) Climate change, global food supply and risk of hunger. Phil Trans R Soc 360: 2125–2138.

Peskett L, Slater R, Stevens C, Dufey A (2007) Biofuels, agriculture and poverty reduction. ODI Natural Resource Perspectives 107, Overseas Development Institute. United Kingdom.

Rosegrant M, Zhu T, Msangi S, Sulser T (2008).Global scenarios for biofuels: Impacts and implications. Rev Agri Econ 30(3): 495–505.

Rosegrant M, Msangi S, Sulser T, Valmonte-Santos R (2006) 'Biofuels and the Global Food Balance', in 2020 Focus 14 – Bioenergy and Agriculture: Promises and Challenges, eds. P. Hazell and R.K. Pachaura, International Food Policy Research Institute: Washington D.C.

Rosegrant MW, Paisner MS, Meijer S, Witcover J (2001) Global food projections to 2020: Emerging trends and alternative futures. Washington, DC: International Food Policy Research Institute.

Runge CF, Senauer B (2007) 'How Biofuels Could Starve the Poor', Foreign Affairs, vol. 86, no. 3, pp. 41–53.

Runge CF, Senauer B (2008) How Ethanol Fuels the Food Crisis, Author update May 28th 2008, Foreign Affairs. http://www.foreignaffairs.org/20080528faupdate87376/c-ford-runge-benjamin-senauer/how-ethanol-fuels-the-food-crisis.html

Schmidhuber J (2006) 'Impact of an increased biomass use on agricultural markets, prices and food security: A longer-term perspective', paper prepared for the International symposium of Notre Europe, 27–29 November, Paris.

Searchinger T, Heimlich R, Houghton RA, et al. (2008) Use of U.S. croplands for biofuels increases greenhouse gases through emissions from land-use change. Sci Express 319, 1238–1240.

Steinfeld H, Gerber P, Wassenaar T, Castel V, Rosales M, de Haan C (2006) Livestock's long shadow: Environmental issues and options. Food and Agricultural Organization of the United Nations (FAO), Rome.

Taheripour F, Hertel T, Tyner W (2008) Biofuels and their by-products: Global economic and environmental implications. Department of Agricultural Economics, Purdue University, West Lafayette, Indiana.

Trostle R (2008) Global Agricultural Supply and Demand; Factors Contributing to the Recent Increase in Food Commodity Prices. WRS-0801. Economic Research Service, US Department of Agriculture.

UN (United Nations) (2007) *World population prospects: The 2006 revision*. New York: United Nations.

West Africa Economic and Monetary Union (UEMOA) (2008) Sustainable bioenergy development in UEMOA member countries. Report written by the United Nations (UN) Foundation for the Hub for Rural Development in West and Central Africa, and the West African Economic and Monetary Union. United Nations Foundation, Washington, DC.

Worldwatch Institute (2006) 'Bio-fuels for Transportation: Global Potential and Implications for Sustainable Agriculture and Energy in the 21st Century'. Extended Summary of Report for the German Federal Ministry of Food Agriculture and Consumer Protection (BMELV). Washington, DC.

Chapter 7
Prospects for Ethanol and Biodiesel, 2008 to 2017 and Impacts on Agriculture and Food

John (Jake) Ferris and Satish Joshi

Abstract The impacts of increased biofuels production on key agricultural variables and consumer prices are analyzed using a multisector econometric model AGMOD. A "baseline" scenario and three alternative crude oil price scenarios are presented. Results indicate that conventional biofuels mandates of the Energy Independence and Security Act of 2007 can be met with moderate increases in crop area and consumer prices. Biodiesel production will increasingly need to draw on non-soybean oil sources. However, except under the "high crude oil price" scenario, ethanol and biodiesel will require a premium over their energy equivalent prices. Production of cellulosic ethanol is likely to be minor through 2017.

7.1 Introduction

Since 2000, the rapid expansion in global biofuel production, accompanied by a dramatic rise in crude oil prices, demand growth in the developing world, dwindling grain stocks and weather anomalies, elevated commodity prices to unprecedented nominal levels in 2008. Following peak levels in early summer, prices, including crude oil, collapsed during the remainder of the year as the world faced a severe financial crisis. The implication of this structural change with agriculture as a source of fuel, as well as food and fiber, over the 2008–2017 period is the focus of this analysis. This chapter analyzes the production of biofuels from traditional (first generation) feedstock and the prospects for the second generation of feedstock from cellulosic sources.

The analytical tool used is AGMOD, an econometric model of US agriculture, developed at Michigan State University, which covers major commodities in the livestock, dairy, poultry, and field crop sectors, including by-product feeds. The international sector focuses on coarse grain, wheat, and oilseeds. The variables

J. Ferris (✉)
Department of Agricultural, Food, and Resource Economics, Michigan State University, East Lansing, MI, USA
e-mail: jakemax33@comcast.net

M. Khanna et al. (eds.), *Handbook of Bioenergy Economics and Policy,*
Natural Resource Management and Policy 33, DOI 10.1007/978-1-4419-0369-3_7,

in the model, of which there are over 1,000, can be classified as exogenous and endogenous. The endogenous variables are those generated within the model and measure behavioral relationships such as how farmers respond to profits and how consumers respond to prices and availability of products. The exogenous variables are those which are external to the model and enter the solution as assumptions, such as is being introduced by the renewable fuel scenarios. Other exogenous variables include population, per capita incomes, energy prices, interest rates, and exchange rates (Ferris, 2005).

Considering the abnormal uncertainties relative to future energy prices, compounded by the global recession which began in 2008, both a "baseline" scenario and three alternative scenarios are presented in an effort to embrace a wide range in possible crude oil prices. Mindful that such projections can only serve as tentative guidelines, the authors suggest that readers view this article as a process to grasp some understanding of possible unfolding future events.

7.2 Assumptions for the Baseline Scenario

Projections of population and real per capita gross domestic product for the United States and foreign nations are from the US Department of Agriculture's (USDA's) "International Macroeconomic Data Set," as were exchange rates (USDA, 2008). Crude oil prices in the projections are for the "composite refiner acquisition cost" as measured by the US Department of Energy (DOE) and derived from the futures quotes on the New York Mercantile Exchange on February 27, 2009 (Table 7.1). As can be noted, crude oil prices are estimated to drop to $45 per barrel in 2009, a level half as high as in 2008, and advance toward the $76 level at the end of the projection period. Crude oil prices are the major determinant of inflation as generated by AGMOD and represented by the implicit price deflator (IPD) for personal consumption expenditures tallied by the US Department of Commerce. The IPD is the deflator of prices, costs, and incomes in this analysis.

The "Energy Independence and Security Act of 2007," (EISA) which became law in December 2007, prescribes mandates called "Renewable Fuels Standard" (RFS) for biofuel use from 2008 to 2022. The specifics are somewhat complex, but in essence the total RFS increases from 34 Giga liters (GL) or 9 billion gallons in 2008 to 136 GL (36 billion gallons) in 2022. Of this, "Conventional Biofuels" refers to ethanol derived from corn starch, which increases from 34 GL in 2008 to 56.8 GL in 2012 and remains at that level. Presumed is that corn ethanol will fill that RFS, although the classification of "Biomass-Based Diesel" is also eligible.

The Act sets the RFS for this biodiesel classification at 1.9 GL for 2009, increasing to a minimum of 3.8 GL by 2012 and beyond. "Biomass-Based Diesel" is also eligible under the classification of "Undifferentiated Advanced Biofuels" to bring the total for biodiesel potential to 17 GL by 2017 and 22.7 GL by 2022. The RFS can be filled by imports as well as from domestic production. In addition, RFSs are prescribed for "Advanced Biofuel except Cellulosic Biofuel" and "Cellulosic Biofuel," the latter increasing from 0.38 GL in 2010 to 20.8 GL by 2017 and 60.5

Table 7.1 Projections of variables related to prices on crude oil, inflation, and assumed production of ethanol and biodiesel in the United States and in foreign nations

Item	2005	2008	2009	2011	2013	2015	2017
Crude oil prices ($/barrel)[1]	50	94	45	61	67	71	76
Implicit price deflator (IPD) (Year 2000 = 1)[2]	1.116	1.215	1.218	1.273	1.321	1.370	1.419
Change in IPD (%)	2.9	3.3	0.2	2.3	1.6	1.8	1.8
US ethanol production (GL)[3]	14.78	34.92	39.75	47.69	52.24	56.78	56.78
Percent of motor gasoline (%)	2.8	7.0	8.0	9.3	10.4	11.7	11.9
US biodiesel production (GL)	0.34	2.60	3.41	5.68	6.44	7.07	7.57
Foreign ethanol production (GL)[4]	8.54	16.81	17.00	21.06	25.38	29.96	34.80
Foreign biodiesel production (GL)[5]	10.87	15.22	17.33	20.41	23.49	26.57	29.66

[1] Refiner acquisition cost, composite of domestic and imported from the US Department of Energy for 2005–2008.
[2] IPD from the US Department of Commerce for 2005–2008.
[3] GL = Giga Liters or 10^9 Liters (1 GL = 264.2 million gallons).
[4] From coarse grain and wheat.
[5] From vegetable oils.

GL by 2022. Presumed is the EISA and other current federal and state legislation will remain intact through 2017. This includes the blenders' tax credits for ethanol and biodiesel and the $0.143/L tariff on ethanol imports. Anticipated is a tariff will be imposed by the European Union on biodiesel imports from the United States. Except for cellulosic biofuel, the assumption is waivers to the RFSs will not be issued and that prices on ethanol and biodiesel will be maintained at a level high enough to generate sufficient profits to meet the mandates.

As indicated in Table 7.1, the projected corn grain-based ethanol production, estimated at 34.9 GL in 2008, will increase to 56.8 GL by 2015 and will remain at that level. As explained in the last section, we do not expect significant production of cellulosic ethanol by 2017, and as a result, cellulosic ethanol is unlikely to provide the needed amounts to reach the total RFS of 90.8 GL by 2017. Hence, ethanol derived from cellulosic feedstock and imports are not analyzed in this paper. Further, by 2012 or 2013, production will reach the 10% "blend wall," but by that time the Environmental Protection Agency (EPA) will likely have increased the allowable blend to about 15%. In addition, the production of "Flex-Fuel" vehicles, which can utilize E-85 (85% ethanol), will likely increase more than enough to provide ample markets for ethanol as projected.

Biodiesel production, at an estimated 2.65 GL in 2008, is projected to be 7.6 GL in 2017, exceeding the energy bill mandates. This is based on existing capacity of about 9.8 GL and the needed profits to meet the mandate to produce at least 3.8 GL.

If biodiesel prices must hold at levels to produce 3.8 GL, the plants representing the unused capacity should at least cover their variable costs. Net exports of biodiesel, registering 54% of domestic biodiesel production in 2008, will likely be reduced as the "splash and dash" program (imports blended with one percent diesel and receiving the $0.264/L subsidy before being exported) was eliminated in October 2008, retroactive to May. Also, exports will be reduced by the anticipated tariff for exports into the European Union.

As for the rest of the world, the projections in Table 7.1 are highly empirical, based upon trends beginning around 2000. The projections for ethanol are only for ethanol produced from coarse grain and wheat, which will be about half the total – the remainder mostly from sugarcane in Brazil.

The essence of the new 5-year Farm bill labeled "Food, Conservation, and Energy Act of 2008" is assumed to continue through 2017. The familiar marketing assistance loans, target prices, counter-cyclical and fixed direct payment features remain in the bill. Only minor changes were made in these features, mostly to increase wheat loans and target prices beginning in 2010. A major new program is called the "Average Crop Revenue Election (ACRE)" program. As an alternative, ACRE addresses the weakness of past programs which have provided price, but not revenue support. In any case, based on projected farm prices, ACRE will not require major modifications of AGMOD projections.

7.3 Global Land Use and Commodity Stocks

The implications of the prospective expansion in the production of biofuels on land use and grain and oilseed stocks are indicated in Table 7.2. Crop yields are slated to increase, in most cases as linear extensions of past trends. In the United States, the combined harvested area of coarse grain, wheat, and soybeans is expected to increase from about 89.7 million hectares (Mha) in 2008 to 93 Mha in 2017. With declining numbers of ruminant livestock, increased feeding efficiency and higher crop yields, area harvested for hay and corn silage and the use of cropland pasture will be declining by as much as 2.8 Mha by 2017.

In addition, the prospective elevated crop prices are likely to draw land out of the Conservation Reserve Program (CRP). As of February 2009, 11.9 Mha were enrolled in the "General Sign-Up." From 2009 to 2017, expiring contracts range from 0.5 Mha to 2.3 Mha each year. Focusing on the CRP assigned to coarse grain, wheat, and soybeans, 1.8 Mha are forecast to leave the program, although CRP land identified with other crops could be a source as well (Table 7.2). In essence, the projected demands on major field crops can be met with less land area than were in harvested field crops, hay, cropland pasture, and the CRP in 2008.[1]

---

[1] Authors' note: Much of the historical data in this analysis were obtained from the "PS&D Online" program of the USDA's Foreign Agriculture Service; from USDA's Economic Research Service, the National Agricultural Statistics Service and the World Agricultural Outlook Board. For prices on biodiesel and related feedstock, the Jacobsen Publishing Co. was a valuable source (Jacobsen Publishing, 2008).

Table 7.2 Projections of the area for major crops in the United States and in foreign nations and for the ending stocks as a percent of utilization on grains and oilseeds

Item	Year						
	2006	2008	2009	2011	2013	2015	2017
Area million ha (Mha)							
United States							
Harvested for coarse grain, wheat, and soybeans	81.6	89.7	88.0	92.1	92.6	92.6	93.2
Harvested for hay	24.6	24.3	23.3	22.2	21.6	22.0	22.1
Corn silage	2.6	2.4	2.3	2.2	2.0	2.0	1.9
Cropland pasture	14.5	14.5	14.3	13.8	13.6	14.0	14.2
Conservation Reserve[1]	9.2	8.9	8.5	8.1	7.7	7.2	6.7
Total	132.6	139.8	136.4	138.3	137.6	137.9	138.1
Foreign							
Coarse grain	272	273	279	285	286	288	291
Wheat	193	194	204	204	203	203	203
Oilseeds	162	156	166	172	178	182	187
Total	627	622	648	661	667	673	682
Ending stocks as a percent of utilization							
US coarse grain	12.0	16.4	13.7	19.1	19.6	12.6	6.6
US soybeans	18.7	7.0	8.6	5.4	7.3	7.4	6.6
US soybean oil	15.0	10.5	11.1	12.5	13.0	13.4	13.6
US wheat	22.3	29.6	22.1	21.4	20.3	17.1	14.0
Foreign coarse grain	13.3	16.1	14.9	14.1	13.1	13.0	13.1
Foreign wheat	19.6	21.4	20.7	19.3	18.0	16.7	15.5
Foreign oilseeds	16.9	16.4	13.6	13.0	13.6	15.3	15.6
Foreign vegetable oils	7.5	7.9	4.9	4.2	4.2	4.2	4.2

[1]Conservation Reserve land allocated to coarse grain, wheat, and soybeans.

Outside the United States, nations will need to increase areas to meet the required demands for food, feed, and fuel even with projected higher yields (Table 7.2). In total, the area for coarse grain, wheat, and oilseeds would expand about 10%, with increases of about 7% for coarse grain, 5% for wheat, and 20% for oilseeds. Foreign acreage of coarse grain and wheat would about equal the peak in 1996. The additional oilseed area in 2017 would represent about 6% above this peak.

7.3.1 US Crops

Ending stocks of US coarse grain, soybeans, and soybean oil will likely continue near "pipeline" or minimum levels over the projection period as measured by percentages of utilization (Table 7.2). This situation is also likely to prevail internationally. An implication is that markets will be vulnerable to shocks, particularly from the effects of weather on crop yields.

Table 7.3 presents the details of the projections on corn, other coarse grains, and the soybean complex in the United States. Harvested corn area is projected to increase from about 32 Mha in 2008 to about 34.9 Mha by 2017 with production reaching 377 million metric tons (Mt). Adding sorghum, oats, and barley, total coarse grain output would be about 390 Mt by 2017. The utilization of coarse grain for livestock feed reflects the substitution of distillers' dried grain and solubles (DDGS) in livestock rations. By 2017, corn processed into ethanol could represent as much as 36% of total production, nearly reaching the amounts fed to domestic livestock. Exports of coarse grain are expected to remain around 50 Mt initially, accelerating toward the end of the projection period as foreign demands increase.

Traditionally, commodity analysts have employed "stock-to-utilization" ratios as the major independent variables in forecasting crop prices with attention to government loan rates if prices are near that level. In AGMOD, corn prices have been a function of stock-to-utilization ratios for US coarse grain and for foreign coarse grain. AGMOD has been modified to also include "break-even" corn prices for ethanol production. A similar modification was introduced for soybean oil prices.

Stock levels and ethanol prices should support corn prices above levels of the past decade averaging between $138/t and $157/t. With general inflation, particularly with energy prices, variable costs per acre will average about $692/ha, about $247/ha above the previous decade. Higher fertilizer and fuel prices than in the past decade are the major factors. Even so, gross margins over variable costs will hold at an elevated level in both nominal and real terms.

Harvested area of soybeans is expected to edge upward from the 30.2 Mha of 2008 to around 32.6 Mha by the end of the projection period with production moving up to the 101 Mt level (Table 7.3). An increased crush will support the expansion in the utilization of soybean oil for biodiesel production as well as for exports. By 2017, biodiesel production would use about a third of the output of soybean oil. The feeding of soybean meal from the expanded crush will increase only modestly as the expanded availability of DDGS will substitute for high protein, as well as energy feeds. Exports of soybean meal could double by 2017.

Soybean stocks are expected to remain relatively low, supporting prices around $330/t to $370/t. Because soybeans require little or no nitrogen fertilizer, the impact of higher energy prices on variable costs is much less than on corn. Like corn, gross margins should hold significantly higher from 2009 to 2017 than in the past decade in both nominal and real terms (Table 7.3).

Considering the elevated gross margins over variable costs projected for the major field crops, farmland prices will continue to benefit. Midwest farmland prices, which reached an average near $10,000/ha in 2008 and early 2009 could advance to the $14,000/ha level and above toward the end of the projection period.

Table 7.3 Projections of selected corn, coarse grain, and soybean variables by crop years plus farmland prices

Item	Year						
	2006	2008	2009	2011	2013	2015	2017
Corn							
Harvested area (Mha)[1]	28.6	31.8	32.2	33.6	33.7	34.2	34.9
Yield (t/ha)	9.35	9.66	9.72	9.99	10.27	10.54	10.82
Production (Mt)[2]	267	307	313	335	346	360	377
Coarse grain							
Production (Mt)	280	326	330	354	362	374	390
Utilization (Mt)							
Feed	148	144	140	143	143	145	142
Ethanol	54	91	106	117	126	131	136
Other domestic	41	40	38	38	38	38	38
Exports	58	48	53	50	59	75	88
Ending stocks (Mt)	36	50	46	67	72	49	27
Farm price ($/t)	120	154	152	133	133	146	169
Variable costs ($/ha)	509	742	606	610	673	754	861
Gross margin ($/ha)[3]	671	800	940	793	767	859	1043
Soybeans							
Harvested area (Mha)	30.2	30.2	30.1	30.9	32.4	32.8	32.6
Yield (t/ha)	2.88	2.66	2.88	2.93	2.98	3.04	3.09
Production (Mt)	87	80	87	91	97	100	101
Crush (Mt)	49	45	52	58	61	62	64
Exports (Mt)	30	31	29	29	31	33	33
Ending stocks (Mt)	16	6	7	5	7	7	7
Farm price ($/t)	236	338	332	338	332	344	376
Variable costs ($/ha)	239	320	296	319	350	387	431
Gross margin ($/ha)[3]	471	608	692	704	673	692	764

Table 7.3 (continued)

Item	Year						
	2006	2008	2009	2011	2013	2015	2017
Soybean oil							
Production (Mt)	9.29	8.53	9.81	11.03	11.55	11.90	12.20
Utilization (Mt)							
Biodiesel (Mt)	1.25	1.45	2.20	3.09	3.48	3.76	3.99
Other (Mt)	7.17	6.67	6.76	6.83	6.92	6.96	6.98
Imports (Mt)	0.02	0.02	0.02	0.02	0.02	0.02	0.02
Exports (Mt)	0.85	0.68	0.91	1.04	1.12	1.17	1.21
Price, Decatur, IL ($/kg)[4]	0.683	0.716	0.796	0.873	0.871	0.897	0.935
Soybean meal							
Production (Mt)	39	35	38	43	44	46	47
Feed utilization (Mt)	31	28	29	29	30	32	32
Exports (Mt)	8	8	9	14	14	14	14
Price, Decatur, IL ($/t)[5]	226	309	310	301	292	310	354
Price of farmland ($/ha)[6]	7,504	9,833	9,980	11,871	12,780	13,474	14,290

[1]Mha= million hectares.
[2]Mt= million metric tons.
[3]Gross margins over variable costs.
[4]Crude, degummed.
[5]48% protein.
[6]Corn belt states.

7.4 Key Biofuel Projections

Vegetable oils, with soybean oil predominant, are the preferred feedstocks for biodiesel production because of the low free fatty acid content. However, beginning in early 2007, a run up in soybean oil prices squeezed profits out of biodiesel facilities using that feedstock. A scramble for cheaper alternatives ensued, mostly for animal fat and yellow grease (recycled fats and oils). The substitution continued in 2008 and, by the end of the year, the processing of non-soybean oil feedstock nearly matched that of soybean oil in biodiesel production. As can be calculated from Tables 7.3 and 7.4, the use of soybean oil and other sources are estimated each to be about the same in the 2008 crop year − 1.45 Mt.

However, because the supplies of animal fat and yellow grease are limited and because they are demanded in other markets (livestock feed and export), only a third of the supplies was assumed to be available as feedstock for biodiesel. Of course, this limit could be raised as cities and states realize the "green" impact of converting yellow grease and even brown (trap) grease into biodiesel.

In any case, additional sources may need to be found to enable the biodiesel industry to expand as forecast. A promising source is corn oil from dry mill ethanol plants extracted at the beginning of the process or from the by-product, DDGS, which contains about 10% corn oil. While not food grade, the product would be acceptable for biodiesel production after the necessary preprocessing to reduce the fatty acid content. In addition, removing corn oil from DDGS improves the handling characteristics and the quality of the feed for dairy rations. Table 7.4 indicates the potential if only 25% of the available corn oil from dry mills is converted to biodiesel toward the end of the projection period.

The price of crude oil and general inflation basically set wholesale gasoline and diesel prices, which, after dropping sharply in 2009, are expected to increase steadily over the 2009–2017 period, but not to the level of 2008 (Table 7.4). These petroleum prices, in turn, predominate in establishing wholesale ethanol (100% ethanol called E-100) and biodiesel (100% biodiesel called B-100) prices in conjunction with the energy program's blenders' tax credit for E-100 ($0.134/L in 2008 and $0.119/L afterward) and B-100 ($0.264/L). Assuming that these tax credits continue through the forecast period, wholesale ethanol and biodiesel prices are projected to hold substantial premiums over their energy equivalents as reflected in the profit margins in Table 7.4. The calculation of the energy equivalents is described in footnote 14 (Table 7.4). Such premiums are needed if biofuels are to meet and exceed the RFSs. For E100, this is a 21 cent/L premium in 2009, 17 cent in 2010, and 14.5 cent afterward. For biodiesel, the premium is set at 15.85 cents/L for 2009–2017. For reference, the premium on ethanol prices averaged $0.11/L and the biodiesel premium averaged $0.203/L in 2008, possibly reflecting minor attention consumers give to retail blends as low as 10% for ethanol and 2 to 20% on biodiesel. In addition, consumers may be willing to pay a premium as a contribution to the environment as they have done in the past.

Also, in Table 7.4 are projected prices on corn and soybean oil on a calendar year basis plus the other alternative feedstocks for biodiesel. Because animal fat and

Table 7.4 Projections of variables related to biofuels

Item	Year						
	2006	2008	2009	2011	2013	2015	2017
Biodiesel feedstock other than soybean oil							
Vegetable oils (Mt)[1]	NA	0.15	0.18	0.19	0.20	0.20	0.20
Animal fat (Mt)[2]	NA	1.25	1.36	1.81	1.55	1.71	1.85
Corn oil from DDGS (Mt)[3]	NA	0.06	0.12	0.32	0.88	0.92	0.96
Wholesale prices ($/L)[4]							
Gasoline[5]	0.52	0.69	0.37	0.44	0.49	0.53	0.57
Ethanol[6]	0.68	0.65	0.49	0.50	0.53	0.55	0.58
Diesel[7]	0.53	0.79	0.37	0.50	0.55	0.60	0.63
Biodiesel[8]	0.75	1.18	0.78	0.87	0.92	0.96	0.99
Feedstock prices (cents/kg)[4]							
Corn, farm	9.0	18.9	15.3	13.9	13.2	14.0	15.9
Soybean oil	53	110	74	86	87	89	92
Tallow[9]	37	75	55	55	55	56	60
White grease[9]	36	74	52	56	56	58	62
Poultry fat[10]	28	61	39	41	40	42	46
Corn oil from DDGS[11]	37	74	53	56	55	57	61
Yellow grease[12]	28	61	43	48	48	50	54
Profit margins (cents/L)[4]							
Ethanol[13]	27.7	3.4	−0.8	0.8	4.5	4.2	2.3
Ethanol (energy)[14]	2.6	−7.7	−21.9	−13.7	−10.0	−10.3	−12.2
Biodiesel							
Soybean oil[13]	10.3	6.3	0.0	−0.3	2.9	4.2	4.0
Soybean oil (energy)[14]	7.7	−14.0	−18.5	−16.1	−12.9	−11.6	−11.8
Tallow	20.9	20.6	2.4	12.7	16.9	18.2	17.2
White grease	22.2	21.9	2.4	11.4	15.3	15.3	13.5

Table 7.4 (continued)

Item	Year						
	2006	2008	2009	2011	2013	2015	2017
Poultry fat	29.1	33.8	17.2	25.6	30.9	32.0	30.4
Corn oil from DDGS	21.1	19.3	10.6	18.8	22.7	23.8	22.7
Yellow grease	16.1	20.3	13.2	19.0	23.5	24.6	23.0
Ethanol feed by-products ($/t)							
Corn gluten feed[15]	78.3	109.2	102.5	93.7	92.6	100.3	118.0
Corn gluten meal[15]	370.5	455.3	399.1	398.0	392.5	411.2	457.6
DDGS[16]	121.3	125.7	118.0	111.4	109.2	116.9	134.5

[1]Mt = million metric tons. Crop year averages.
[2]Includes yellow grease.
[3]Includes corn oil extracted from ethanol plants before processing.
[4]Calendar year averages.
[5]Refiner prices for resale (DOE).
[6]F.O.B.,Omaha, NE.
[7]No. 2, refiner prices for resale (DOE).
[8]Upper Midwest (Jacobsen Publishing Co.).
[9]Bleachable fancy tallow and choice white grease, Chicago (Jacobsen).
[10]Stabilized, Alabama/Georgia (Jacobsen).
[11]A synthetic price derived as the average for tallow and white grease prices (Jacobsen).
[12] (Jacobsen).
[13]Costs include feedstock, direct processing, depreciation, and a nominal return on investment for a new 190-ML ethanol plant or a 38-ML biodiesel plant.
[14]Assumes that ethanol and biodiesel are priced at their energy value relative to petroleum plus the blenders' tax credit. This would be two-thirds of retail gasoline prices plus 11.9 cents for ethanol and 92% of retail diesel prices plus $0.264 for biodiesel, both adjusted back to the wholesale level by the marketing spread.
[15]Illinois points (USDA, ERS 2009).
[16]Lawrenceburg, IN.

yellow grease substitute for energy feeds like corn in livestock rations, their prices are strongly correlated with prices on corn as well as soybean oil.

The cost estimates for ethanol and biodiesel production, as described in footnote 13 (Table 7.4), were based on relatively small plants from surveys and process models (Shapouri and Gallagher, 2005; Haas et al., 2006; Canakci and Van Gerpen, 2001). Costs over time after these studies were completed were estimated from input price changes. Profits would tend to be higher for larger plants due to returns to scale, but the trends would be largely intact as shown in Table 7.4.

While the prices on biodiesel from soybean oil were established to provide nominal profits, the use of animal fat, DDGS corn oil, and yellow grease would provide very attractive profit margins. This was the case even though extra preprocessing costs relative to soybean oil were included and the product discounted. The problems for the cheaper feedstocks have been consistent quality and performance in cold weather. However, by blending in these feedstocks with soybean oil and other virgin oil, the dual problem of economics and quality control may be effectively addressed. On the other hand, even as technology closes the quality gaps between biodiesel made from soybean oil and the alternative feedstock, prices on animal fat, DDGS corn oil, and yellow grease will likely bid up to levels that will reduce the profits from these feedstocks substantially.

Price forecasts for the livestock feed by-products of ethanol production – corn gluten feed and meal from wet mills and DDGS from dry mills – involved their conversion to energy and protein equivalents (Table 7.4). The feed prices were derived from synthetic energy and protein prices based on corn and soybean prices by a procedure outlined by Ferris (2006). In addition, prices on DDGS will be under some pressure because of increased supplies. Nearly all the increase in ethanol production over the forecast period is expected to be from dry mills.

7.4.1 Impacts on Livestock

The implications of the expansion in biofuels for the US livestock industry are outlined in Table 7.5. The cyclical decline in beef cow numbers which began in 2007 is expected to continue through the midpoint of the forecast period with beef production holding relatively steady as productivity increases. The pork industry is also slated for a cut in production in response to unprecedented losses in 2008. By 2014, pork production should have recovered to the 2008 level. Broiler output is expected to be the major growth sector in the meat complex with turkey fairly stable. Growth is projected for both egg and milk production.

The impact of higher feed costs and general inflation is reflected in the projection set on livestock prices in Table 7.5. As with feed prices, livestock prices will be holding at a higher level than in the recent past. In general, production will adjust to bring real gross margins over feed costs closer to long-term averages than experienced in 2008.

Table 7.5 Projections of livestock variables, calendar years

Item	Year						
	2006	2008	2009	2011	2013	2015	2017
Livestock production							
Beef (Mt)[1]	12.03	12.10	11.89	11.76	11.54	11.56	11.63
Pork (Mt)	9.56	10.60	10.43	10.07	10.51	10.97	11.06
Broiler (Mt)	15.93	16.54	16.23	17.03	17.88	18.68	19.42
Turkey (Mt)	2.54	2.80	2.70	2.66	2.67	2.69	2.71
Eggs (Mil. Doz)	7,610	7,514	7,529	7,735	7,919	8,100	8,255
Milk (Mt)	82.46	86.05	85.75	86.76	86.67	91.71	94.92
Livestock prices							
Choice steers ($/kg)[2]	1.88	2.03	1.90	2.05	2.17	2.26	2.34
Barrows and gilts ($/kg)[3]	1.04	1.05	1.07	1.21	1.20	1.13	1.24
Broilers ($/kg)[4]	1.42	1.76	1.82	1.85	1.91	1.97	2.08
Turkeys ($/kg)[5]	1.70	1.93	1.84	1.96	2.02	2.08	2.16
Eggs ($/doz)[6]	0.72	1.28	1.21	1.08	1.09	1.10	1.13
Milk, average farm ($/kg)	0.28	0.40	0.29	0.36	0.40	0.39	0.38

[1] Mt = million metric tons.
[2] Nebraska, Direct, 1,100–1,300 lbs (500–590 kg).
[3] National Base, Live equiv. 51–52% lean.
[4] Wholesale, 12-city average.
[5] 8–16 lbs (3.6–7.3 kg) hens, Eastern Region.
[6] Grade A large, New York, volume buyers.

7.4.2 Consumer Prices

The future combined impacts of the elevated energy and farm prices on inflation in consumer prices can be observed in Table 7.6. While the annual increase from 2.8% on all items in the Consumer Price Index (CPI) in calendar 2007 to 3.8% in 2008 may have caused concern, double digit inflations were common in the 1970s and early 1980s. The average annual inflation rate between 1970 and 1982 was nearly 8%. With virtually no inflation forecast for 2009, the rate is projected to accelerate moderately before converging to about 2% in the latter part of the forecast period.

On food, inflation at 5.5% in 2008 is projected to drop to 1.0% in 2009, averaging about 2% in 2010–2017, but less than on all items. The CPI on energy, increasing 14% in 2008, is expected to drop sharply in 2009 and stabilize in 2010, exceeding the inflation on all items in 2011–2017. Clearly, both energy and food are the volatile sectors in the CPI. The weighting for food is about 14% and energy about 9%. These weights will need to be revised upward.

Energy prices affect food prices as they impact both farm production costs and marketing costs. In 2005–2007, the farm-to-retail spread represented 78% of the retail price of the so-called Market Basket of US farm foods (USDA, 2009). Retail

Table 7.6 Projections of inflation rates for consumer price indices and selected foods[1]

	Year						
Item	2006	2008	2009	2011	2013	2015	2017
General indices (% change)							
All items	3.2	3.8	0.1	2.7	1.7	2.1	2.0
Food	2.4	5.5	1.0	2.4	1.6	1.8	1.8
Food at home	1.7	6.4	1.1	2.4	1.5	1.7	1.8
Food away from home	3.1	4.4	0.9	2.4	1.7	1.9	1.8
Energy	11.2	14.0	−24.4	7.5	3.6	2.9	2.6
Selected food items (% change)							
Beef	0.9	4.5	−0.9	3.4	1.9	2.2	2.0
Pork	−0.2	2.3	−0.2	3.2	1.4	1.3	1.7
Broilers	−2.5	5.1	2.1	3.3	1.3	1.9	2.1
Turkeys	1.1	4.7	−0.4	2.6	1.3	1.6	1.7
Eggs	4.9	14.0	1.0	2.5	0.4	1.1	1.6
Dairy products	−0.6	8.0	−4.2	3.1	1.9	1.2	1.5
Cereals and bakery	1.8	10.3	5.4	1.8	1.1	1.6	1.9
Fats and oils	0.2	13.8	−4.3	2.5	1.3	1.7	1.7

[1]U.S city average, US Department of Labor, Bureau of Labor Statistics.

price changes on selected food items in Table 7.6 reflect both the variation in farm prices and the secular increases in the farm-to-retail price spread. The marketing spread dominates for cereals and fats and oils, which has represented about 93 and 80% of the retail price, respectively. However, in 2008, the increase in the calendar year average price of wheat and soybean oil of about 40% contributed noticeably to the increased prices at retail. For beef, pork, and dairy products, increased retail prices were due mostly to the impact of inflation on marketing costs. Special situations relate to anticipated changes in retail prices on selected foods in 2009 and 2010, with annual inflation between 1 and 3% in most cases afterward.

7.5 Alternative Scenarios

The dominance of crude oil prices in establishing the markets on gasoline, diesel, and their biofuel counterparts is evident along with feedstock prices (Tables 7.1 and 7.4). Unfortunately, crude oil prices are very difficult to forecast as is evident from the performance in recent years of major models designed to generate long-range energy projections. For that reason, three alternative scenarios were introduced into AGMOD based upon the "Low, Reference, and High" projections of crude oil prices by the US Department of Energy (USDOE, December 2008). These projections along with the baseline (futures) set presented in this paper are illustrated in Fig. 7.1

As can be noted, the DOE's Reference projection is substantially higher than the crude oil prices generated from the NYMEX's contracts as of February 27, 2009 with the low holding near to the level of early 2009. The possibility of crude oil

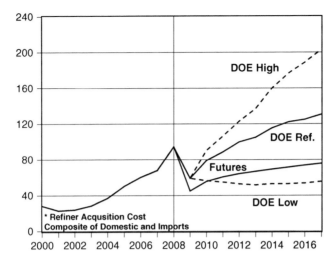

Fig 7.1 Annual average crude oil prices, 2000 to 2008 and projected to 2017 by futures and the DOE ($/Barrel).*

prices reaching $200 per barrel by 2017 was suggested, a figure frequently quoted as realistic in mid-2008. To present the implications of these alternative crude oil projections, four scenarios were developed with the results displayed in Tables 7.7, 7.8, and 7.9 as averages for 2010–2017.

The assumptions for the "DOE Low" scenario were similar to the baseline (futures) described earlier, except that the premium over the energy value for ethanol was set at 17.2 cents/L, 2.7 cents/L higher than in the baseline. The premium on biodiesel was set at 21.1 cents/L, 5.3 cents/L higher than in the baseline. No change was made in the projections for ethanol or biodiesel production.

In the DOE Reference scenario, the profit expectations for ethanol production would be encouraging enough to use the current (2009) excess capacity of over 18.9 GL (including plants under construction) estimated by the Renewable Fuels Association (RFA, 2009). In addition, another 11.3 GL of capacity was projected for the 2010–2017 period, bringing total ethanol production (mostly from corn starch) to over 75 GL by 2017. With the increased production, the ethanol premium over the energy value was reduced to 7.9 cents/L and, by 2013, the blenders' tax credit was reduced from 11.9 cents/L to 3.9 cents/L.

For biodiesel in the "DOE Reference" scenario, with the current (2008) capacity estimated by the National Biodiesel Board at 9.8 GL, production would increase rapidly into 2011 before leveling off, reaching 11.4 GL by 2017 (NBB, 2008). With the increased production, the biodiesel premium over the energy value would drop to 5.3 cents/L and by 2014 the blenders' tax credit would be cut from $0.26/L to $0.13/L.

In the "DOE High" scenario, production would reach 95 GL by 2017, ethanol prices would drop to the energy value by 2014, and the blenders' tax credit would be

Table 7.7 Impacts on biofuels from alternative scenarios[1]

Item	Scenario[2]			
	DOE low	Futures	DOE reference	DOE high
	Average for 2010–2017			
Crude oil prices[3]	54	67	108	148
Ethanol				
Production (GL)	52.52	52.52	61.16	66.78
Gasoline prices ($/L)	0.39	0.49	0.81	1.12
Ethanol prices ($/L)	0.49	0.54	0.63	0.72
Corn prices, farm ($/t)				
From AGMOD	138.98	141.73	160.24	179.13
Ethanol break even	143.70	152.76	174.41	191.34
Processing costs ($/L)[4]	0.15	0.16	0.21	0.25
DDGS prices ($/t)	113.56	114.66	130.10	143.33
Profit margins				
Projected ($/L)	0.01	0.03	0.04	0.07
At energy value ($/L)	−0.17	−0.12	−0.03	0.04
Biodiesel				
Production (GL)	6.44	6.44	9.89	10.60
Diesel prices ($/L)	0.45	0.56	0.90	1.22
Biodiesel prices ($/L)	0.86	0.92	1.06	1.23
Soybean oil prices ($/kg)	0.84	0.87	0.96	1.04
Processing costs ($/L)[5]	0.12	0.12	0.14	0.16
Profit margins				
Soybean oil				
Projected ($/L)	0.01	0.03	0.07	0.16
At energy value ($/L)	−0.20	−0.13	0.01	0.10
Tallow ($/L)	0.13	0.16	0.22	0.31
White grease ($/L)	0.11	0.13	0.19	0.25
Poultry fat ($/L)	0.25	0.29	0.36	0.45
Corn oil from DDGS ($/L)	0.19	0.22	0.27	0.35
Yellow grease ($/L)	0.19	0.22	0.28	0.36

[1]Check footnotes to Table 7.4.
[2]The scenarios represent the low, reference, and high crude oil price projections of the US Department of Energy's Energy Information Administration (Early Release, December 2008) and the derivation from the crude oil futures on the NYMEX as of February 27, 2009.
[3]Refiner acquisition cost, domestic and imported composite.
[4]Processing costs (not including feedstock) less returns from DDGS.
[5]Processing costs (not including feedstock) for soybean biodiesel less returns from glycerol.

eliminated. For biodiesel, production would reach 13.2 GL by 2017, the premium would remain at 5.3 cents/L, and the blenders' tax credit would be eliminated by 2013.

Of course, these scenarios are quite subjective and were developed by observing the profit margins evolving. Alternative scenarios could be established by modifying feedstock costs as well as or in addition to ethanol and biodiesel prices.

Table 7.8 Impacts on selected crop and livestock variables from alternative scenarios, average for 2010–2017[1]

| | Scenario[2] | | | |
| | DOE low | Futures | DOE reference | DOE high |
Item	Average for 2010–2017			
Crude oil prices[3]	54	67	108	148
Crops[4]				
Soybean prices, farm ($/t)	337	344	384	431
Soybean meal prices ($/t)	307	311	350	400
Wheat prices, farm ($/t)	192	196	224	253
Variable costs				
Corn ($/ha)	679	706	798	886
Soybeans ($/ha)	353	363	395	427
Gross margins, crops ($/ha)[5]	674	684	765	891
Price of farmland ($/ha)	12,610	12,879	13,756	14,509
Total acreage, 2017 (Mha)[6]	93	93	96	98
Livestock[7]				
Production				
Beef (Mt)	11.64	11.64	11.57	11.55
Pork (Mt)	10.62	10.61	10.38	10.24
Broiler (Mt)	18.06	18.06	18.02	18.00
Turkey (Mt)	2.68	2.68	2.66	2.65
Eggs (Mil. Doz)	7,963	7,961	7,935	7,925
Milk (Mt)	89.63	89.54	88.91	88.51
Prices				
Choice steers ($/kg)	2.16	2.18	2.27	2.35
Barrows and gilts ($/kg)	1.17	1.19	1.28	1.37
Broilers ($/kg)	1.91	1.93	2.03	2.12
Turkeys ($/kg)	2.01	2.03	2.12	2.19
Eggs ($/doz)	1.08	1.09	1.14	1.19
Milk, farm ($/kg)	0.37	0.38	0.39	0.40

[1] Check footnotes to Tables 7.3, 7.4, and 7.5.
[2] Check Footnote 2 in Table 7.7.
[3] Refiner acquisition cost, domestic and imported composite.
[4] Crop year averages.
[5] Average gross margins over variable costs for corn, soybeans, and wheat, weighted by harvested area, and includes returns from government payments as well as from the market.
[6] Harvested areas of coarse grain, wheat, and soybeans.
[7] Calendar year averages.

7.5.1 Impact of the Alternative Scenarios

In each of Tables 7.7, 7.8, and 7.9, the average crude oil price for 2010–2017 is indicated for the four scenarios. Rather than presenting an extensive commentary on each table, only a few observations will be made. Most of the results need little explanation. In Table 7.7, one might have expected higher feedstock prices in the DOE Reference and High alternatives. However, over an 8-year span, the

Table 7.9 Impacts on the consumer price indices, US city average (1982–1984 = 1.000) from alternative scenarios, average for 2010–2017

| | Scenario[1] | | | |
| | DOE low | Futures | DOE reference | DOE high |
Item	Average for 2010–2017			
Crude oil prices[2]	54	67	108	148
General indices				
All items	2.365	2.389	2.473	2.552
Food	2.346	2.367	2.438	2.505
Food at home	2.330	2.352	2.434	2.509
Food away from home	2.368	2.385	2.444	2.499
Energy	1.857	2.080	2.800	3.481
Selected food items				
Beef	2.428	2.456	2.560	2.652
Pork	2.015	2.035	2.119	2.196
Broiler	2.216	2.238	2.323	2.399
Turkey	2.080	2.100	2.165	2.226
Eggs	2.174	2.196	2.288	2.368
Dairy products	2.217	2.240	2.316	2.386
Cereals and bakery	2.257	2.781	2.883	2.974
Fats and oils	2.038	2.062	2.136	2.203

[1]Check Footnote 2 in Table 7.7.
[2]Refiner acquisition cost, domestic and imported composite.

higher prices early in the period generate responses which increase production and carryovers toward the end. Certainly, as mentioned before, the profit margins for biodiesel produced from animal fat, yellow grease, and corn oil from DDGS will be diminished as prices on these feedstocks will be bid up. In addition, opportunities will be attractive to blend soybean oil and other virgin vegetable oils with these alternative feedstocks to achieve the needed quality and reduce the free fatty acid content.

Table 7.8 provides more detail on the effects on the soybean complex and wheat. Also indicated is how much higher energy prices would impact variable costs of production on corn and soybeans. In AGMOD, the major determinant of the total acreage harvested of the major field crops of corn, sorghum, barley, oats, wheat, and soybeans is the aggregate gross margins on these crops. By 2017, these higher gross margins under the DOE High scenario would increase these acres by 4.5 Mha over the baseline. Also, farmland prices in the Midwest would reach nearly $14,500/ha with high energy prices vs. $12,800/ha in the baseline.

The effects on the livestock sector might be somewhat less than might be expected. Production under the higher price scenarios would be slightly lower with prices more noticeably higher due to inelastic demands. As with crops, livestock producers adjust output over time and prices converge to equilibrium levels.

To summarize the results between the DOE High scenario and the baseline (futures) in Table 7.9, the CPI for all items would be 6.8% higher in the high case.

On food consumed at home, prices would be 6.7% higher compared to only 4.8% for food consumed away from home. This is because farm prices will increase more percentage-wise than the marketing spread, which is closely tied to general inflation. In the CPI, the weight for food consumed away from home is 43% of all food. The elevation in retail prices for selected food items would range from 6% for turkey to 8% for beef.

7.6 Prospects for Cellulosic Ethanol

The renewable fuel standards under the EISA, 2007 set forth a phase-in for renewable fuel volumes beginning with 34 GL in 2008 and growing to 136 GL by 2022. The Act further defines and establishes separate volumetric targets for conventional biofuels, advanced biofuels, cellulosic biofuels, and biomass-based diesel. Conventional biofuel is defined as ethanol derived from corn starch that achieves at least 20% greenhouse gas (GHG) emissions reduction compared to baseline life-cycle GHG emissions (for plants coming on line after 2008). Advanced biofuel is derived from renewable biomass other than corn starch and achieves a 50% GHG emissions reduction. Cellulosic biofuel is renewable fuel derived from any cellulose, hemicellulose, or lignin that is derived from renewable biomass and achieves a 60% GHG emission reduction.

The conventional biofuel volumes are limited to 56.8 GL and advanced biofuel volumes are mandated to be 79.5 GL, including 60.5 GL of cellulosic ethanol by year 2022. The cellulosic ethanol mandate starts in 2010 at 0.38 GL, rising to 20.8 GL by 2017.

However, we assume in our projections that production of cellulosic ethanol will be negligible through 2017. Despite aggressive targets set under EISA 2007, commercial feasibility of cellulosic ethanol on a large scale remains a formidable challenge for a number of reasons (Collins, 2007). Based on our projected gross margins for coarse grains and soybeans, significant switching of land from conventional crops to growing energy crops such as switchgrass is unlikely. Demand for renewable electricity is likely to limit forest biomass supply for ethanol conversion. Estimated biomass feedstock costs under current productivity conditions are considerably higher than feedstock costs assumed in previous studies of economic competitiveness of cellulosic ethanol. Cellulose conversion technologies continue to face major uncertainties including the fundamental question of which is the best pathway for conversion of biomass to ethanol, thermochemical or biochemical or a combination of the two as proposed by Coskata (2008). In addition, problems related to system integration, commercial scale up, and overall process optimization remain unresolved. Projected capital requirements for cellulosic ethanol plants are much higher than for corn-ethanol dry mills. The estimated capital costs of annual capacity of ethanol range from $0.75/L (Aden et al., 2002) to $1.44/L (McAloon et al., 2000) for cellulosic plants compared to a range of $0.28/L to $0.80/L for corn-ethanol plants and $0.05/L to $0.26/L for expansion of existing plants (Shapouri and Gallagher, 2005). Significant new investments will also be necessary in establishing appropriate biomass supply chains including harvesting and storage infrastructure. Several potential supply chain configurations are possible including centralized processing at large biorefineries (Aden et al., 2002), a system of regional biomass preprocessing centers (Carolan et al., 2007), and on-farm processing.

To help bring cellulosic ethanol to market, DOE is investing over $385 million for six biorefinery projects with a total investment of more than $1.2 billion (US Department of Energy, 2007). These plants use a variety of cellulosic feedstock such as urban yard and wood waste, wheat and barley straw, corn stover, switchgrass, wood residues, and woody energy crops. By 2012, when fully operational, these biorefineries are expected to produce about 0.57 GL of cellulosic ethanol per year. Future development of the cellulosic ethanol industry critically depends on technical and commercial success of these DOE-funded plants. Although commercial-scale ethanol plants have been proposed by Verenium-BP (136 ML/year), Range fuels (75 ML/year), DuPont-Danisco, and Mascoma (150 ML/year), we project that cellulosic-ethanol production will be minor through 2017 in view of current depressed oil prices and the crisis in the credit markets.

Following a similar logic, the Annual Energy Outlook 2008 from DOE assumes that total cellulosic-ethanol production will be limited to 0.57 GL from these six DOE projects by 2012; and cellulosic ethanol will contribute less than 4% of total fuel ethanol produced in the United States by the year 2017. In fact, the Annual Energy Outlook 2008 assumes that imported cane ethanol will count toward meeting the advanced biofuel mandate; and even by 2022, cellulosic ethanol will contribute only 27.3 GL compared to the EISA mandate of 60.6 GL (US Department of Energy, 2008).

Given the uncertainty about whether the new RFS schedule can be achieved, EISA 2007 contains a general waiver based on technical, economic, or environmental feasibility; and the EPA administrator is also provided some flexibility in setting GHG limits. Further, the cellulosic biofuel mandate includes a credit program that is activated in years when the mandated level of cellulosic biofuel is judged by the EPA Administrator as unlikely to be met. For all the fuel mandates, if there is a 20% deficit in more than two consecutive years or a 50% deficit in any one year, regulatory adjustment mechanisms are provided to lower the mandated levels from that point forward. This rule, which could be enacted by the EPA Administrator, would modify all applicable volumes (including the overall and advanced biofuel totals) for all subsequent years (EIA, 2008). We expect that this regulatory flexibility will be exploited to meet/alter the provisions of EISA 2007. However, we do not have forecasts on how the policies will evolve.

References

Aden A, Ruth M, Ibsen K, et al. (2002) *Lignocellulosic biomass to ethanol process design and economics utilizing co-current dilute acid prehydrolysis for corn stover.* NREL/TP-510-32438. Golden, CO: US Department of Energy, National Renewable Energy Laboratory. http://www1.eere.energy.gov/biomass/pdfs/32438.pdf Accessed 11 September 2007.

Canakci M, Van Gerpen J (2001) *A pilot plant to produce biodiesel from high free fatty acid feedstocks*.Paper No. 016049. 2001 ASAE Annual International Meeting, Sacramento, CA, July 30-August 1, 2001. http://asea.frymulti.com/request.asp?JID=5&AID=4209&CID=sca2001&T=2 Accessed 11 September 2007

Carolan J, Joshi S, Dale B (2007) Technical and financial feasibility analysis of distributed bio-processing using regional biomass pre-processing centers. *J Agric Food Industrial Org*, 5 (2) http://www.bepress.com/jafio/vol5/iss2/art10/

Collins K (2007) *Statement of Keith Collins, Chief Economist, U.S. Department of Agriculture before the U.S. Senate Committee on Agriculture, Nutrition and Forestry*, January 10, 2007. Washington, DC: Office of the Chief Economist, US Department of Agriculture, http://www.usda.gov/oce/newsroom/congressional_testimony/Collins_011007.pdf Accessed 20 June 2008.

Coskata (2008) http://www.coskata.com /Accessed 20 June 2008.

Ferris J (2005) *Agricultural prices and commodity market analysis*. 2nd ed. East Lansing, MI: Michigan State University Press: 136–137.

Ferris J (2006) *Modeling the US domestic livestock feed sector in a period of rapidly expanding by-product feed supplies from ethanol production*. Staff Paper 2006-34. East Lansing, MI: Michigan State University, Department of Agricultural Economics.

Haas M, McAloon A, Yee W, Foglia T (2006) A process model to estimate biodiesel production costs. *Bioresource Technology* 97 (4):671–678.

Jacobsen Publishing Co (2009) *Biodiesel Bulletin. Fats and Oils Bulletin. Grain & Food – The Feed Bulletin*. Various issues and archives. 1123 West Washington, Chicago, IL

McAloon A, Taylor F, Yee W, Ibsen K, Wooley R (2000) *Determining the cost of producing ethanol from corn starch and lignocellulosic feedstocks*. NREL/TP-580-28893. Golden, CO: US Department of Agriculture and US Department of Energy, National Renewable Energy Laboratory. http://www.ethanol-gec.org/information/briefing/16.pdf Accessed 20 June 2008.

National Biodiesel Board (2008) *Commercial Biodiesel Production Plants*, www.nbb.org. Accessed 20 September 2008.

Renewable Fuels Association (2009) *2009 Ethanol Industry Outlook*.February 2009.

Shapouri H, Gallagher P (2005) *USDA's 2002 Ethanol cost-of-production survey*. Agricultural Economic Report Number 841. Washington, DC: US Department of Agriculture, Office of Energy Policy and New Uses. http://www.usda.gov/oce/reports/energy/USDA_2002_ETHANOL.pdf Accessed 20 June 2008.

US Department of Agriculture, Economic Research Service (2009) *Statistical Indicators*. Table 8. January. http://www.ers.usda.gov/Publications/Agoutlook/AOTables /

US Department of Agriculture, Economic Research Service (2008) *International macroeconomic data set*. http://www.ers.usda.gov/data/macroeconomics/, Accessed 20 December 2008.

US Department of Agriculture, Foreign Agriculture Service (2008) *Production, Supply and Distribution Online*. http://www.fas.usda.gov/psdonline/psdhome.aspx Accessed 20 December 2008.

US Department of Energy (2007) *DOE selects six cellulosic ethanol plants for up to $385 million in federal funding*. Press Release 4827. Washington, DC: US Department of Energy, February 28, 2007. http://www.energy.gov/news/4827.htm Accessed 11 September 2007

US Department of Energy, Energy Information Administration (2008) *Annual energy outlook 2008 with projections to 2030*. Report # DOE/EIA-0383(2008) Washington, DC: US Department of Energy, Energy Information Administration. http://www.eia.doe.gov/oiaf/aeo/pdf/0383(2008).pdf Accessed 28 June 2008

US Department of Energy, Energy Information Administration (2008) *Annual energy outlook 2009*. #DOE/EIA – 0383(2009) Early Release.

Chapter 8
The Global Bioenergy Expansion: How Large Are the Food−Fuel Trade-Offs?

Jacinto F. Fabiosa, John C. Beghin, Fengxia Dong, Amani Elobeid, Simla Tokgoz, and Tun-Hsiang Yu

Abstract We summarize a large set of recent simulations and policy analyses based on FAPRI's world multimarket, partial-equilibrium models. We first quantify and project the emergence of biofuel markets in US and world agriculture for the coming decade. Then, we perturb the models with incremental shocks in US and world ethanol consumption in deviation from this projected emergence to assess their effects on world agricultural and food markets. Various food−biofuel trade-offs are quantified and examined. Increases in food prices are moderate for the US ethanol expansion and even smaller for the ethanol expansion outside the United States, which is based on sugarcane feedstock, and which has little feedback on other markets. With the US expansion, the high protection in the US ethanol market limits potential adjustments in the world ethanol markets and increases the demand for feedstock within the United States. Changes in US grain and oilseed market prices propagate to world markets, as the United States is a large exporter in these markets. With changes in world prices, land allocation in the rest of the world responds to the new relative prices as in the United States but with smaller magnitudes because price transmission to local markets is less than full.

J.F. Fabiosa (✉)
Iowa State University, Ames, IA, USA
e-mail: jfabiosa@iastate.edu

Authors: All authors are affiliated with Iowa State University unless noted otherwise
- Jacinto F. Fabiosa is a scientist and co-director of FAPRI-ISU
- John C. Beghin is a Marlin Cole Professor of International Agricultural Economics
- Fengxia Dong is an associate scientist
- Amani Elobeid is an associate scientist
- Simla Tokgoz was an associate scientist and is now a research fellow at IFPRI
- Tun-Hsiang Yu was an associate scientist and is now an assistant professor at the University of Tennessee in Knoxville

M. Khanna et al. (eds.), *Handbook of Bioenergy Economics and Policy*,
Natural Resource Management and Policy 33, DOI 10.1007/978-1-4419-0369-3_8,
© Springer Science+Business Media, LLC 2010

8.1 Introduction

This chapter summarizes a large set of recent simulations and policy analyses focusing on US and global biofuel markets. These were undertaken at the Center for Agricultural and Rural Development at Iowa State University and based on the world multimarket, partial-equilibrium agricultural models of the Food and Agricultural Policy Research Institute (FAPRI). We first quantify and project the emergence of biofuel markets in US and world agriculture for the coming decade using the *FAPRI 2008 U.S. and World Agricultural Outlook* and earlier editions. Then, we perturb the models with incremental shocks in US and world ethanol consumption in deviation from the projected emergence to assess their effects on world agricultural and food markets. Various food−biofuel trade-offs are quantified and examined. We discuss how trade and energy policies condition these trade-offs, in both the United States and the world.

The world FAPRI models[1] capture the biological, technical, policy, and economic relationships among key variables within a particular commodity and across commodities and countries or regions. They are based on historical data analysis, current academic research, and a reliance on accepted economic, agronomic, and biological relationships in agricultural production and markets. The analysis incorporates major trade-offs among bioenergy, feedstock, feed, and food production and consumption arising with the emergence of biofuels. The analysis accounts for grain- and sugar-based ethanol, biodiesel, and the potential for cellulosic ethanol production. See Appendix A, available from the authors for further details on the model description.

The food−biofuel trade-offs appear first in the competing demand for feedstock crops among food demand, feed demand, and energy demand. These competing demands increase the prices of feedstock crops and hence the prices of feed and food items intensive in these crops and intensive in feed, namely meat and dairy products. Second, the change in prices of feedstock crops translates into supply effects. Land use is attracted to profitable feedstock crops and moves away from other crops. As a result, the prices of the nonfeedstock crops increase and have a second-round effect on food prices for products intensive in these other crops, namely oilseeds and products. Another important feedback effect originates in grain-based ethanol production, which generates 30% co-product called dried distillers grain (DDG), which can be used to substitute corn at a rate close to one-for-one, alleviating the pressure for more land (see Appendix B, available from the authors, for more details on this substitution). The use of DDG in animal feed varies depending on the type of livestock; it is highest in the beef cattle sector at a maximum inclusion rate of 40% (for

---

[1] These sets of models are jointly developed by a consortium, with the international models developed by FAPRI at Iowa State University and the US models by FAPRI at the University of Missouri. The individual commodity models or the full set of models have been employed for numerous studies, such as Abler et al. (2008), Elobeid et al. (2007), Fabiosa et al. (2007), Elobeid and Beghin (2006), and Fabiosa et al. (2005). The FAPRI models have been used by several organizations to develop baselines as well as for policy analysis.

wet distillers grain), followed by the dairy sector at 20%, the pork sector at 20%, and the poultry sector at 10% (Fabiosa 2008a). A major concern of livestock and poultry producers in the adoption of DDG as a feed ingredient is its quality, including stability of its nutritional content, storability, and ease of transport (Fabiosa 2008b).

In Section 2, we examine the projected evolution of production, consumption, trade, and land allocation under biofuel emergence by type of crop for key countries growing feedstock for ethanol (corn, barley, wheat, sugarcane, and other grains) and for major crops competing with feedstock for land resources, such as oilseeds. We incorporate feedback effects on prices for meat, dairy, oil, and food products intensive in these crops, and on live-animal production. Section 2 is based on recent issues of the *FAPRI U.S. and World Agricultural Outlook*.

We then report on two major analyses that identified the effects of growing energy demand for agricultural inputs. We decompose the global biofuels emergence in terms of domestic (US) and rest-of-the-world components. Using the analysis of Tokgoz et al. (2007), we look at the implications of a domestic (US) ethanol expansion driven by higher oil prices combined with extensive adoption of flexible-fuel vehicles (FFVs) using E-85 fuel. In this analysis, we compare the long-term equilibrium imposed in 2016/2017 under a baseline established for the analysis with the long-term equilibrium under a scenario with higher oil prices and no bottleneck in FFVs and E-85 markets. Long-term equilibrium means that on the supply side, profit margins for ethanol producers have been exhausted, no incentives exist to enter or exit the ethanol industry, and returns in the dairy and livestock industry are normal, while at the same time on the demand side, the price of ethanol approaches its energy value, removing any incentive to invest in FFVs.[2] In a second analysis by Fabiosa et al. (2007), we shocked the models with exogenous expansion in ethanol demand in Brazil, China, the European Union, and India. The scenario captures the essence of the global expansion outside the United States. We compute shock multipliers for land allocation, production, consumption, trade, and prices for the various crops, food items, and countries of interest. The multipliers, reported as the average of 10 annual multipliers between 2007/2008 and 2016/2017, show how sensitive (or not) these variables are to the growing demand for ethanol in foreign countries with sizeable ethanol markets and other countries growing feedstock crops. We also highlight the movement of area away from major crops that compete with major feedstock crops used for ethanol production.

These investigations include all major policy distortions affecting relevant markets and international feedback effects through world commodity prices and trade. Because of the high US tariffs on ethanol, higher US demand for ethanol essentially translates into a US ethanol production expansion. This production expansion has strong global effects on land allocation and output because the prices of coarse grains, the major feedstock in the United States, transmit significant shocks

---

[2]See Tokgoz et al. (2007), pages 3–5, and Elobeid et al. (2007) for further discussion of these conditions and implications. The two scenarios (expansions inside and outside the United States) are, respectively, based on dedicated baselines. Changes in assumptions occur between these baselines although their underlying modeling approach is the same.

worldwide. These price effects eventually trickle down to all crops, meat and dairy prices, vegetable oil prices, and other food prices worldwide. In contrast, expansion in ethanol use and production in the rest of the world chiefly affects sugarcane area and production, the major feedstock in Brazil and other countries, and to a lesser extent, in other sugar-producing countries. Land uses and food prices other than those for sugar show little change in most countries.

The impact of the US expansion on US food prices is significant but moderate overall because agriculture's share of food cost is small for many food items. Following large increases in corn prices and their direct effect of higher feed costs, US food prices would increase by more than 1% over baseline levels. Beef, pork, and poultry prices would rise by more than 4% and egg prices would rise by about 7% (Tokgoz et al. 2008). The impact on world food prices of this expansion is likely to be larger wherever agricultural inputs represent a larger share of the food cost for households consuming food that is less processed and closer to the farm-gate.

Finally, we look at the effect of conditioning exogenous factors of the identified trade-offs, such as fossil energy prices, and policy distortions (US tariffs on ethanol and the US ethanol tax credit). The impact of altering these policy distortions is assessed.

8.2 Stylized Facts on the Global Emergence of Biofuels

8.2.1 Biofuels in the United States

US ethanol production increased dramatically in the last few years, from 2.1 billion gallons in 2002 to 6.5 billion gallons in 2007 (EIA 2008a). Consequently, the demand for corn used in ethanol production has increased accordingly, exceeding the level of corn exports in 2007. Higher demand for corn translated into higher corn prices and higher corn acreage. In 2007, corn area increased by more than 15 million acres in response to rising corn prices, which have doubled in the past couple of years (USDA-NASS 2008). The increased area dedicated to corn came at the expense of competing crops such as soybeans and wheat. This led to lower production and therefore higher prices for these crops. Hence, the ethanol expansion in the United States has resulted in higher agricultural commodity prices and land being bid away from competing crops to corn. With the minimum target levels of use of various biofuels required by the Energy Independence and Security Act of 2007 (EISA), corn production is expected to continue to increase in subsequent years to meet growing demand for fuel, feed, and food. The higher demand for ethanol is also met by higher imports from countries like Brazil despite high US border tariffs. The US biodiesel industry is relatively small and the demand for biodiesel is primarily driven by the mandates issued in the EISA.

8.2.2 Biofuels in the World

Prior to 2006, Brazil was the major producer and consumer of ethanol. However, Brazil has produced less ethanol than the United States in the past 3 years. Brazil's production increased from 4.1 billion gallons in 2005 to 5.2 billion gallons in 2007 (USDA-FAS 2008a). Countries such as China and India are emerging as significant producers of ethanol, as more countries begin to promote ethanol as an alternative fuel, mainly through mandates and/or directives.[3] As countries increase their production of ethanol, the higher supply leads to a decline in the world ethanol price. However, with increased use of "advanced biofuels" under provisions of the EISA in the United States, including imported sugar-based ethanol, the world ethanol price is expected to increase. The EU-27 has the most mature biodiesel industry in the world market. The biofuel targets in each member state push up the production for biodiesel; however, domestic demand still needs to rely on imports from emerging producers, such as Argentina, Brazil, or Southeast Asia.

8.2.3 Comparison Among FAPRI Outlooks: Catching Up with Reality and Policy Changes

Table 8.1 presents the evolution of the ethanol sector pre- and post-EISA 2007 using the current (2008) and previous (2006 and 2007)*FAPRI U.S. and World Agricultural Outlook*.[4] The 2006 FAPRI Outlook projected that US corn-based ethanol production and consumption would reach 8.1 and 8.3 billion gallons, respectively, by 2015, an increase of about 80% over 2006 levels. Ethanol imports were projected to increase by 67%. As the ethanol sector continued to expand under the 2005 US energy bill, the 2007 FAPRI Outlook presented higher production, consumption, and trade projections with production and consumption of ethanol reaching well over 12 billion gallons by 2015 and net imports projected to be 2.5 times higher than what they were in the 2006 FAPRI Outlook. With the implementation of the EISA at the end of 2007, these projections were adjusted again in the 2008 FAPRI Outlook such that US ethanol production would reach 15.5 billion gallons and consumption would reach 16.6 billion gallons by 2015, more than three times their 2006 levels. Under the EISA of 2007, ethanol net imports are projected to reach 1.2 billion gallons by 2015 compared to only 0.3 billion gallons projected by the 2007 FAPRI Outlook. Part of these changes has been driven by unanticipated increases in fossil fuel prices, which stimulated substitution toward ethanol. Furthermore, ethanol imports from Brazil to the United States are expected to increase significantly since, under the EISA, Brazilian ethanol qualifies under the "advanced biofuel" category. Hence, the increase in US imports of ethanol is primarily policy driven.

---

[3] Although the Chinese government has stopped approving new grain-based ethanol plants because of food security concerns, the Chinese Renewable Energy Plan mandates an almost tenfold increase in fuel ethanol production, to 3.3 billion gallons (mostly from non-grain feedstock) by 2020 (USDA-FAS 2008b).

[4] For each year, the*FAPRI U.S. and World Agricultural Outlook* provides 10-year projections. For comparison among Outlooks, we chose the years 2006, 2009, 2012, and 2015, although the first year and last years of the projection differ across the three Outlooks.

Table 8.1 Evolution of the biofuel market: comparision of 2006, 2007, and 2008 FAPRI outlooks for ethanol

	2008 FAPRI outlook				2007 FAPRI outlook				2006 FAPRI outlook			
	2006*	2009	2012	2015	2006*	2009	2012	2015	2006	2009	2012	2015
Ethanol prices					(U.S. dollars per gallon)							
Anhydrous ethanol price, Brazil**	1.92	1.41	1.25	1.41	1.80	1.55	1.43	1.36	1.21	1.16	1.24	1.31
Ethanol, FOB Omaha	2.58	1.90	1.78	2.13	2.58	1.71	1.63	1.58	1.89	1.81	1.66	1.73
Biodiesel prices												
Central Europe FOB price**	3.34	4.40	5.06	5.57	–	–	–	–	–	–	–	–
Biodiesel plant	3.12	3.87	4.67	5.01	–	–	–	–	–	–	–	–
U.S. Production and consumption					(Million Gallons)							
Ethanol												
Production	4,884	11,274	12,283	15,458	4,856	11,501	12,290	12,436	4,592	6,438	7,704	8,146
Consumption	5,436	11,554	12,668	16,594	5,370	11,684	12,594	12,750	4,635	6,501	7,797	8,261
Net trade	–686	–402	–397	–1178	–679	–288	–306	–322	–74	–98	–111	–124
Biodiesel												
Production***	268	638	1,081	1,109	385	578	534	472	–	–	–	–
Consumption	246	403	950	1000	–	–	–	–	–	–	–	–
Net trade	21	235	131	109	–	–	–	–	–	–	–	–
Brazilian production and consumption												
Ethanol												
Production	4,536	6,076	7,058	8.756	4,763	5,386	6,201	7,153	5038	5,667	6,108	6,554
Consumption	3,696	4,946	5,500	6,094	3,848	4,606	5192	5,954	4,325	4,762	5,059	5,352
Net trade	905	1,127	1,558	2,664	928	779	1,007	1,198	712	905	1,050	1,203
EU production and consumption					(Million Gallons)							
Ethanol												
Production	1,244	1,359	1,607	1,899	864	1,031	1,218	1,398	601	752	819	831
Consumptions	1,145	1,440	1,785	2,164	935	1,175	1,409	1,628	604	810	904	951
Net trade	99	–82	–180	–268	–71	–145	–192	–232	–3	–59	–86	–121

Table 8.1 (continued)

	2008 FAPRI outlook				2007 FAPRI outlook				2006 FAPRI outlook			
	2006*	2009	2012	2015	2006*	2009	2012	2015	2006	2009	2012	2015
Biodiesel												
Production	1,548	1,401	1,924	2,286	5,504	6,526	6,639	7,161	—	—	—	—
Consumption	1,633	2,131	2,313	2,724	—	—	—	—	—	—	—	—
Net trade	−29.72	−688.8	−348.3	−397.6	—	—	—	—	—	—	—	—
Rest of the world net trade												
Ethanol												
Canada	−124	−192	−261	—	—	—	−72	−121	—	—	—	—
China	267	86	26	24	42	−8	—	—	—	—	—	—
India	0	52	−67	−158	−118	−152	−179	−193	—	—	—	—
Japan	−165	−224	−261	−290	−171	−222	−258	−292	−155	−193	−226	−258
South Korea	−66	−96	−116	−132	−75	−96	−116	−135	—	—	—	—
Rest of the world	−344	−337	−370	−401	23	11	−5	−25	—	—	—	—

*Historical numbers not projections.

**Represents the world price.

***U.S. biodiesel production is in marketing year (Oct–Sept) in the 2007 FAPRI Outlook.

Note: Different data sources were used in the 2008 FAPRI Outlook, which explains the significant difference in numbers between outlooks for some countries.

According to the 2006 FAPRI Outlook, Brazilian sugarcane-based ethanol production and consumption were projected to increase to 6.6 billion gallons and 5.4 billion gallons, respectively, by 2015, an increase of 30% for production and 24% for consumption when compared to 2006 levels. Projections were higher for 2015 in the 2007 FAPRI Outlook, with ethanol production at 7.2 billion gallons and ethanol consumption at almost 6 billion gallons, a reflection of higher FFV use in Brazil. The provisions of the EISA of 2007 resulted in significantly higher projections in the 2008 FAPRI Outlook. With higher demand from the United States, ethanol net exports in Brazil were projected to almost triple to 2.7 billion gallons by 2015 (compared to 1.2 billion gallons projected in the 2007 FAPRI Outlook), as domestic production was to double compared to that in 2006, reaching 8.8 billion gallons. This additional increase of 1.5 billion gallons of ethanol between the 2007 and 2008 FAPRI Outlooks would translate to well over 850,000 additional hectares of sugarcane.[5] In the most current Outlook, Brazilian ethanol consumption is projected to reach 6.1 billion gallons, 65% higher than 2006 levels.

Subsequent projections for the EU ethanol sector also increased over time. Ethanol production, consumption, and net trade increased significantly between the 2006 and 2008 FAPRI Outlooks. A comparison of the two most recent Outlooks also reflects higher production and consumption of ethanol in India and China and higher import demand from Japan, South Korea, and the aggregate of "Other Countries."

As a result of the continued higher demand for ethanol from the United States, Brazil, and other countries, projections for the world ethanol price increased from $1.31 per gallon by 2015 in the 2006 FAPRI Outlook to $1.41 per gallon for the same year in the 2008 FAPRI Outlook. US ethanol price projections also increased, from $1.73 per gallon to $2.13 per gallon for the same period between the two Outlooks.

8.3 Land Allocation Effects of Biofuel Expansion

8.3.1 US Expansion

To gauge the US expansion effects, we use the scenario with no bottleneck in the ethanol market and a $10 increase in the oil price, as in Tokgoz et al. (2007). This scenario implies a large shock on the US ethanol market, but with a modest increase in ethanol imports into the United States. Hence, the direct feedstock effect is on the US corn market and US crops competing with corn for land, namely, oilseeds and other crops. These shocks propagate worldwide through relative world prices. Land devoted to corn and other coarse grains increases, whereas land devoted to wheat and oilseeds decreases. This movement into coarse grains and out of oilseeds occurs worldwide. The impact multipliers in percent changes for area from ethanol expansion are presented in Table 8.2 for major crops and countries. Corn area in the

---

[5]This is calculated based on 23 gallons of ethanol per ton of sugarcane and 75 tons of sugarcane per hectare.

Table 8.2 Impact multipliers for area from expansion scenarios (in percent change 100* %ΔX/%Δethanol use)

Countries	US ethanol demand expansion (Tokgoz et al. 2007) (long-run equilibrium)						
	Corn	Wheat	Sorghum	Sugarcane	Soybean	Rapeseed	Crops*
World	8.1	−0.3	2.0	0.0	−1.7	−1.2	1.7
US	23.4	−10.1	3.9	3.5	−14.6	2.4	2.6
Brazil	6.0	−0.6	−	−1.1	6.6	−	5.2
EU-25	0.8	0.5	−	−	1.7	−2.6	0.3
China	2.8	1.4	−	0.0	0.4	0.3	1.5
India	7.1	1.3	1.8	0.3	2.5	−2.4	1.7
Argentina	13.6	−0.8	2.8	0.2	−1.2	−	0.1
Australia	2.3	0.3	0.9	0.3	−	−0.1	0.4
Canada	2.6	0.3	−	−	1.7	−0.8	0.2
Mexico	2.1	5.5	6.7	0.7	−	−	2.8
South Africa	6.1	0.0	−0.2	0.5	−	−	5.3
Countries	Global ethanol demand expansion (Fabiosa et al. 2007) (10-year average)						
	Corn	Wheat	Sorghum	Sugarcane	Soybean	Rapeseed	Crops*
World	0.3	0.0	0.1	13.8	−0.1	0.0	0.1
US	0.9	−0.1	0.2	0.3	−0.6	−0.1	0.1
Brazil	0.3	0.0	−	44.4	0.2	−	5.8
EU-25	0.4	0.4	−	−	0.0	−0.1	0.4
China	0.1	0.1	−	0.3	0.0	0.0	0.1
India	0.4	0.0	0.1	1.7	0.1	−0.1	0.2
Argentina	0.6	−0.1	0.1	1.5	−0.1	−	0.0
Australia	0.1	−0.2	0.0	1.8	−	0.0	−0.1
Canada	0.1	0.0	−	−	0.0	−0.1	0.0
Mexico	0.1	0.2	0.3	0.1		−	0.1
South Africa	0.3	−	0.0	2.5	−	−	0.4

*Crops include corn, wheat, sugarcane, sugar beet, sorghum, soybean, rapeseed, sunflower, peanuts, and barley.

United States has a multiplier of 23.4, which means that a doubling of ethanol use in the United States would increase area devoted to corn by 23.4%. Corn area in Argentina and Brazil has a multiplier of 13.6 and 6, respectively. World corn area has a multiplier of 8.1. Area devoted to other coarse grains increases in aggregate in the United States and the world, but with much smaller multiplier values than those of corn (e.g., 4 for US and Argentine barley, 0.4 for world barley). The US oat area actually decreases (multiplier of −5.9). The US wheat area decreases considerably (multiplier = −10.1) as land moves to corn production. In other countries, the changes are moderate, summing up to a small decrease in world wheat area (−0.3). The most noticeable change is in the United States. This is explained by the fact that outside the United States, relative prices change moderately (most nominal prices went up by related proportions). Because area allocation is driven by relative prices, the changes are moderate.

The amount of land devoted to oilseeds falls in the United States and in aggregate. The US soybean area falls substantially (multiplier of −14.6). Soybean area in Brazil and Canada increases (multipliers of 6.6 and 1.7). In aggregate, world soybean area decreases slightly (multiplier of –1.7). Area devoted to other oilseeds is also affected but to a lesser extent than soybean area, with the notable exception of sunflower acreage in Argentina, which falls substantially (multiplier of –7.9). Land devoted to sugar increases slightly for US sugarcane area (multiplier of 3.5) but changes very little in any other country.

Table 8.2 also presents multipliers for total crop area for world and major producers. In response to a US ethanol expansion, world crop area increases, with a multiplier of 1.7. Most of the increase in world crop area is through a world corn area increase, which has a multiplier of 8.1. Brazil and South Africa respond the most, with multipliers of 5.2 and 5.3, respectively. They are followed by Mexico, the United States, and India.

8.3.2 Global Emergence Scenario

We use scenario 2 of Fabiosa et al. (2007), in which ethanol consumption in the rest of the world is increased by 10%, to gauge the effect of an expansion of ethanol use on land use in Brazil, the EU, China, India, and the rest of the world (non-US). The impact is mostly felt in land allocated to feedstock in these countries, and with an overwhelming impact on sugar crop area in Brazil and to a lesser extent in India. Further sugar-crop area in other sugar-producing countries increases slightly as the world price of sugar is impacted positively. Because few crops compete with sugar crops in the land allocation, these sugar-related changes have little impact beyond sugar crops. The US ethanol market is insulated from the world ethanol market by trade restrictions (two tariffs and some tariff rate quotas). Hence, the US ethanol market is nearly unaffected, and so too the corn market and land devoted to corn and competing crops remain nearly unchanged. The shock imposed on the EU ethanol market has some effect on EU grain markets, but small effects on world grain markets and resulting land allocations. This moderate impact is explained by the modest size of EU biofuels in world grain use. Land effects in the United States are even smaller. Worldwide, sugarcane land area increases with a multiplier of 13.8, but world sugar output falls slightly as expected (multiplier of –1.2). The impacts on most other crops and sectors are modest (multiplier values near 0 to 2 with the exception of Brazil, with a multiplier of 44.4).

In response to a global ethanol demand expansion, Brazil's total crop area responds the most, with a multiplier of 5.8, whereas most of the expansion is through sugarcane area, with a multiplier of 44.4. Following Brazil are the EU-25 and South Africa, both of which have a multiplier of 0.4 for total crop area. Total world crop area expands very modestly, with a multiplier of 0.1.

The world crop area multiplier in response to a global (non-US) ethanol expansion is very small relative to a US ethanol expansion. This is also the case for other countries, with the exception of Brazil. Since the United States is a major exporter of grains and oilseeds, any change in US corn demand impacts world markets considerably. This leads to indirect land use changes in the world in response to a US ethanol market expansion, and inclusion of these changes is crucial in estimating greenhouse gas emissions and savings from ethanol, as discussed in Searchinger et al. (2008).

8.4 Trade-Offs Among Feed, Feed Crops, and Bioenergy

We begin with an important stylized fact: With the actual expansion of US ethanol production, mostly using a dry milling process, DDG production and use have increased. Livestock feeders have found a way to incorporate DDG into their feed rations. DDG enters the rations of ruminant animals and replaces corn mostly and soybean meal only to a limited extent. With large US and international markets for DDG in ruminants, the DDG market price reflects its feed value in ruminant rations as a replacement for corn. That is why DDG prices closely track corn prices in

Table 8.3 Impact multipliers for feed use from expansion scenarios (in percent change 100* %ΔX/%Δethanol use)

Countries	US ethanol demand expansion (Tokgoz et al. 2007) (long-run equilibrium)						
	Corn use for ethanol	DDG	Corn feed	Sorghum feed	Soybean meal	Rapeseed meal	Sunflower meal
World	116.3	124.8	−6.5	−0.3	−2.1	−1.0	−0.5
US	118.8	70.9	−15.1	32.1	−11.0	−18.5	2.6
Brazil	0.0	0.0	−6.7	0.0	−4.6	–	–
EU-25	0.0	756.1	0.9	0.0	0.0	−0.1	−0.1
China	−2.7	1030.5	−1.9	0.0	0.6	1.1	0.6
India	0.0	0.0	−7.9	−7.2	−0.2	−0.4	–
Argentina	0.0	0.0	2.4	−2.4	−3.1	–	−2.7
Australia	0.0	0.0	−2.6	−1.6	–	−0.4	–
Canada	0.0	1098.9	−5.3	0.0	−3.6	−1.6	–
Mexico	0.0	463.4	−0.6	−4.2	–	–	–
South Africa	0.0	0.0	−5.5	−23.7	–	–	–
Countries	Global ethanol demand expansion (Fabiosa et al. 2007) (10-year average)						
	Corn use for ethanol	DDG	Corn feed	Sorghum feed	Soybean meal	Rapeseed meal	Sunflower meal
World	5.4	4.2	−0.3	0.1	−0.1	0.0	0.0
US	4.2	4.5	−0.7	1.0	−0.3	−0.3	−0.2
Brazil	0.0	0.0	−0.3	0.0	−0.2	–	–
EU-25	34.0	1.9	−0.1	0.0	0.0	0.0	−0.1
China	75.3	0.0	−0.1	0.0	−0.1	−0.1	0.0
India	0.0	0.0	−0.3	−0.3	0.0	0.0	–
Argentina	0.0	0.0	0.0	−0.1	−0.2	–	−0.2
Australia	0.0	0.0	−0.1	−0.1	–	−0.2	–
Canada	0.0	0.9	−0.2	0.0	−0.2	−0.2	–
Mexico	0.0	0.7	−0.1	−0.1	–	–	–
South Africa	0.0	0.0	−0.2	−0.7	–	–	–

the market, especially since 2006, with the implication that rations, with or without DDG instead of corn, exhibit similar costs.

With the US ethanol demand shock, the derived demand for corn in ethanol production increases by more than one in the United States and worldwide. In response to an increase in US corn ethanol use, corn feed use declines for most countries. In the United States and Mexico, DDG feed use increases more than does sorghum feed use. DDG feed use is quite responsive in all countries, as indicated by high multipliers. For the global ethanol demand shock, Chinese and EU-25 demands for corn for ethanol production are the most responsive. However, US demand responds the most for the increase in DDG feed use and the decline in corn feed use in response to higher prices. Corn use for feed declines in all countries, whereas sorghum feed use declines for all but the United States.

The US ethanol demand expansion introduces shocks in the protein meal market, primarily through the reduction of soybean production and to some extent as increasing DDG production replaces the share of protein meal in the feed ration. In the United States, the use of soybean meal for feed declines considerably (−11), as seen in Table 8.3. Soybean meal use for feed in Brazil, Canada, and Argentina also declines, but to a lesser extent. Rapeseed meal use drops significantly in the United States (−18.5), while other countries are not much affected by US ethanol expansion. Sunflower meal use decreases slightly in the world and in some individual countries. Ethanol demand expansion outside the United States does not create enough shocks to move the protein meal markets because it has a small impact on oilseed markets. In addition, the ethanol production expansion does not generate co-products that can be used to substitute for protein meal in feed rations.

8.5 Trade-Offs Among Food, Food Crops, and Bioenergy

8.5.1 Meat and Dairy Consumption

The impacts of the two scenarios are shown sequentially in Table 8.4 for beef, pork, and poultry, and for nonfat dry milk, cheese, butter, and whole milk powder consumption. The table covers selected key countries. Detailed results on other countries are available from the authors. As explained in the previous section, expansion of the US biofuel sector exerts an upward pressure on the prices of all feed ingredients. Facing higher costs of production, livestock and poultry producers reduce their supplies, causing meat prices to increase as well. There is a differential in the magnitude of price changes by meat type because the share of feed cost in total cost, as well as the ability to use biofuel by-products as substitute feed ingredient varies by animal type, favoring ruminant over monogastric animals. Even within the ruminant category, differences in production practices such as the use of grains for supplemental cattle feeding in some countries (e.g., United States and Canada) vs. pure pasture-based production in others (e.g., Australia and Brazil) will explain differences among country-specific impacts. In general, consumers lower their demand for meat products in response to the higher prices of all meat products.

Table 8.4 Impact multipliers for meat and dairy consumption (In percent change 100*
%ΔX/%Δethanol use)

Countries	US ethanol demand expansion (Tokgoz et al. 2007) (long-run equilibrium)						
	Beef cons.	Pork cons.	Broiler cons.	NFD cons.	Cheese cons.	Butter cons.	WMP cons.
World	−0.3	−0.9	−1.4	−0.7	−0.3	−0.1	−0.2
US	−1.9	−1.7	−3.4	−0.5	−0.5	−1.2	−
Brazil	0.0	−0.9	−1.0	−0.8	−0.3	−0.1	−0.7
EU-27	−0.1	−0.3	−0.5	−0.4	−0.2	0.1	−0.3
China	0.0	−0.8	−1.0	−1.3	−0.1	−0.1	0.1
India	0.4	−	−1.6	−0.6	−	−0.1	−
Argentina	−0.6	−0.6	−1.1	−0.9	−0.3	−0.1	−0.5
Australia	−0.3	−0.8	−2.2	−1.1	−0.4	0.0	−0.3
Canada	−0.7	−1.6	1.2	0.0	0.0	1.1	0.0
Mexico	0.4	−1.1	−0.2	−1.4	−0.4	−0.1	0.0
South Africa	0.8	−	−0.3	−	−	−	−

Countries	Global ethanol demand expansion (Fabiosa et al. 2007) (10-year average)						
	Beef cons.	Pork cons.	Broiler cons.	NFD cons.	Cheese cons.	Butter cons.	WMP cons.
World	0.4	−2.8	−7.1	0.0	0.0	0.0	0.0
US	−0.3	−4.7	−17.0	0.0	0.0	−0.1	−
Brazil	0.6	−2.3	−4.2	0.0	0.0	0.0	0.0
EU-27	−0.8	−3.3	−6.7	0.0	0.0	0.0	0.0
China	0.6	−2.2	−2.8	−0.1	0.0	0.0	0.0
India	−0.4	−	−8.0	0.0	−	0.0	−
Argentina	−1.5	−1.6	−4.1	−0.1	0.0	0.0	0.0
Australia	−0.5	−1.7	−7.7	0.0	0.0	0.0	0.0
Canada	0.0	−8.6	4.6	0.0	0.0	0.0	0.0
Mexico	2.7	−5.0	−0.6	−0.1	0.0	0.0	0.0
South Africa	1.7	−	0.8	−	−	−	−

Note: NFD indicates nonfat dry milk and WMP indicates whole milk powder.

The change in the price of pork and poultry is two to three times larger than
the change in the price of beef. This differential is the main driver in the consumption response, whereby poultry and pork declined the most in many countries and
beef declined the least. In fact, beef consumption increased in some countries, as
the effect of the larger price increases in pork and broiler induced substitution in
favor of beef, which dominated the response to the smaller increase in beef price.
Moreover, where entry and exit in the livestock sector are allowed to impose on the
long-run equilibrium condition, the magnitudes of price changes and their corresponding impacts on consumption adjustments are much larger compared to the case
in which a long-run equilibrium is not imposed. The impact of the global ethanol
shock outside of the United States is very small, since grain markets are moderately
influenced by the expansion of sugarcane-based ethanol.

In US dairy markets, increased feed prices force milk producers to switch to
some lower-cost but less efficient feed and consequently push down dairy yield per

cow. And higher production costs force producers to reduce cow numbers. Together, these two factors reduce total milk output. Consequently, tight milk supplies constrain production of dairy products and put an upward pressure on dairy prices. Higher prices of dairy products induce lower consumption, but with magnitudes varying by country. The expansion of ethanol use in the United States has the greatest effect on world nonfat dry milk (NFD) consumption with a multiplier of –0.7, as the United States is the biggest exporter of NFD. In the US domestic market, butter consumption is affected the most, with a multiplier of –1.2, a moderate change.

The global expansion of ethanol use in the rest of the world has similar effects to those under the expansion in the United States, but the effects are even smaller, not only for each country but also for the world (e.g., the multiplier for NFD world consumption is –0.7 under the US expansion of ethanol use vs. negligible under expansion of ethanol use in the rest of the world). The main reason for this result is the low impact of this global scenario on grain-based feed costs.

8.5.2 Vegetable Oils

As shown in Table 8.5, the US ethanol demand expansion scenario affects vegetable oil consumption slightly in most countries. Soybean oil use for food generally declines, except in the United States. Palm oil use for food increases significantly in the EU-25 (12.8). This scenario assumed that biodiesel would not expand further than planned in the baseline, as margins are negative in biodiesel production. By design, the global scenario focuses exclusively on ethanol production because biodiesel worldwide is unprofitable given the high vegetable oil prices. Hence, the global ethanol demand shock has very limited impacts on vegetable oil consumption.

8.5.3 Sugar

The impact of a US ethanol expansion on world sugar consumption is small, as shown in Table 8.5. Global sugar consumption falls by a negligible amount (multiplier of –0.1). Sugar consumption has a multiplier of –0.3 in both India and South Africa, while multipliers for the rest of the countries range between –0.1 in China and –0.2 in Australia. On the other hand, sugar consumption in the United States increases, with a multiplier of 1.8, as a result of an increase in the price of high-fructose corn syrup (HFCS). Since the expansion of ethanol occurs predominantly in the United States, most of the impact occurs in the United States and we see less of a response in the rest of the world.

In the global scenario, the expansion of ethanol use in Brazil, the European Union, China, India, and the rest of the world results in a small increase in US sugar consumption (multiplier of 0.2) while it results in a decline in sugar consumption in the rest of the world, ranging from –0.5 in China to –2 in India. Global sugar consumption declines by 1.1%, which is 10 times higher than the impact from the US ethanol expansion. This is because with the expansion occurring in countries other

Table 8.5 Impact multipliers for grain, sugar, and oil food use (in percent change 100*
%ΔX/%Δethanol use)

	US ethanol demand expansion (Tokgoz et al. 2007) (long-run equilibrium)						
Countries	Wheat	Barley	Corn	Soybean oil	Rapeseed oil	Sugar	Palm oil
World	−0.8	0.0	−3.6	0.0	0.0	−0.1	0.0
US	−2.1	−2.3	−3.9	0.5	5.9	1.8	−
Brazil	0.4	4.8	−3.7	−2.1	−	−0.1	−
EU-25	0.0	−0.7	−1.6	−3.6	0.1	0.0	12.8
China	0.4	0.1	−3.0	−1.8	−0.3	−0.1	0.0
India	−5.0	0.0	−3.6	−2.0	−0.3	−0.3	0.1
Argentina	−2.2	−5.5	−4.6	−2.4	−	−0.1	−
Australia	−1.4	0.0	−7.6	−	0.3	−0.2	−
Canada	0.1	−2.4	−4.9	−2.1	−0.7	−0.1	−
Mexico	0.6	14.9	−0.3	−	−	−0.1	−
South Africa	−	0.0	−8.8	−	−	−0.3	−
	Global ethanol demand expansion (Fabiosa et al. 2007) (10-year average)						
Countries	Wheat	Barley	Corn	Soybean oil	Rapeseed oil	Sugar	Palm oil
World	0.3	0.7	−0.3	−0.1	0.0	−1.1	0.0
US	−0.1	−0.1	−1.3	−0.2	−0.1	0.2	−
Brazil	0.0	−0.2	−0.1	0.0	−	−0.9	−
EU-25	−0.4	−0.2	0.1	−0.1	0.0	0.0	0.0
China	0.0	−0.1	−0.3	0.0	0.0	−0.5	0.0
India	−0.2	0.0	−0.2	0.0	0.0	−2.0	0.0
Argentina	−0.1	−0.2	−0.3	0.0	−	−1.0	−
Australia	−0.1	0.0	−0.4	−	0.0	−1.0	−
Canada	0.0	−0.1	−0.2	0.0	0.0	−0.9	−
Mexico	0.0	−0.2	0.0	−	−	0.0	−
South Africa	−	0.0	−0.3	−	−	−1.7	−

than the United States, we see a relatively larger impact of the shock in countries
like Brazil and less of an impact on the United States when compared to the effects
of the US ethanol expansion.

However, the overall impact of the global expansion is relatively small in magnitude in comparison to some of the other commodities, and this is for two reasons.
First, in sugarcane-producing countries, the acreage response to price changes is
limited, especially in the short run, because of the biology of the slow growth of
sugarcane and the fact that several annual crops can be harvested from one planting
of sugarcane. Second, Brazil, one of the major sugarcane producers, has the potential to expand area significantly. Sugarcane area in Brazil increased by almost 9% in
2007 and has averaged an annual increase of over 6% in the last 5 years. This helps
explain the relatively small impact on sugar resulting from an expansion of global
(non-US) ethanol use.

8.5.4 Grains

Table 8.5 presents the impact multipliers for grains. In the US expansion scenario, wheat replaces corn in food use in Brazil, China, Canada, and Mexico, whereas barley replaces corn for food use in Brazil, China, Mexico, and South Africa. World barley food use increases slightly, whereas world wheat and corn food use declines. In the second scenario, world wheat food use and barley food use increase, whereas world corn use decreases with a global ethanol demand shock. These effects are moderate in both scenarios, even more so for the second scenario, for which multipliers are often one order of magnitude smaller than in the US shock scenario.

8.6 Policies and Exogenous Factors Conditioning the Trade-Offs

The foregoing discussion illustrates that the expansion of the biofuels sector has had significant impacts on world agricultural and food markets. There are multiple factors that caused this expansion, such as domestic policies that support and promote the expansion of the biofuels sector as a supplier of fuel needs for transportation purposes and higher crude oil prices. This has important consequences for the US and world agricultural sectors since ethanol is mostly produced from corn in the United States, and the United States is a major exporter of agricultural commodities. In this context, a clearer understanding of the fundamentals of the ethanol market and analysis of the impacts of some potential policy changes might be helpful for policymakers and other stakeholders.

The volatility of domestic ethanol prices in the United States and the recent increases in crop and food prices in US and world markets have led to discussions of eliminating the tariffs on US ethanol imports. One study that contributed to this discussion was by Elobeid and Tokgoz (2008), which analyzed the impact of removing US trade barriers and the federal tax credit on ethanol markets. Table 8.6 presents the impact of the trade and tax credit removals on ethanol markets both in the United States and Brazil, which is the major exporter of ethanol. The results are reported as the average of the annual percentage changes (2006–2015) between the baseline and the respective scenarios. This analysis shows that US trade barriers have been effective in protecting the US ethanol industry and in keeping domestic prices strong in most countries of the world, except when domestic ethanol prices are extremely high. Under current policy and with the caveat on high prices, there is separability of the US ethanol market from world markets. With trade liberalization, the ethanol market deepens, making it less susceptible to price volatility.[6] The effect of trade liberalization extends beyond ethanol markets, affecting crop markets. The results also show that the impact of removal of the tax credit overrides the impact of the tariff

---

[6]This statement holds ceteris paribus, i.e., factors such as crude oil prices may have a larger impact on ethanol prices, which override supply-side forces in the ethanol market. This is indicated in Tokgoz et al. (2008), wherein a moderate increase in crude oil prices has a significant impact on ethanol prices.

Table 8.6 Impact of removal of US trade barriers and federal tax credit on US and Brazilian ethanol markets (average percent change between baseline and scenario)

Average 2006–2015	Removal of US trade barriers			Removal of US trade barriers and federal tax credit		
Prices	World ethanol price 23.9%	US ethanol price −13.6%	US corn price −1.5%	World ethanol price 16.5%	US ethanol price −18.4%	US corn price −2.1%
US	Ethanol production −7.2%	Ethanol consumption 3.8%	Ethanol net imports 199.0%	Ethanol production −9.9%	Ethanol consumption −2.1%	Ethanol net imports 137.0%
Brazil	Ethanol production 9.1%	Ethanol consumption −3.3%	Ethanol net exports 64.0%	Ethanol production 6.3%	Ethanol consumption −2.3%	Ethanol net exports 44.0%

Adapted from Elobeid and Tokgoz (2008).

removal. The removal of trade distortions lowers the US domestic ethanol price by an average of 13.6%, which results in a decline in US ethanol production and an increase in consumption when compared to the status quo. Consequently, US net ethanol imports increase significantly. The resulting higher world ethanol price leads to an increase in ethanol production and a decrease in total ethanol consumption in Brazil, causing net exports to increase relative to the baseline.

According to Elobeid and Tokgoz (2008), the effect of the removal of trade distortions extends beyond the ethanol market, affecting corn and other crop markets and their by-products. The US corn price decreases by 1.5% on average with the decline in demand for corn used in ethanol production. This affects the prices of other crops in the United States, as well as the area allocation among them since area allocation depends on relative net returns. The US ethanol protection has exacerbated the food–biofuel trade-offs in the United States and beyond.

The removal of the US federal tax credit of 51¢ per gallon for refiners blending ethanol leads to a reduction in the US refiners' and final consumers' demand for ethanol. Thus, this scenario shows a lower increase in the world ethanol price relative to ethanol trade liberalization (16.5% vs. 23.9%). The tax credit acts as consumption subsidy for US ethanol consumers. This is based on the assumption that the tax credit is passed on completely from the blenders to the final consumers. The removal of the tax credit overrides the impact of the tariff removal.

The recent surge in crop and food prices has generated much debate about the impact of the recent policies of the United States, EU-27, and Brazil that support the biofuels sector and how much these policies were responsible for the increases in food prices. Although expansion of the biofuels sector is one of the reasons for the increase in crop and food prices, it is not the only reason for this increase. Energy policies such as the EISA of 2007 and the energy bill of 2005 have contributed to the expansion of the biofuels sector in the United States, but they are not the only reason why the US ethanol sector has grown. Higher crude oil prices increased gasoline prices, which in turn made ethanol a good alternative as a fuel for transportation and contributed to this demand increase.

US refiners' acquisition cost of imported crude oil increased 78% between May 2006 and May 2008 (EIA 2008b). This led to higher gasoline and diesel fuel prices since they are derived from crude oil. This increase has also contributed to the increase in retail food prices through higher transportation, refrigeration, and production costs. For example, the ocean freight rate from the US Gulf to China increased 194% between May 2006 and May 2008. The ocean freight rate from the US Gulf to the European Union increased 333% between May 2006 and May 2008, adding to the cost of internationally traded products (IGC various). Costs of inputs such as fertilizers and irrigation have increased with higher crude oil prices as well.

Tokgoz et al. (2008) look at the impact of higher energy prices on the US ethanol sector and crop prices. They show that since the ethanol sector has become integrated into the agricultural sector, the agricultural sector's susceptibility to volatility from the energy prices has increased considerably. With the emergence of biofuels, crude oil prices have a much more direct impact on the US agricultural sector compared to the pre-biofuels era when it mostly affected the cost of production. The expansion of FFVs is a crucial element in this integration.

8.7 Conclusions

Favorable policies such as the EISA of 2007, the energy bill of 2005, and high crude oil prices have largely contributed to the emergence and expansion of the biofuels sector in the United States and in several other countries including Brazil, the EU, India, China, and Japan. This study contributes to the current debate on the food–fuel trade-offs by providing a systematic and quantitative analysis on how biofuel emergence and expansion affect US and world food and agricultural markets under the most recent policy environments. It also provides information on the impacts of some potential policy changes, which might be helpful for policymakers and other stakeholders. In particular, we quantify various food–biofuel trade-offs. To do this, we summarize major analyses based on FAPRI's world multimarket, partial-equilibrium agricultural models. Two scenarios on the expansion of biofuels are compared to a baseline situation to assess their effects on world agricultural and food markets.

This study highlighted that the impact of the emergence and expansion of biofuels on world agricultural markets and food–biofuel trade-offs is influenced by a host of factors. These include

- Policy regimes such as border protection and domestic support
- Biofuels conversion technology, including the different feedstock used such as corn in the United States, sugarcane in Brazil, and wheat and other grains in the EU

- Crop and livestock production technology such as less flexibility in land allocation between sugarcane and other crops, and supplemental grain feeding in cattle
  production in North America vs. mostly pasture-based cattle production in South
  America and Oceania
- Market structure, including the size of the biofuel market in respective countries,
  as well as the market shares of impacted countries in the world export market.

In particular, where biofuel growth occurs, that is, in the United States or in the
rest of the world, will determine the market outcome. Biofuel growth originating in
the United States results in larger and more widespread impacts. This is because the
energy conversion technology in the United States uses mostly corn as feedstock
in a dry mill process, producing substantial amounts of co-products that are good
substitutes, primarily for corn, in feed rations for livestock production that practices supplemental feeding of grains. Also, there is more flexible land substitution,
especially between corn and soybeans, in the major producing area of the Midwest,
significantly impacting production and available exportable surpluses of commodities where it is a major supplier in the world market. The United States accounts for
62% of world corn exports, 42% of soybeans, 25% of wheat, 29% of pork, 36% of
poultry, and 28% of nonfat dry milk. Land allocation adjustments in the rest of the
world follow the same direction as in the United States but at much smaller magnitudes, as most prices changed by related proportions. In contrast, the dominance of
sugarcane feedstock in Brazil's conversion technology, the lack of co-products useful for livestock production, the less flexible changes in allocation for land planted
with sugarcane, and the policy-induced insulation of the biggest ethanol market in
the world − the United States − mute the impact of biofuel expansion originating
from the rest of the world and mitigate its spillover effects.

Production of DDG in the US conversion technology favors ruminants, as livestock producers can use (lower cost) wet distillers grain at higher inclusion rates than
can producers using rations for monogastric animals that use dry distillers grain at
lower rates of inclusion. Moreover, with feed cost of ruminants accounting for only
29% of production costs compared to 55% in monogastric animals, the price impacts
on beef are much smaller than those on pork and poultry. This favors consumers in
the Americas, who have higher beef consumption, rather than consumers in Europe
and Asia, who have higher pork consumption.

Overall effects on food prices are moderate for the US expansion since agricultural commodities make up a small share of the cost of food in the United States.[7]
Moreover, with only a small fraction of their income allocated to food, the impacts
for consumers in the United States are modest as well. In contrast, because households in developing countries spend a substantial proportion of their income on food,
consumers in these countries may feel the impact of higher prices more deeply.

---

[7]The conclusion that the biofuel expansion has had some impact on food prices but that most of
the recent dramatic increase in prices is a result of other factors including high crude oil prices and
income growth has been put forth by other studies, including OECD (2008) and Trostle (2008).

References

Abler D, Beghin J, Blandford D, and Elobeid A (2008) Changing the U.S. sugar program into a standard crop program: Consequences under NAFTA and Doha. *Rev Agr Econ* 30(1):82–102.

Elobeid A and Beghin J (2006) Multilateral trade and agricultural policy reforms in sugar markets. *J Agr Econ* 57(1):23–48.

Elobeid A, Tokgoz S (2008) Removal of US ethanol domestic and trade distortions: Impact on US and Brazilian ethanol markets. *Amer J Agr Econ* 90(4):918–932

Elobeid A, Tokgoz S, Hayes DJ, Babcock BA, Hart CE (2007) The long-run impact of corn-based ethanol on the grain, oilseed, and livestock sectors with implications for biotech crops. *AgBioForum* 10(1):11–18

Energy Information Administration (EIA) (2008a) Fuel ethanol overview. Historical Energy Data, Monthly Energy Review, Renewable Energy. Energy Information Administration, http://www.eia.doe.gov/emeu/mer/pdf/pages/sec10_7.pdf Accessed June 2008

Energy Information Administration (EIA) (2008b) Monthly Energy Review. State and US Historical Data. http://www.eia.doe.gov/overview_hd.html sAccessed June 2008

Fabiosa JF (2008a) Distillers dried grain product innovation and its impact on adoption, inclusion, substitution, and displacement rates in a finishing hog ration. CARD Working Paper 08-WP 478, Center for Agricultural and Rural Development, Iowa State University

Fabiosa JF (2008b) Not all DDGS are created equal: An illustration of nutrient-profile-based pricing to incentivize quality. CARD Working Paper 08-WP 481, Center for Agricultural and Rural Development, Iowa State University

Fabiosa JF, Beghin JC, Dong F, Elobeid A, Tokgoz S, and Yu T (2007) Land allocation effects of the global ethanol surge: Predictions from the international FAPRI model. Economics Department Staff General Research Paper No 12877, Iowa State University, September forthcoming in Land Economics

Fabiosa J, Beghin J, De Cara S, et al. (2005) The Doha Round of the WTO and Agricultural Markets Liberalization: Impacts on Developing Economies. *Rev of Agr Econ* 27(3):317–335.

International Grains Council (IGC) Grain market report. Various issues. http://www.igc.org.uk/en/Default.aspx Accessed June 2008

Organization for Economic Cooperation and Development (OECD) (2008) Economic assessment of biofuel support policies. Directorate for Trade and Agriculture. http://www.wilsoncenter.org/news/docs/brazil.oecd.biofuel.support.policy.pdf Accessed February 2009

Searchinger T, Heimlich R, Houghton R, et al. (2008) Factoring greenhouse gas emissions from land use change into biofuel calculations.*Science* (February 29):1238–1240.

Tokgoz S,Elobeid A,Fabiosa J, et al.(2008)Bottlenecks, drought, and oil price spikes: Impact on US ethanol and agricultural sectors.*Rev Agr Econ* 30(4):604–622.

Tokgoz S, Elobeid A, Fabiosa J, et al. (2007) Emerging biofuels: Outlook of effects on US grain, oilseed, and livestock markets. Staff Report 07-SR 101, Center for Agricultural and Rural Development, Iowa State University, July (revised)

Trostle R (2008) Global agricultural supply and demand: Factors contributing to the recent increase in food commodity prices. USDA Economic Research Service Report WRS-0801, July, http://www.ers.usda.gov/Publications/WRS0801/WRS0801.pdf Accessed February 2009

USDA Foreign Agricultural Service (USDA-FAS) (2008a) Brazil: biofuels annual report 2008. Global Agriculture Information Network Report Number BR8013, July 22, http://www.fas.usda.gov/gainfiles/200807/146295224.pdf Accessed July 2008

USDA Foreign Agricultural Service (USDA-FAS) (2008b) People's Republic of China: Biofuels annual report 2008. Global Agriculture Information Network Report Number CH8052, June 26, http://www.fas.usda.gov/gainfiles/200806/146295020.pdf Accessed July 2008

USDA National Agricultural Statistics Service (USDA-NASS) (2008) Quick stats: U.S. and all states data – crops. USDA-NASS.http://www.nass.usda.gov/QuickStats/Create_Federal_All.jsp#top Accessed June 2008

Chapter 9
Demand Behavior and Commodity Price Volatility Under Evolving Biofuel Markets and Policies

Seth Meyer and Wyatt Thompson

Abstract Rising petroleum prices, growing incomes in developing nations, and changes in energy policy around the world are having a significant effect on agricultural markets as biofuels begin to play a more substantial role in meeting the world's energy needs. Some see biofuels as a means to reduce carbon emissions, increase energy independence, and raise farm income. Others note that the new demand for commodities to produce biofuels can lead to higher food prices, potentially undermining food security, and bring about unintended environmental consequences from expanding crop acreage that may result in greater carbon emissions. While there are a large number of studies that examine the influence of biofuels and biofuel policy on the prices of agricultural commodities, one poorly understood impact of greater biofuel production is how these changes have affected commodity price volatility. Price volatility is important for food security, farm income, policy formation, and investment behavior. We examine how greater biofuel production, whether motivated by policy or market prices, might change commodity price volatility in the future.

9.1 Assumptions About Long-Run Ethanol Market Behavior

Commodity markets in the United States have traditionally been influenced by energy markets through effects of energy prices on agricultural commodity supply. Rising or falling oil, natural gas, and other energy prices cause corresponding changes in production costs, including fertilizer, as well as in transportation and processing.[1] In the past, corn and soybean oil use as feedstocks in the production of

---

S. Meyer (✉)
Food and Agricultural Policy Research Institute (FAPRI), University of Missouri, Columbia, MO, USA
e-mail: meyerse@missouri.edu

[1] Fuel, electricity and oil accounted for 6% of total farm costs in 2000–2006 and fertilizer and lime account for as much (USDA ERS (www.ers.usda.gov/Data/FarmIncome/FinfidmuXls.htm). Fuel, electricity and transportation account for almost 10% of the margin between farm and consumer prices (USDA ERS (www.ers.usda.gov/data/FarmToConsumer/marketingbill.htm).

M. Khanna et al. (eds.), *Handbook of Bioenergy Economics and Policy*,
Natural Resource Management and Policy 33, DOI 10.1007/978-1-4419-0369-3_9,
© Springer Science+Business Media, LLC 2010

biofuels remained a small niche demand exerting little influence on commodity pricing. Changes in market conditions, technology, petroleum prices, and policies for both energy and agriculture brought about an explosion in feedstock demand. These changes radically altered the relationship between the two sectors. Some authors suggest that this creates a direct link from petroleum price, through gasoline and ethanol markets, to corn prices (see, for example, Elobeid et al. 2006). Other authors suggest a more modest, but still important, link from energy markets to agricultural markets (Westhoff et al. 2008). As discussed below, the more general conclusion is that growing biofuel processing in the United States has brought about greater integration between energy and agricultural markets. We find reasons to expect that this link extends the reach of energy market volatility into agriculture markets.

The long-run view of biofuel demand for feedstocks takes this demand to be perfectly elastic on the basis of the very small share of such fuels in motor fuel markets. This approach has been applied in the case of corn demand by ethanol processors (Tyner 2007; Tyner and Taheripour 2008). The underlying assumption of this approach is, given the small role of ethanol in the wider motor fuel market, a large change in the ethanol quantity relative to the base value may have a very small effect on motor fuel prices. On the other hand, this argument holds that movements in the large motor fuel market will cause changes in ethanol volumes that are so large that, in effect, the gasoline price will drive the ethanol price. The argument goes further to suggest that the ethanol use of corn will drive the corn market in turn, so that there will be a chain of linkages from the petroleum price through to the corn price. This one-way avenue of impact leads authors to conclude that there is a perfectly elastic demand for corn at the breakeven price at which ethanol processors could make normal profits from their operations. That is to say, any change in the relative prices of corn and ethanol and producers will purchase increasing quantities of corn to produce ethanol until the price of corn is pushed up and the price of ethanol pushed down and all "excess" profits are squeezed from the marketplace, forcing corn and ethanol prices to move in lockstep over time. An example of a calculation of this type starts from a petroleum price and estimates a gasoline price based on that petroleum price using historical relationships, adding any markup for octane or markdown for lower energy content in ethanol relative to gasoline, to calculate the price at which ethanol sells given the gasoline price. This ethanol price is tied to a breakeven corn price for ethanol processors based on assumed costs of ethanol production net of distillers grains, the feed coproduct in ethanol production, and technical coefficients. This representation leads to a direct link from petroleum prices to corn prices. While volatility in petroleum markets may not be addressed directly, these calculations are typically repeated over a range of petroleum prices to draw a line of corn breakeven prices corresponding to those petroleum prices. The representation of demand for feedstocks as perfectly elastic depends on a long-run equilibrium condition where entry or exit in the ethanol processing sector quickly eliminates economic profits or losses (relative to normal risk-adjusted returns to capital). This representation has trouble explaining observed fluctuations over a shorter time frame. Factors relating to supply and demand are constrained in the short run. As recent events show, it takes time for ethanol processors to adjust to high profits or loses. Westhoff et al. (2007) describe how market events led to a spike in ethanol processor returns in 2006 that could not be satiated by a sudden run-up in quantities supplied. Market participants cannot respond immediately and perfectly to changes in breakeven calculations.

Ethanol production capacity, distribution infrastructure, and consumer adoption pose short-run constraints on increased ethanol production and use. The transportation, blending, and ability to dispense various ethanol blends pose distribution obstacles as does the number of flex fuel vehicles, those able to use up to 85% blends, which exist in the fleet. Tokgoz et al. (2007) emphasized that their representation of ethanol processor demand for corn is a long-run one. Even so, by introducing some restrictions on the number of flex fuel vehicles, and consequently on E85 use, their representation is less long run in this sense than Elobeid et al. (2006) who imposed no limits on E85 use. These studies also differ in their treatment of supply-side factors, with Elobeid et al. using a breakeven corn price to drive results, whereas Tokgoz et al. seem to solve for market-clearing prices that assume imperfectly elastic responsiveness. In practice, Birur et al. (2008) also assume finite constant elasticities of substitution between motor fuels, as well as for the parameters governing biofuel production.

An exclusively long-run view seems potentially misleading in discussions of ethanol market variability and its effects on commodity market variability. For example, Gallagher et al. (2000) considered a shock to demand, namely a then-hypothetical increase in the use of ethanol as an additive, and differentiated between the short and long runs by whether or not production capacity is fixed. They then observed that the short-run effect is a run-up in profitability of existing plants with a movement to normal returns as capacity expands. Westhoff et al. (2007) consider the market events surrounding the eventual regulatory change that led fuel blenders throughout the United States to use ethanol as an additive in place of MTBE.[2] They find that MTBE replacement lead to a surge in prices and profitability in 2006. Westhoff et al. (2008) show that effects of biofuel policy on consumer use are highly dependent on the current context, including petroleum prices. More generally, limited ethanol-processing capacity alters the short-run relationship between ethanol and corn prices. In a given period, there may not be sufficient capacity to process large additional quantities of corn into ethanol even if the petroleum price increases sharply. A large crop could still mean low corn prices without much of the effect being passed on to ethanol markets if production was already near or at capacity. Questions of fixed factors in the short run affect market price variability as limits to shifting quantities of ethanol production or consumption lead to greater price effects than would be observed in long-run analysis where such factors are not constrained.

---

[2]Methyl tertiary butyl ether (MTBE) serves as a lead-free additive to gasoline to raise its octane. Motor fuels with MTBE burn differently and generate fewer emissions of certain types, leading to its historical eligibility to serve as an additive that alters fuels so that they meet regulatory standards in certain areas or at certain times. However, MTBE can have other harmful effects on human health if leaked into groundwater, so it was replaced by ethanol by 2006.

9.2 Changing Market Relationships

Ethanol demand can be represented as a derived demand for this motor fuel as an input into transportation services. Characterizations of ethanol demand as a single category may lead to incorrect analysis of price variability and market links. Ethanol demand is segmented into involuntary and voluntary uses that have different characteristics and distinct responses to changes in relative prices. Involuntary uses leave the consumer no decision whether to include ethanol in their motor fuel purchasing decisions. Involuntary use includes the gas reformulation requirements in California and the state wide 10% ethanol blend requirement in the state of Minnesota. Involuntary demand expanded when ethanol replaced MTBE as the common fuel additive, as discussed below. Such demand is highly inelastic with respect to ethanol and petroleum prices (Fig. 9.1a).

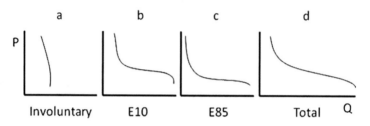

Fig. 9.1 Demand representation of categories of ethanol demand

Consumers may choose to purchase ethanol in most other locations, and it is this choice which drives the voluntary uses. The two common forms of voluntary ethanol consumption are E10 and E85 blends. E10 can be consumed by most standard vehicles in the United States with no modification. At high ethanol prices relative to gasoline prices, demand may be highly inelastic as consumers purchase ethanol not primarily for its energy content, but for other perceived characteristics, such as its contribution to energy independence, reduction of greenhouse gases, and support for domestic farmers. There are several reasons to suppose that these consumers represent a minority of voluntary ethanol demand and that the marginal or next consumers will be more sensitive to relative ethanol and gasoline prices. Historical ethanol consumption before the recent run-up in petroleum prices was small as most consumers opted not to buy ethanol at a premium over gasoline. Recently, the premium that was paid for ethanol in earlier years seems to have been eroded (de Gorter and Just 2007). Current use easily exceeds estimates of the amount of involuntary use of ethanol on the order of 3 or 4 billion gallons, such as those of Thompson et al. (2008b). At a lower relative price, ethanol becomes competitive for its value as a motor fuel. Demand becomes much more elastic as the bulk of consumers are willing to adopt this alternative motor fuel. At some even lower price, demand once again becomes less responsive as all consumers who are willing to purchase E10 have done so saturating this market (Fig. 9.1b).

E85 adoption is constrained by the need for consumers to own or buy a special flex fuel vehicle and for there to be area fueling stations which dispense the product. As with voluntary E10 demand, some consumers have been willing to purchase E85 blends at prices relative to gasoline which are above their energy value. This demand may be quite inelastic with respect to prices as consumers seek other benefits to ethanol use as outlined above. Because E85 represents a basic energy substitute for gasoline, rather than also providing the octane boost of E10, the price of E85 must be lower relative to the gasoline price before E85 is widely substituted for gasoline. Thus, the kink in demand, the relative price level at which it becomes much more elastic, is expected to be lower than that for E10 (Fig. 9.1c). The limit to E85 expansion within a short-run period is not known, so it is unclear how this demand expands for lower and lower E85 price relative to the gasoline price. Demand could again become inelastic in the short run if all available flex fuel vehicle owners who would purchase E85 and are able to find dispensers were doing so. In the long run, this would be eased by additional flex fuel vehicles and infrastructure.

Total demand for ethanol is the sum of involuntary use (as an additive and as mandated by states), voluntary E10 use, and voluntary E85 use. The result is a total demand curve for ethanol that varies greatly in elasticity (Fig. 9.1d). For example, at a high ethanol-to-gasoline price ratio, demand may be largely involuntary uses and insensitive to price. At a price ratio close to the energy content of ethanol, voluntary uses may respond quickly to any price changes. In the event that the ethanol-to-gasoline price ratio were very low, the E10 market would become saturated and further E85 expansion might be costly, implying little scope for further adjustment no matter the price change.

Supply may also be constrained in the short run through limits on existing ethanol production capacity. In the long run, additional capacity can be constructed while, in the short run, increases in production are achieved only through more intensive use of existing facilities. However, increasing output from existing capacity is costly and subject to an absolute maximum at some point. Imports represent an important source of short-run supply, albeit one that is subject indirectly to many of the same constraints as US production, but are largely ignored in the present discussion.

The replacement of MTBE by ethanol in the motor fuel supply completed in 2006 illustrates the role of short-run production constraints on market outcomes (Fig. 9.2). Ethanol supply, S^e, is reflected by a convex upward sloping curve reflecting the increasing costs as utilization rates increase. This supply is intimately linked to ethanol processor demand in the feedstock markets, D^f. This relationship is a derived demand; as the agent who demands corn and supplies ethanol is actually the ethanol processor in both cases the feedstock demand curve is derived from the same underlying technological and economic conditions as ethanol supply. Excess supply in feedstock markets is the difference between the supplies of corn and all other corn demands, including exports, at each price, and this is represented by ES_0 (in Fig. 9.2). Intuitively, as the price ethanol producers are willing to pay rises, the corn market is willing to supply more corn for this purpose because producers sell more at a higher price – which is unlikely in the short run – or other demands

Fig. 9.2 Displacement of
MTBE by ethanol

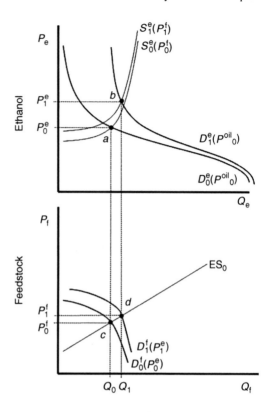

are outbid by the ethanol processors. Prior to the replacement of MTBE, the initial equilibrium is at point a in the ethanol market and at point c in the feedstock market. In this representation, quantities of supply and demand balance in both the ethanol market and feedstock markets, and the quantity of ethanol supply is consistent with the quantity of feedstock demand given the technical relationship binding them. As blenders replace MTBE, ethanol demand increases to D_1^e and becomes more inelastic. The derived demand for feedstocks also moves D_1^f as processors are more willing to buy corn when ethanol prices rise. The increased feedstock demand modestly raises feedstock prices, from P_0^f to P_1^f, as ethanol is only 14% of total corn demand, including exports, in the 2005/2006 marketing year.[3] The higher corn price discourages ethanol processors from making ethanol, so ethanol supply shifts back.

In summary, the sharp increase in involuntary ethanol demand associated with MTBE replacement gives evidence that ethanol can disconnect from gasoline prices and be a source of price variability in the feedstock market. Ethanol demand pushes

---

[3]This does not account for other feed grains or the feed coproduct from wet and dry mill ethanol production.

against capacity constraints, the inelastic portion of the ethanol supply curve in 2006, forcing ethanol prices to move substantially. However, the effect of this particular change on the price of corn was by no means as large because the quantity of corn processed remained a small, but growing, share of the total market. At the new equilibrium, points b in the ethanol market and d in the feedstock market, rents accrue to owners of existing ethanol capacity and a strict long-run relationship between feedstock prices (corn) and petroleum prices is broken until capacity can fully adjust.

9.3 Volatility of Markets

Over certain ranges or prices and given time to adjust, demand for feedstocks to produce biofuels may be highly elastic and stabilize corn prices with respect to shocks that originate in agricultural markets, but the level at which it stabilizes the price is contingent on the price of petroleum and therefore subject to its fluctuations. While common perception may be that petroleum prices are more volatile than other commodity prices, Plourde and Watkins (1998) concluded in a comparison of crude oil prices and other raw commodity prices, that although oil prices were more volatile than the other commodities examined over the 1985–1994 period, "results do not show volatility in oil prices as a clear outlier." (p 260) The lone representative agricultural commodity in the study was wheat, however. Regnier (2007) examined a much larger number of commodities over 5-year periods starting in 1946 and ending in 2005. The conclusion was oil prices were less volatile than most other raw commodities prior to 1986 and, while oil price volatility had increased after 1986 and exceeded volatility of most other commodities, farm product price volatility has kept pace. Given the already volatile nature of agricultural commodities, it is important to examine what the link between petroleum and commodity prices adds to this volatility (Fig. 9.3).

The supply and demand representation reflects a market where capacity is less of a constraint and ethanol demand is elastic at the starting point. The initial market equilibrium is at a in the ethanol market and c in the feedstock market with equilibrium prices at P_0^e and P_0^f, respectively. An increase in oil prices, from P_0^{oil} to P_1^{oil}, shifts the ethanol demand curve out, D_0^e to D_1^e. As ethanol prices rise, feedstock demand shifts out, D_0^f to D_1^f, and the quantity of ethanol supply increases. Expanding ethanol quantity increases the feedstock demand. The consequent rise in feedstock price in turn pulls the ethanol supply curve back somewhat. A new equilibrium is reached at b in the ethanol market and at d in the feedstock market, and equilibrium quantities in both markets increase, from Q_0 to Q_1. The resulting size of the price movement in the feedstock market, from P_0^f to P_1^f, depends on the size of the oil shock, P_0^{oil} to P_1^{oil}, as well as on the elasticities of supply and demand in the ethanol market. The extent to which the petroleum price affects the corn market depends on the presence of any short-run constraints, and thus the net effect of the

Fig. 9.3 Oil price increase

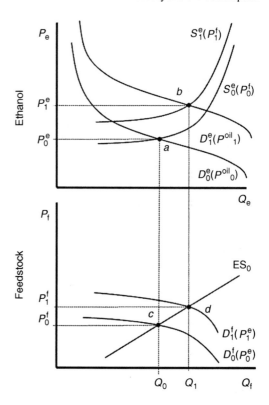

impact of petroleum market volatility on commodity price volatility in the short run is uncertain.

9.4 Energy Policy and Its Influence on Commodity Price Volatility

Historically, the primary policy mechanism influencing the agriculture sector has been the farm bill, the key legislation guiding agricultural policies that is rewritten every 5–7 years. From its inception in 1949, it focused on stabilizing farm income and supporting rural economies. Since the 1980s, a period where farm policies emphasized supply management through acreage controls and buffer stock programs, a transition to more market-oriented policies was furthered in the 1996 Federal Agriculture Improvement and Reform (FAIR) Act. The legislation maintained the long-standing goal of supporting farm income, but represented a large step toward market orientation. This transition and its effect on commodity farm price volatility was a source of considerable debate and anxiety with intense analysis leading up to its passage and beyond. Shortly after passage of the FAIR Act,

Westcott (1998) found that volatility derived from changes in policies to lower public stocks may be offset by improved market responsiveness due to increased supply flexibility allowed under the farm bill. Collins and Glauber (1998) were also skeptical of the potential for increased price volatility, noting that the disposition of large government stock-holding was subject to diverse policy objectives and not always responsive to market forces, reducing their stabilizing effect. Years later, the effects of market-oriented policy changes remain a source of debate. While admitting the inability to assign the change to the 1996 FAIR Act, Zulauf and Blue (2003) found that implied volatility for select commodities had increased since its passage. Regnier (2007), although focusing on a comparison of volatility in oil and other commodity prices, showed an increase in volatility beginning in the transition years prior to enactment of the 1996 FAIR Act. While interventionist supply management policies have not returned, the subsequent 2002 and 2008 farm bills pursued the objective of stabilizing farm income, but not commodity prices in the market place. Meanwhile, policies outside of traditional agriculture legislation have begun to play an increasingly important role.

A national blenders credit for every gallon of biofuel used has been in existence since the 1970s (Duffield and Collins 2006). However, limitations in the technology to convert corn to ethanol, prevailing corn and petroleum prices, and consumer acceptance kept biofuel at a small scale. The value of the blenders credit fluctuated over the years and in 2004 was set to $0.51 per gallon of ethanol blended with gasoline and reduced in 2009 to $0.45. All ethanol used, even imported ethanol, receives the blenders credit. The credit's application to imported biofuels has been used to justify a specific tariff of $0.54. A small ad valorem tariff is also applied. The Energy Policy Act (EPA) of 2005 introduced the first national renewable fuel standard equivalent to 4.0 billion gallons in 2006 growing to 7.5 billion gallons by 2012. The Energy Independence and Security Act (EISA) of 2007 set additional layered mandates extending out to 2022 for four types of renewable fuels being used in the motor fuel supply, broken out primarily along feedstock type and green house gas reduction characteristics. These mandates represent a new and as yet untested policy.

These mandates obligate fuel blenders to blend a minimum quantity of renewable fuels. At high enough oil prices or low enough feedstock prices, the mandates have little effect as ethanol buyers and sellers would choose quantities above the mandated amount. In this case, the equilibrium quantities exceed the mandates, and they are not binding. However, at lower oil prices or higher feedstock prices, the mandates may become binding. In this case, the fuel blenders must pay enough to producers or importers to obtain the mandated volume of biofuels and must sell ethanol-blended fuels at a price low enough to induce the consumers to purchase the specified quantities of biofuels.

The mandate largely disconnects the link biofuels generate between commodity markets from energy price movements when binding. Feedstock demand for corn is no longer sensitive to ethanol and commodity prices. The price effect of any significant crop loss or supply shock is amplified, because this element of feedstock demand does not adjust to help the market absorb the shock. The effect of mandates on commodity price volatility in the event of a supply shock such as a short crop as a result of unfavorable growing conditions can be represented graphically (Fig. 9.4). Take the case without any mandate on the left-hand side (panel A). The initial equilibrium is at a in the ethanol market and at point c in the feedstock market. In the event of a supply shock, such as a poor planting or growing conditions, the excess supply in feedstock markets shifts back, from ES_0 to ES_1. In the absence of a mandate, the ethanol supply curve would shift back as feedstock prices rise and ethanol prices would also rise. The no mandate case would result in equilibrium, point b in the ethanol market and d in the feedstock market, with prices rising because of the supply shock, from P_0^e to P_1^e in the ethanol market and from P_0^f to P_1^f in the feedstock market. Quantities of feedstocks and ethanol fall, from Q_0 to Q_1 with this reduction softening the impact of the supply shock on the feedstock price increase.

If a mandate were imposed at the original quantity, Q_0, then ethanol and feedstock markets would be unable to respond to the smaller feedstock supplies as shown on the right-hand side (Fig. 9.4, panel B). In effect, feedstock demand would become perfectly inelastic.[4] At the mandated quantity, the feedstock market would reach a

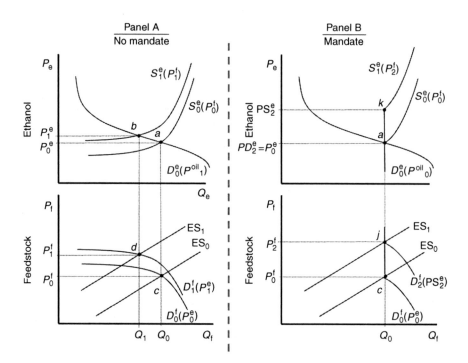

Fig. 9.4 Feedstock supply shock, ethanol mandate

---

[4]There are likely to be provisions in the regulations that allow blenders some flexibility in the timing of their obligation, potentially reducing the inelasticity resulting from the mandate. EISA provisions to waive mandates also represent a source of flexibility.

new equilibrium, j, after the supply shock, resulting in feedstock prices higher than the unconstrained case, $P_1^f < P_2^f$. A mandate, when binding, is likely to increase feedstock price volatility caused by changes in the feedstock market.

The price dynamics in the ethanol market will be affected by a mandate as well. In the absence of a mandate, a feedstock supply shock causes ethanol quantities to fall, from Q_0 to Q_1, as prices rise, from P_0^e to P_1^e. When the mandate is imposed, the supply curve crosses the mandated quantity at a point, k, that implies a higher price. Ethanol producers face higher costs because of rising feedstock prices, so their marginal costs are higher. So ethanol producers must be offered a higher price, PS_2^e, to provide sufficient supplies to meet the mandate despite the higher feedstock costs. However, the demand curve still crosses the mandated quantity at the original point, a, and the original price. Consumers must be offered the same price as before, $P_0^e = PD_2^e$, to consume the same mandated quantity, Q_0. Because the mandates are imposed on blenders and not on consumers, the cost of meeting the mandate, $PS_2^e - PD_2^e$, initially must be paid by blenders who are likely to pass the cost on to motor fuel users in some way (Thompson et al. 2008a). While the effects of a mandate can be dramatic, they are only relevant if binding. At high enough oil prices or low enough feedstock prices, the mandates are not binding and have little effect as the market will clear at quantities above the mandated amount.[5]

While the net effect of closer links between energy and agriculture markets on commodity price volatility is unclear, it appears probable that the imposition of mandates through energy policy may increase volatility in commodity markets in the short run.

9.5 Calculating Volatility

The FAPRI–MU model has a long history of use in examining the market impacts of policy and other factors on US agricultural commodity markets and is useful in examining effects on price levels and volatility from these factors over several years. The model covers major crop and livestock markets and is a nonspatial, partial equilibrium model where world markets are represented by reduced form equations (Westhoff et al. 2006). The model also contains a representation of the biofuels sector (Thompson et al. 2008a). The model can be simulated in a partial stochastic framework using draws on empirical distributions of a subset of exogenous variables which maintain historical correlations. Thus, historical deviations in yields, petroleum prices, unexplained variation in several demand categories, and other key determining factors are reproduced. While the model reflects a realistic range of probabilities, it cannot be assumed to contain all conceivable outcomes given that ranges are defined based on historical data and only a subset of external factors are

---

[5]The choice here to impose the mandate at the quantity of the initial no-mandate equilibrium is solely for exposition purposes. There is no reason to expect that the mandated quantity will be exactly equal to the quantity the market would settle on without a mandate.

varied. With these distributions, the model is simulated with 500 different stochastic input sets to assess the effects of biofuel policies under varying market conditions (Westhoff et al. 2008). Here, these results are used to examine how biofuels have changed commodity price volatility over a wide range of possible market outcomes.

One scenario simulates biofuel and agricultural markets over the next 10 years in the absence of blenders' credits, import tariffs, and national mandates. A second scenario extends blenders credits, both for ethanol and biodiesel, and the specific ethanol import tariff, but not the mandates, through the end of the 10-year period. The tax credit subsidizes the blending of ethanol with other fuels, shifting out demand and increasing the size of the industry.[6] Comparing these two scenarios illustrates how the size of the biofuel industry, with its varying elasticity, affects the links between petroleum and corn price volatility. A third scenario adds the national mandates as specified in the EISA to the scenario extending the blenders' credits and tariffs for the 10-year period. Results of this scenario explore how the addition of a mandate affects the link between petroleum and corn price volatility.

Sorting the 500 simulations by the petroleum price and calculating the average corn price reveal how petroleum price affects commodity price behavior under the three scenarios (Fig. 9.5). The blenders' credits increases the size of the ethanol industry and raises price levels at all petroleum prices as compared to the case

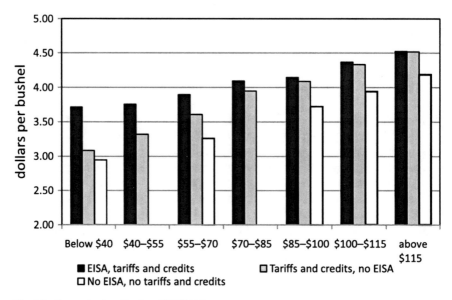

Fig. 9.5 Corn price by oil price, 2015/2016 crop year

---

[6]The ethanol tariff is maintained as though it is attached to the credit. The potential for import surges if these policies are not tied together is not discussed here.

without credits or EISA mandates. The blenders credit also results in ethanol prices that grow more rapidly as oil prices increase over the range explored here.

When the mandate provisions in the EISA are added along with credits, a disconnect between petroleum and ethanol prices becomes apparent. The mandate is less likely to be binding at high petroleum prices, so the corn price for any given petroleum price is roughly the same as the case of credits without any mandate provisions. However, at low petroleum prices, the mandate is more likely to be binding and supports ethanol production so corn prices are higher than in other scenarios. Thus, for a range of petroleum prices outlined by historical variations, the mandate reduces the range of corn price variation caused by petroleum price fluctuations, all else constant.

However, all else is not constant. Examining corn prices over the distribution of corn yields, the blenders credit raises corn prices at every value of yield relative to the case of no credit or EISA mandates by a roughly constant amount (Fig. 9.6). When the mandate is added, the corn prices rise faster as corn yields fall. Analytical results show that the EISA mandate increases corn price volatility due to corn supply shocks. The low-yield case corresponds to a feedstock supply shock, from a drought as discussed earlier where the reduction of corn supplies in the presence of a binding mandate cause greater increases in corn prices than in the case without a mandate. Here, the shocks to corn yield are also seen to result in higher corn price with the EISA mandate than without. Given the opposing effects on commodity price volatility highlighted in Figs. 9.5 and 9.6, a more comprehensive examination of the net effects on volatility seems warranted.

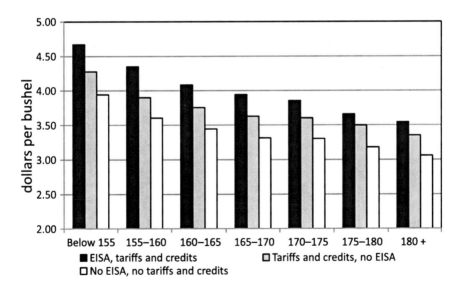

Fig. 9.6 Corn price by corn yield, 2015/2016 crop year

A measure of volatility can be calculated as the standard deviation of the difference of the log of current and lag prices (Regnier 2007); (Pindyck 2004); (Plourde and Watkins 1998). This equation takes the form

$$r_s = \log(p_t) - \log(p_{t-1})$$

where p is the price of corn, t is the year, and s is the number of years for which a difference is calculated, in this case 7 (the calculation is over the 2011/2012 to 2017/2018 crop year). The standard deviation of each of the 500 simulations is calculated and then averaged.

$$\bar{\bar{\sigma}} = \frac{1}{500} \sum_{j=1}^{500} \sqrt{\frac{1}{s-1} \sum_{1}^{s} (r_{js} - \bar{r}_j)^2}$$

The calculations are done over a wide range of market conditions and do not reflect a specific market situation, but rather a global measure of volatility that spans multiple regimes. In some solutions ethanol capacity may be limited, in some oil prices will be high, and others yields may be below the trend, all three may be true or none may be.

The relative changes in the standard deviations of the difference in log prices ($\bar{\bar{\sigma}}$) indicate the effects of these policies (Table 9.1). For corn, the standard deviation when no policies are in place is roughly equivalent to that when credits are added. Thus, the absolute level of volatility associated with a given pattern of petroleum prices is largely unaffected by the introduction or removal of credits, our proxy for industry size. Soybean farm prices, as well as soybean oil market prices, show a decline in volatility from implementation of the credits and tariffs, which includes a blenders credit for biodiesel, as the credit has a significant impact on the size of the biodiesel industry (Westhoff et al. 2008). With the imposition of a mandate, volatility returns in both soybean and soybean oil prices. Wheat prices, not commonly used as a biofuel feedstock, show little direct effect from changes in biofuel policies. The imposition of tax credits and tariffs and then the imposition of the mandate

Table 9.1 Standard deviation of the log difference in prices, 2011–2017, 500 stochastic solutions

	No EISA, no tariff or credits	Tariffs and credits	EISA, tariffs, and credits
Corn farm price	0.2038	0.2044	0.2185
Wheat farm price	0.1560	0.1543	0.1555
Soybean farm price	0.3057	0.2876	0.3268
Soybean oil market price	0.2212	0.1487	0.2083
Ethanol rack price	0.1220	0.1134	0.1083

appear to reduce the volatility in the rack, or wholesale, ethanol price. The analysis of changes is over a wide range of contextual assumptions, including yields and petroleum prices, and the cross effects in demand and supply mitigate some of the impact. Their inclusion into the model is necessary to a give realistic examination of the effects of biofuel policy on commodity price volatility. The results suggest the net effect of the imposition of tax credits and tariffs on the price volatility of biofuel feedstock crops, such as corn and soybeans, is indeterminate while the addition of the mandate likely increases the price volatility of those commodities.

9.6 Conclusion

The effect that national policies have on commodity price volatility has long been a concern. Analysis of farm bills often considered how commodity price volatility may change as a result of these policies, even distinguishing between producer and consumer effects. The growth in biofuel production and implementation of more expansive biofuel policies have brought in new players and introduced new forces into old market links. Decision makers unaccustomed to these markets and policies may not yet fully understand the closer relationship between the energy and agriculture sectors.

The net effect of petroleum price and biofuel policies on the corn price volatility is highly dependent on the market context. Preliminary analysis suggests that a larger biofuel industry and thus larger demand for commodities as feedstocks might increase volatility modestly. Preliminary results suggest that biofuel use mandates increase corn price volatility if they are binding, and they are more likely to be binding if oil prices are low or feedstock prices are high. Political actors and market participants must develop familiarity with these new and strengthening relationships in the future to understand market volatility and gauge the consequences for farm income, food security, and biofuel investments.

References

Birur D, Hertel T, and Tyner W (2008) "The Biofuels Boom: Implications for World Food Markets." Paper prepared for presentation at the Food Economy Conference sponsored by the Dutch Ministry of Agriculture, as presented at the International Agricultural Trade Research Forum on January 9, 2008.

Collins KJ and Glauber JW (1998) "Will Policy Changes Usher in a New Era of Increased Agricultural Market Variability?" Choices(Second Quarter): 26–29.Elobeid, Amani, Simla Tokgoz, Dermot Hayes, Bruce Babcock, and Chad Hart. "The Long-Run Impact of Corn-Based Ethanol on the Grain, Oilseed, and Livestock Sectors: A Preliminary Assessment." Center for Agricultural and Rural Development (CARD) Briefing Paper 06-BP 49. November 2006.

De Gorter H and Just J (2007) "The Economics of a Biofuel Consumption Mandate and Excise Tax Exemption: an Empirical Example of the U.S. Ethanol Policy." Working Paper of the Department of Applied Economics and Management, Cornell University. October.

Duffield JA and Collins K (2006) "Evolution of Renewable Energy Policy." Choices, p. 9–14, 21(1), Quarter 1.

Elobeid A, Tokgoz S, Hayes D, Babcock B, and Hart C (2006) "The Long-Run Impact of Corn-Based Ethanol on the Grain, Oilseed, and Livestock Sectors: A Preliminary Assessment." Center for Agricultural and Rural Development (CARD) Briefing Paper 06-BP 49. November 2006.

Gallagher P, Otto D, and Dikeman M (2000) "Effects of an Oxygen Requirement for Fuel in Midwest Ethanol Markets and Local Economies." Rev of Agr Econ. 22(2), 292–311, 2000.

Pindyck RS (2004) "Volatility and commodity price dynamics." J of Futures Markets 24(11): 1029–1047.

Plourde A and Watkins GC (1998) "Crude oil prices between 1985 and 1994: how volatile in relation to other commodities?" Resources Energy Econ 20(3): 245–262.

Regnier E (2007) "Oil and energy price volatility." Energy Econ 29(3): 405–427.

Thompson W, Meyer S, and Westhoff P (2008a) "Policy risk for the biofuels industry" proceedings of the farm foundation "Risk, Infrastructure and Industry Evolution" Oak Brook, IL: Farm Foundation, June.

Thompson W, Meyer S, and Westhoff P (2008B) "State Support for Ethanol Use and State Demand for Ethanol Produced in the Midwest." FAPRI-MU Report #11-08.

Tokgoz S, Elobeid A, Fabiosa J, et al. (2007) "Emerging Biofuels: Outlook of Effects on U.S. Grain Oilseed, and Livestock Markets." Center for Agricultural and Rural Development Staff Report 07-SR 101. May 2007.

Tyner W (2007) "U.S. Ethanol Policy – Possibilities for the Future." Purdue Extension report ID 342-W.

Tyner W and Taheripour F (2008) "Policy options for integrated energy and agricultural markets." Rev Agr Econ 30(3): 387–396.

Westcott PC (1998) "Implications of U.S. policy changes for corn price variability." Rev Agr Econ 20(2): 422–434.

Westhoff P, Thompson W, Kruse J, and Meyer S (2007) "Ethanol Transforms Agricultural Markets in the USA." Eurochoices 6(1): 14–21.

Westhoff P, Thompson W, and Meyer S (2008) "Biofuels: Impact of Selected Farm Bill Provisions and other Biofuel Policy Options" FAPRI-MU Report #06-08.

Westhoff P, Brown S and Hart C (2006) "When Point Estimates Miss the Point: Stochastic Modeling of WTO Restrictions." J Int Agr Trade Dev 2:87–107.

Zulauf CR and Blue EN (2003) "Has the Market's Estimate of Crop Price Variability Increased since the 1996 Farm Bill?" Rev Agr Econ 25(1): 145.

Part III
Designing the Infrastructure for Biofuels

Chapter 10
Optimizing the Biofuels Infrastructure: Transportation Networks and Biorefinery Locations in Illinois

Seungmo Kang, Hayri Önal, Yanfeng Ouyang, Jürgen Scheffran, and Ü. Deniz Tursun

Abstract Growing biofuel mandates pose considerable challenges to the infrastructure needed across all stages of the supply chain − from crop production, feedstock harvesting, storage, transportation, and processing to biofuel distribution and use. This chapter focuses on the biofuel transportation and distribution network infrastructure, using Illinois as a case study. Building on an optimal land use allocation model for feedstock production, a mathematical programming model is used to determine optimal locations and capacities of biorefineries, delivery of bioenergy crops to biorefineries, and processing and distribution of ethanol and co-products (DDGS). The model aims to minimize total system costs in a multiyear planning horizon for the period of 2007–2022. Certain locations may be more suitable for corn and corn stover-based ethanol plants, others more for producing ethanol using perennial grasses (*miscanthus*)

10.1 Introduction

Ethanol is the major renewable fuel in the United States and blended in half of the nation's gasoline consumption, mostly at 10% rate. Driven by the demand for fuel additives and biofuel mandates, ethanol production increased from about 6.1 billion liters (BL, equivalent to 1.6 billion gallons, BG) in 2000 to 24.6 BL (6.5 BG) in 2007. In 2007 alone the number of operating ethanol plants was increased from 110 to 180, with an annual production capacity of 42 BL. Twenty-one additional refineries were under construction at the end of 2008, expanding the total capacity further by 6.1 BL (RFA 2008). According to the 2007 Energy Independence and Security Act (EISA) Renewable Fuel Standard (RFS), at least 136 BL of renewable fuel

S. Kang (✉)
Energy Biosciences Institute, University of Illinois, Urbana-Champaign, IL, USA
e-mail: skang2@illinois.edu

Authors' names are listed in alphabetical order, seniority of authorship is not assigned.

M. Khanna et al. (eds.), *Handbook of Bioenergy Economics and Policy*, Natural Resource Management and Policy 33, DOI 10.1007/978-1-4419-0369-3_10, © Springer Science+Business Media, LLC 2010

must be blended into motor-vehicle fuel by 2022, including 57 BL of conventional ethanol and 79 BL of advanced biofuels (U.S. EPA 2007). The latter comprises 61 BL of cellulosic ethanol from corn stover, perennial grasses, and woody biomass and 18 BL of biofuels from undifferentiated sources.

These mandates pose enormous challenges to the infrastructure needed (GAO 2007). Considerable investments are required to overcome the technical and economic barriers at all stages of the production and supply chain — from crop production, feedstock harvesting, storage, transportation, and processing to biofuel distribution and use (CRS 2007). Complex issues must be addressed on a regional, national, and international level to determine (i) the infrastructure requirements most appropriate to environmental and economic conditions; (ii) an optimal design of biorefineries for an efficient conversion of biomass materials into bioproducts through various processes; and (iii) the transportation and distribution networks of feedstocks and bioproducts. Due consideration must be given to balancing centralization and decentralization, bioproduct markets and trade, water demand and availability, energy balance and efficiency, land use, greenhouse gas emissions and other environmental concerns (Worldwatch 2006).

This chapter focuses on the biofuel transportation and distribution infrastructure, specifically the flow of materials, and biorefinery location and size issues. After a brief discussion of some key infrastructure components, we will introduce a mathematical model to determine the optimal biorefinery location in the transportation network, considering Illinois as a case study. This is part of a comprehensive study on the biofuel supply chain problem that spans from bioenergy inputs production to processing and consumption of ethanol and by-products. The model determines the optimal construction and expansion plan for biofuel refineries in a multiyear time horizon to minimize the total production and transportation costs.

10.2 The Biorefinery — Connecting Feedstocks and Bioproducts

A cornerstone in the emerging bioeconomy is the integrated biorefinery that takes all organic feedstocks as input and applies multiple biomass conversion processes to produce a variety of bioproducts, including fuels, power, and chemicals from biomass (Kamm et al. 2005). Due to the wide range of options, it is a challenge to design a biorefinery to optimally convert biomass into bioproducts in a way that is adaptive to market demands and specific regional conditions. Next-generation biorefineries are expected to utilize plant cell-wall matter derived from waste material (such as corn stover) and dedicated energy crops (such as switchgrass and *miscanthus*). Current biorefineries are largely based on a single input (corn, sugarcane, or plant oil) and a single output (ethanol or biodiesel), often generating co-products (distillers, grains, bagasse). Driven by government subsidies and mandates, the capacity of ethanol plants has been increasing. The current standard capacity is 379 million liters (100 million gallons) per year. Some smaller-scale plants have been expanded to take advantage of economies of size (Brown et al. 2007).

Ethanol becomes more competitive if distillers dried grains and solubles (DDGS) and higher valued chemicals are produced as co-products. Currently, DDGS are used as animal feed, while most of the bagasse from sugarcane production in Brazil is burned for power generation. It has been estimated that DDGS production in the United States could increase rapidly, from 15 million tons in 2006/2007 to more than 35 million tons in 2009/2010, and eventually to about 43 million tons by 2013/2014 (Babcock et al. 2008). This expansion could go together with utilizing the fiber and oil content of DDGS to increase the efficiency and profitability of ethanol plants and reduce energy requirements and greenhouse gas (GHG) emissions (Singh et al. 2004, Kim et al. 2008). While DDGS dried to 10% moisture can be stored and shipped across the country and into export markets, wet distillers grains with solubles (WDGS) has a wet moisture content of 65–70% (Dooley et al. 2008) and is usually shipped in a 80-km circle.

According to Brown et al. (2007), the capital cost of building a dry grind ethanol plant has increased in recent years, "ranging from $1.50 to $2.00 per gallon, so the capital required to build a 100 million gallon dry grind ethanol plant can be in the $200 million range with rail access included." Economic analyses of dry grind ethanol in the United States and Europe estimate production costs to be between $0.80 and $1.36 per gallon ($0.21 and $0.36 per liter) depending upon various factors, especially feedstock costs (Wallace et al. 2005, Wooley et al. 1999). Compared to a project investment of $76 million for a corn ethanol plant with the capacity of 189 million liters per year, the National Renewable Energy Laboratory estimated the total investment for a cellulosic ethanol plant with similar capacity to be about $250 million (NREL 2007). Besides the fixed investment cost disadvantage, the cost of producing cellulosic ethanol is also higher, estimated as twice the cost of corn-based ethanol.

10.3 Biomass Transportation Networks and Biorefinery Locations

Transportation of feedstocks to biorefineries and distribution of bioproducts to the consumers are important cost factors in a regional biofuel assessment. Field harvested corn and cellulosic biomass have a low-energy density compared to solid fossil fuel such as coal and require feedstocks to be collected over large areas and transported to biorefineries. Since biorefineries are more likely to be profitable closer to production and demand centers, finding the balance between the supply in rural areas and the demand in urban areas is a task for optimization.

An efficient distribution network would combine cost-effective and sustainable transport of biomass, biofuels, and biogas through roads, rails, waterways, and pipelines with properly located storage and conversion plants (Fig. 10.1). Locating and sizing of plants and refineries for power, fuel, chemical, and food production have to address trade-offs between large facilities that take advantage of economies of size and decentralized network nodes closer to producers and consumers. Ethanol

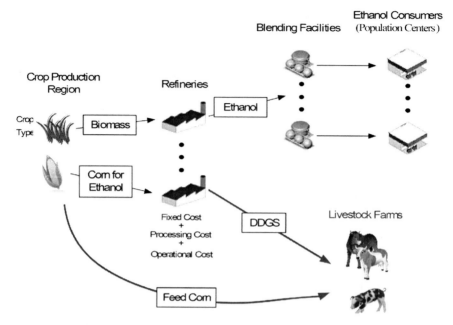

Fig. 10.1 Biomass transportation, biorefinery locations, and biofuel distribution

moves from biorefineries to petroleum terminals where it is blended with gasoline. An inadequate blending and distribution infrastructure may become an impediment to the marketing of ethanol as industry expands. To address these constraints, capital investment "in tanks, rail siding and cars, barges, piping, blending, metering, terminals, etc., is needed to sustain the expected national diffusion of demand." (Brown et al. 2007) Each of the transportation modes has its own characteristics, making it appropriate in a particular environment.

10.3.1 Road/Highway

Truck transportation is appropriate for short distances; for large distances, costs may become a disadvantage. An important factor is access to interstate highways. Trucks carry most of the feedstock to nearby refineries, sometimes leading to heavy traffic congestion and hence possible community resistance. For instance, a biorefinery that produces 450 megawatts of bioenergy per year requires 2.1 million tons of dry biomass annually, corresponding to about one truck delivery every 4 minutes (Mahmudi and Flynn 2006). Some studies suggest that a truck-and-train intermodal transport system is more economical (Mahmudi et al. 2005). There is a minimum shipping distance for rail transport though, below which gains due to lower costs per km are offset by incremental fixed costs.

10.3.2 Railroad

Rail transshipment may be preferable for large volumes and long distances or if road congestion precludes efficient truck delivery. Currently, most of the ethanol moved from plants to blending terminals is carried by trains. Since railroad cars are larger than trucks, it normally takes longer to fill these units. With an expansion of ethanol shipment to the East and West Coasts, rail (and barge when possible) transportation plays a major role and requires an infrastructure system to handle larger, more efficient units (Brown et al. 2007).

10.3.3 Waterways

Water transportation is much cheaper and thus provides a significant advantage compared to rail and truck for longer distances and larger volumes. A river barge can hold about 14 times more ethanol than one rail car and around 50 times more than a truck, a ship twice that amount (Brown et al. 2007), although this implies longer waiting time until the barge is filled, usually several days. River transportation is vulnerable to winter weather and extreme water levels, which could delay delivery for weeks. Despite all these, barges deliver large amounts of ethanol from the Midwest through the Inland Waterway System to the Gulf Coast and further to population centers along the East and West Coasts, and foreign countries.

10.3.4 Pipelines

Compared to the other modes of transportation, pipelines have a significant advantage as they allow large flows of materials in liquid or gas form to be transported over long distances and at low cost without congesting traffic systems or polluting the environment. While pipelines are standard for petroleum and natural gas, they have not played a role yet for ethanol transport because ethanol is corrosive, mixes with water and ethanol-blended gasoline tends to separate in pipelines. Existing pipelines are not configured to carry biofuels from the Midwest to the coasts. While some studies suggest that shipping ethanol via pipeline could be feasible (API 2008), no major investments have been made yet (CRS 2007).

The distribution system for ethanol (and biodiesel) is more complicated and more costly than for petroleum fuels. While nationwide petroleum transportation from refineries to fueling stations is estimated to cost about 0.79 to 0.32 cents per liter, the overall cost of transporting ethanol is estimated to be between 3.43 and 4.75 cents per liter, depending on the distance traveled and the mode of transportation (GAO 2007:23). The increase in ethanol production has contributed to shortages in regional transportation capacity. For example, the number of ethanol carloads tripled between 2001 and 2006, which increased further by 30% in 2007 (Brat and Machalaba 2007). The freight rail system has limited capacity and may not be able to meet the growing demand. Replacing, maintaining, and upgrading the existing aging rail infrastructure is extremely costly (GAO 2007).

Another major limitation in the biofuels industry is the so-called blending wall. Currently, the amount of ethanol permitted to blend with gasoline is set at 10% for conventional vehicles, because fuel economy diminishes at higher blending rates. Flex fuel vehicles (FFVs) can use any gasoline−ethanol mix efficiently, including E-85 (85% ethanol, 15% gasoline). In early 2007, about 1% of fueling stations in the United States, primarily in the Midwest, offered E-85 or high blends of biodiesel. However, significant growth in the number of E-85 stations and FFVs requires considerable investment.

10.4 The Optimal Biomass Transportation and Biorefinery Location Problem

As the previous discussion explained, transportation and distribution are important factors in the biofuel supply chain, making it essential to find the best transportation network and location of biorefineries. To support decision-making on infrastructure design, analytical modeling tools are needed to simulate the optimal supply chain to biorefineries and beyond.

For at least 30 years, economists have studied the economic feasibility of corn residues as auxiliary fuel in coal-fired power plants by using mathematical programming models, including costs for farm production, transportation, and processing and handling (English et al. 1981; Nienow et al. 2000). Sokhansanj et al. (2006) used a logistics model for an integrated supply analysis of the collection, storage, and transportation of corn and cellulosic biomass supply. Using time-dependent discrete event simulation and queuing analysis that represent the entire network of material flows from the field to a biorefinery, they predicted the number and size of equipment needed to meet the biorefinery demand for feedstocks. Tembo et al. (2003) have developed a multiregion, multiperiod, mixed-integer programming model, encompassing production of alternative feedstocks, transportation and processing to assess the most economical structure of the biomass-to-ethanol industry. Building on this work, Mapemba (2005) and Mapemba et al. (2007) have estimated the cost to deliver feedstock to a biorefinery in the range of $26 to $58 per ton, depending upon the size of the biorefinery, the number of harvest days, and the harvest frequency. The results demonstrated how the increase in biorefinery capacity would affect the expected delivery cost.

Estimating the cost of delivered biomass, quality of biomass supplied, emissions during collection, energy input, and maturity of supply system technologies, Kumar et al. (2006) provided a ranking of biomass-collection systems. For a given capacity, rail transport of biomass was identified as the best option, followed by truck transport and pipeline transport. Rail transshipment may also be preferable in cases where road congestion precludes truck delivery. Mahmudi and Flynn (2006) analyzed the conditions under which a truck-and-train intermodal transport system would be the most economical.

The problem is to locate the processing facilities so as to minimize the total transportation cost and the plant cost adjusted by the returns from co-product sales, such as DDGS. This type of facility-location problem has been solved using integer-programming (Daskin 1995, Drezner 1995), which searches for an optimal network configuration (Fuller et al. 1976; Hilger et al. 1977) and selects biorefinery locations from a set of candidates, considering prespecified criteria such as water availability or distance to the transportation network (Peluso et al. 1998). A mathematical programming model by Kaylen et al. (2000) analyzes economic feasibility of producing ethanol from lignocellulosic feedstocks at minimal cost, distinguishing between capital cost, operating cost, feedstock cost, and transportation cost. As plant capacity increases, marginal operating cost declines due to economies of scale, but the transportation cost increases because feedstock will have to be shipped from longer distances.

Improving the feedstock logistics, use of DDGS and high-valued co-products and finding the best mix of biomass-conversion technologies can lead to significant cost reductions for the future biorefinery industry (Hess et al. 2007; Wright and Brown 2007). Integrated models incorporate a number of factors simultaneously (feedstock, farm type, biorefinery site and size, technology, market conditions). For instance, Eathington and Swenson (2007) have developed a GIS-based decision tool for the selection of optimal site, size, and technology of ethanol plants to assess different policy and economic scenarios, including biofuels-related job impacts, local demand, and growth of the industry. Integrating this analysis into a modeling framework such as POLYSYS allows to support long-term policy decisions (De La Torre Ugarte and Ray 2000).

Building on our research on biofuels and land use in Illinois (Khanna et al. 2008; Scheffran and Bendor 2009), we are expanding this analysis to optimization of the feedstock and ethanol transportation network together with biorefinery location and size. An earlier version of this model (Tursun et al. 2009) addressed the problem of optimal biomass and ethanol deliveries to and from biorefineries using linear mixed-integer programming. Here, we expand this model to also accommodate DDGS deliveries to demand locations where it is used as livestock feed.

10.5 Model Description

A large-scale mixed-integer programming model has been used to determine the optimal land use for feedstock production in the State of Illinois (for details see Chapter 17; Khanna et al. (2009) in this book). The model we develop here is based on the results of that study.

Based on transportation costs from production sources of the feedstock by the road/railroad network, certain locations may be more suitable for corn-based ethanol plants, while others may be more suitable for producing ethanol using corn stover or perennial grasses (specifically *Miscanthus*; switchgrass was not found to be economical in Illinois conditions). The model aims to minimize the total system costs for transportation and processing of feedstock, transportation of ethanol from refineries to blending terminals, cost of shipping ethanol from blending terminals to demand destinations, capital investment in refineries, and transportation of the co-product DDGS to livestock producing areas for the period of 2007–2022. For readability, we present details of the mathematical model used in our analysis in the Appendix. In this section, we provide a brief description of the model structure and data requirements.

Illinois is used as a test bed in this case study because of the input data availability and the State's important role in renewable energy production. Each of the 102 counties in Illinois is considered as a supply region (where one or more of the bioenergy crops can be produced) and a demand destination to which ethanol is to be delivered. Also, each county is assumed to be a candidate plant location where a corn-based or a cellulosic biorefinery, or both, can be built in any given year. Cellulosic biorefineries are assumed to process both corn stover and *Miscanthus*. The existing corn ethanol plants' locations and capacities, and the locations of blending terminals are specified exogenously.

Production and distribution stages of the biofuel supply chain involve several integrated decision layers that must be addressed simultaneously (see Fig. 10.1). These include (1) the type of processing facilities, their capacities, years in which they are built, and locations; (2) the amount of raw materials (corn, corn stover, and *Miscanthus*) transported from production regions to biorefineries; (3) the amount of ethanol deliveries to existing blending facilities (terminals) and from there to demand destinations; and (4) the amount of DDGS shipments from refineries to livestock production regions.

This is a multilayered network design and transshipment problem with unconstrained network flows (see, for example, Dantzig and Thapa 2003) including yes/no type facility location selection and capacity decisions. We formulate the problem as a mixed-integer linear programming model where all the transportation and distribution decisions and refinery capacities are defined as nonnegative variables while the decision to build a biorefinery in any candidate location in any given year is defined as a binary variable.

Currently, DDGS is used as a substitute for dry matter feed for livestock such as cattle and hogs (for simplicity, here we include corn as the only feed grain). The maximum blending rates of DDGS in dry matter are specified for different types of livestock. It is assumed that livestock farmers choose an optimum feed ration including DDGS and feed corn (within the specified limits) based on the regional corn prices and transportation cost of DDGS from the refineries. The total amount of DDGS shipped from a biorefinery to the livestock-producing regions is restricted to the DDGS produced at that facility and the total amount of corn fed to all livestock categories in a given production region is restricted to the available feed corn in that region and the nearby regions. The total feed demand of each livestock producer region is calculated a priori based on the existing number of livestock in that region and the average annual feed requirement per head by livestock type. The livestock numbers in each region, thus the feed demand, are assumed to be constant throughout the planning horizon.

10.6 Model Specification and Data

The model requires several sets of input data. Since the fermentation processes of cellulose and glucose vary significantly from each other, we consider two different types of facilities, namely corn-based and cellulosic ethanol plants, which differ in terms of their fixed costs, processing costs, and other operational costs. Fixed conversion factors determine the amount of ethanol produced by each type of facility processing and converting corn or biomass into ethanol. The current technology and related conversion factors are assumed to be unchanged throughout the planning horizon. Other model inputs include regional supplies of bioenergy crops, costs of transportation from production regions to plant locations, blending facilities and demand locations, and the costs for different types of ethanol plants. Procedures for data generation are explained below.

10.6.1 Supply Input of Bioenergy

A key data set needed in the facility location model involves the spatial and temporal distribution of bioenergy input supplies, i.e., the amounts of corn and cellulosic biomass supplied by each county in each year of the planning horizon. These were generated by using a spatial and temporal resource allocation and market-equilibrium model whose methodological details and main findings are presented in Chapter 17 of this book. The model developed and employed in the present analysis uses the results of the economic model as input data.

10.6.2 Multimodal Transportation Network and Cost Matrix

The main criteria for qualifying candidate locations of biorefineries include access to the transportation mode and availability of sufficient water resources for ethanol processing. Most counties in Illinois have access to railroad and highways within county boundaries, and water is also widely available from major surface waters and aquifers. Hence, all counties are assumed to be a candidate site for future biorefineries. As an approximation, we treat the centroid of each county both as an origin and destination of all types of freight. Transportation costs for corn, biomass, and ethanol are calculated based on the highway and railroad network provided by the Bureau of Transportation Statistics National Transportation Atlas Database. Using centroid connectors linking the centroids to their nearest node of the highway and railroad networks, the Dijkstra's shortest path algorithm (Dijkstra, 1959) is used to determine the minimum network distance for each pair of centroids.

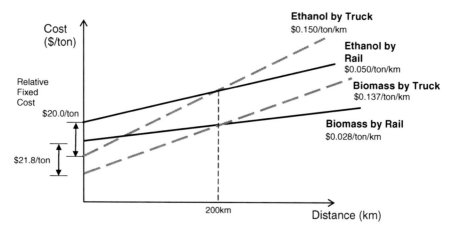

Fig. 10.2 Transportation cost structure

As shown in Fig. 10.2, railroad transportation has significantly lower per-km variable cost, but transshipment through railroad usually incurs a higher fixed cost (Mahmudi and Flynn 2006) because of extra handling of the load (e.g., hauling, storage, and unloading within the railroad terminal, for transshipment between truck and railcars). We assumed a fixed cost for railroad transportation and calculated the per-bushel-km delivery costs for corn and cellulosic biomass for both truck and rail transportation based on Sokhansanj et al. (2006). Similarly, we calculated the ethanol transportation costs per liter-km as suggested by Morrow et al. (2006).

10.6.3 Cost Structure of Biorefineries

Corn-based and cellulosic ethanol plants are associated with different cost structures. Refinery costs for each type of plant are divided into three main components: (i) annualized fixed cost, which includes the cost of land allocated to the refinery physical structure (based on farmland prices and the size of required land), and the costs of construction and machinery investment; (ii) processing cost, which is proportional to the capacity utilized (i.e., the amount of corn or cellulosic feedstock processed); and (iii) other costs related to operational expenses, such as labor and administrative expenses, which are linked to the refinery capacity. The cost parameters for corn-based refineries are generated by the "Dry Mill Simulator" component of Farm Analysis and Solution Tools (FAST) developed by Ellinger (2008) at the University of Illinois. These costs are based on the simulated performance of a 100-million gallon capacity corn ethanol plant. As the costs of cellulosic biorefineries we use the estimates by Wallace et al. (2005) for a 95-million liter capacity plant.

10.6.4 Ethanol Demand

As mentioned earlier, each Illinois county is assumed as a demand center. The mandated target for corn ethanol increases monotonically in the first half of 2007–2022 and then remains constant, whereas the cellulosic ethanol target constantly increases throughout the planning horizon. We assumed that Illinois will produce and blend a specified portion of the national ethanol target based on the State's current share (19%). For simplicity, we assume that this share is the same for both corn-based and cellulosic ethanol and the biofuels from undifferentiated sources in the RFS mandates (19 billion liters) will also come from cellulosic sources. Based on the populations of individual counties and the State's share in the RFS mandate, the ethanol demand for each county is specified for each year of the planning horizon.

10.6.5 Livestock Feed Demand

DDGS can substitute for corn as dry matter feed for livestock up to a specified limit depending on the livestock type. Babcock et al. (2008) review various studies regarding the blending possibilities of DDGS. It is generally agreed that a DDGS ratio of about 20% in dry matter is most productive for beef or dairy cattle, while DDGS can be used up to 30% for hogs. In this study, we employed those ratios in livestock feed rations. We determined the maximum amount of DDGS that can be used in each livestock-producer region based on the daily feed amount of dry matter per head. The livestock inventory in Illinois is obtained from the USDA Census of Agriculture.

10.7 Model Results

The results presented below for the State of Illinois serve as an exemplary and preliminary case for a comprehensive modeling effort, which aims to address similar policy issues and prospects for the entire US ethanol industry. Figure 10.3 shows the projected regional production of corn, corn stover, and *Miscanthus*.[1] The model results are driven mainly by these exogenous input data. For space restriction here we present the results mostly for 2022.

The optimum corn-based and cellulosic ethanol refinery locations are shown in Figs. 10.4 and 10.5, respectively, along with the locations of the existing blending facilities and the top 10 major demand areas. Figure 10.4 reveals an expected result: the corn-based biorefineries are located in those counties where (i) corn for ethanol is available at relatively large amounts and (ii) the plant location is close enough to the major demand locations. To meet the increased demand of ethanol over the

---

[1]Note that the color of each county indicates the total amount of biomass production, not the density.

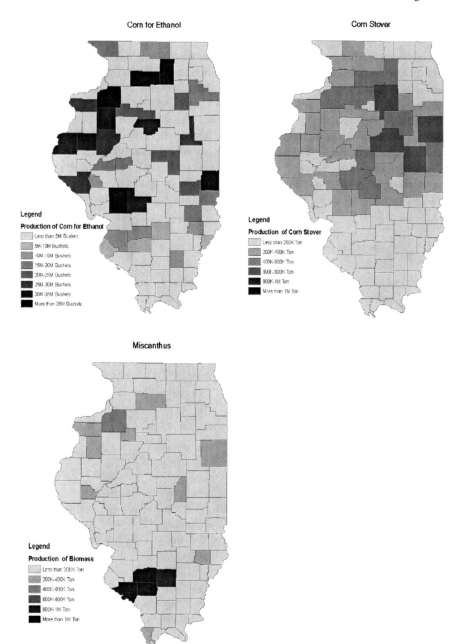

Fig. 10.3 Spatial distribution of projected (2022) biofuel production inputs in Illinois

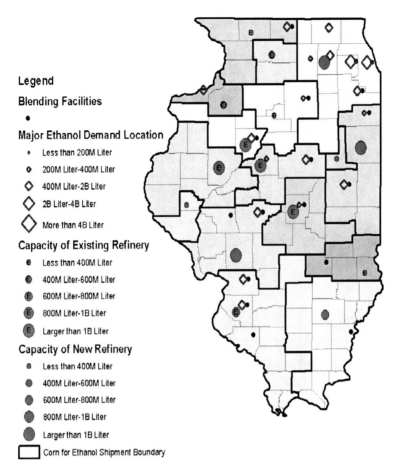

Legend

Blending Facilities
- •

Major Ethanol Demand Location
- ◦ Less than 200M Liter
- ◊ 200M Liter-400M Liter
- ◇ 400M Liter-2B Liter
- ◇ 2B Liter-4B Liter
- ◇ More than 4B Liter

Capacity of Existing Refinery
- ⊜ Less than 400M Liter
- ⊜ 400M Liter-600M Liter
- ⊜ 600M Liter-800M Liter
- ⊜ 800M Liter-1B Liter
- ⊜ Larger than 1B Liter

Capacity of New Refinery
- ⦁ Less than 400M Liter
- ⦁ 400M Liter-600M Liter
- ⦁ 600M Liter-800M Liter
- ⦁ 800M Liter-1B Liter
- ⦁ Larger than 1B Liter
- ☐ Corn for Ethanol Shipment Boundary

Fig. 10.4 Optimal location of corn-based ethanol refineries in 2022

planning horizon, many of the existing corn ethanol refineries would expand their capacities. The situation for cellulosic biorefineries is similar. Although most of the larger cellulosic refineries are located in central Illinois, where much of the corn stover is produced, several biorefineries would be built in the southern counties, supplying relatively large amounts of *Miscanthus*. Besides the regional raw material availabilities, exact locations and sizes of the biorefineries are driven by the cost trade-off between transportation of raw material from production regions to processing plants and of ethanol from refineries to major demand centers.

According to the model solution, within the first 3 years of the planning horizon the number of corn-based biorefineries would grow from 10 in 2007 to 13 in 2012.[2] The refinery capacity for those counties is the total capacity of all refineries

---

[2]The number of refineries is actually the number of counties having at least one refinery. Some counties have multiple refineries, such as Peoria and Tazewell.

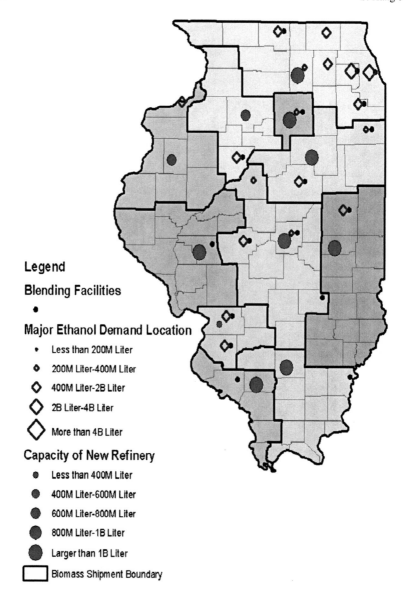

Fig. 10.5 Optimal location of cellulosic ethanol refineries in 2022

located in the county. After 2012, only one additional corn-based refinery is built (in 2015, see Fig. 10.6). Six of the existing refineries increase their capacities (shown by the concentric circles in Fig. 10.4), which are relatively small refineries. The three largest refineries (located in Central Illinois, namely in Macon, Peoria, and Tazewell counties) maintain their processing capacity. In order to satisfy the rising demand in the northern and northeastern counties, several new refineries have to be built in

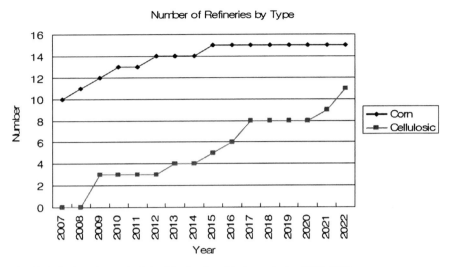

Fig. 10.6 Projected growth of the Illinois ethanol industry during 2007–2022

that region (Ford, Iroquois, Kane, and Putnam counties, all near the greater Chicago metropolitan area) while two large corn-based refineries are built in the southwest (Macopin and St. Clair counties, near St. Louis). Two of these new refineries have the maximum capacity that we specified exogenously (1.1 billion liters per year based on the actual sizes of existing plants; the largest corn ethanol plant in 2007 has the annual capacity of 1.03 billion liters – ADM's plant in Macon county). The average corn-based refinery capacity rises from 462 million liters (ML) in 2007 to 712 ML in 2022.

In contrast, the number of cellulosic biorefineries increases steadily from zero (no plant exists in 2007) to a total of 11 plants in 2022 (Figs. 10.5 and 10.7). This corresponds to the increasing trend in the RFS targets for cellulosic ethanol production. In 2022, the smallest cellulosic plant has an annual production capacity of 454 ML while the average plant size is 908 ML, higher than the average corn-based ethanol plant size. About 36 percent of the cellulosic ethanol comes from corn stover. Five of those cellulosic plants, all located in southern counties (Bond, Jefferson, Perry, Richland, and Washington), hit the maximum capacity limit (1.1 billion liters per year), while in the north two large-scale cellulosic plants are built at near maximum capacity (in La Salle and Livingston counties, with 1,106 and 995 ML, respectively).

Table 10.1 displays some summary statistics (minimum, average, and maximum) for the sizes and operational details of the corn-based and cellulosic biorefineries in 2022. According to the model result, there is no significant difference between the average distances corn and biomass are transported from production regions to processing plants, while the transportation distance of corn is slightly longer than that of cellulosic biomass. The procurement areas of individual corn-based and cellulosic biorefineries are shown in Figs. 10.4 and 10.5, respectively. These maps show

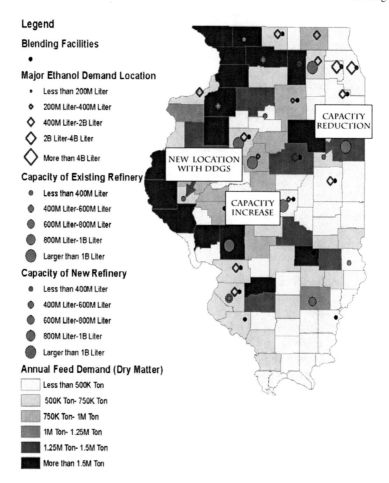

Fig. 10.7 Capacity and location of corn refineries with livestock feed demand

only illustrative examples of transportation of corn and biomass, which are shipped from a certain region to multiple refineries. The relationship between the size of the procurement area and the plant size depends on the amount of corn and biomass supply availability in the areas surrounding a given plant. For instance, some large corn ethanol plants in central counties have relatively smaller procurement areas compared to the plants located in southern counties because of the relative abundance of corn for ethanol in those areas.

The baseline scenario does not have the DDGS layer. Figure 10.7 illustrates the impact of DDGS use to meet the regional livestock feed demands on the locations and capacities of corn refineries. When the cost of DDGS transportation to the livestock areas is included in the objective function, some plant locations were altered and the capacity of refineries near the high-feed demand areas would increase

Table 10.1 Transportation distances and capacities of biorefineries in 2022

		Min	Average	Max
Transport	Corn	18.5	89.1	266.3
distance (km)	Biomass	14.2	84.1	193.4
Capacity	Corn ethanol	182	712	1,136
(million liters/year)	Cellulosic ethanol	454	908	1,136
Number of plants	Corn ethanol		15	
	Cellulosic ethanol		11	

Table 10.2 Increased DDGS use between the current-DDGS-price scenario and half-price scenario (Unit:Ton)

Corn refineries	Livestock farms	Year 13	Year 14	Year 15	Year 16
Crawford	Edgar	11,489.8	11,489.8	11,489.8	−466.3
Crawford	Effingham	93,525.1	93,525.1		93,525.1
Macon	Edgar	466.4	466.3	466.3	466.3
Macoupin	Fayette		32,010.6	32,010.6	32,010.6
St.Clair	Alexander	3,122.8	3,122.8	3,122.8	3,122.8
St.Clair	Jackson	23,831.1	23,831.1		23,831.1
St.Clair	Union	23,906.3	23,906.3	23,906.3	23,906.3
Wayne	Hardin	2,915.7	5,616.3		5,616.3
Wayne	Johnson	19,985.7	19,985.7		19,985.7
Wayne	Massac	16,391.2	16,391.2		16,391.2
Wayne	Pope	6,417.9	6,417.9		6,417.9
Wayne	Pulaski	5,670.2	5,670.2	5,670.2	5,670.2
Wayne	Williamson	14,369.1	14,369.1	14,369.1	14,369.1
Total % change		**5.1**	**6.0**	**2.0**	**5.7**

over time, compared to the scenario where DDGS delivery cost was not considered. For example, a refinery located in Schuyler County would be moved to an adjacent county (shown by the arrow) that is closer to a major livestock-producer region.

Table 10.2 represents results for two possible scenarios regarding future DDGS prices. In the first scenario, the DDGS price is assumed to remain at the current price level. In the second scenario, the DDGS price is decreased to half of its current level while everything else remains the same. There is no significant difference in the refinery capacities and locations in the two solutions, indicating that the impact of DDGS price on the objective function is relatively small. However, the DDGS purchased by livestock farms would be increased by 2–6% in the half-price scenario.

10.8 Summary and Conclusions

This article discusses current challenges for bioenergy infrastructure, critical issues of the biofuel supply chain, and the literature on biomass transportation and refinery location. These considerations demonstrate the need for analytical tools to optimize the bioenergy infrastructure and supply chain. A mathematical programming model is developed to determine the optimal locations and capacities of biorefineries, delivery of bioenergy crops to biorefineries, and processing and distribution of ethanol and co-products produced in those facilities.

Our results show that biorefineries must be located near those counties with large crop supply and/or large ethanol demand. Many of the corn ethanol refineries would expand their capacities to meet the mandated demand for ethanol over the planning horizon, 2007–2022. With four new corn ethanol plants built and six plants expanding their capacities (some reaching the maximum capacity of 1.1 billion liters) early in that period, the corn-based biorefineries would produce enough ethanol to meet the share of Illinois in the RFS mandate (19% ~27 billion liters). The number of cellulosic biorefineries increases steadily from zero to 11 plants in 2022 with an average plant size of 908 million liters, higher than the average corn-based ethanol plant size (681 million liters). Cellulosic refineries would be located in central Illinois, where much of the corn stover is produced, and in Southern Illinois, where *Miscanthus* is produced at larger amounts. Average shipping distances for corn and biomass from production regions to biorefineries are comparable (usually between 80 and 90 km). Procurement areas of corn ethanol plants are somewhat smaller in corn-abundant Central Illinois than in Southern Illinois. Including the distribution cost of DDGS in the model changes the locations of some plants and increases the capacity of corn refineries near high livestock feed demand areas.

The model development and application to the State of Illinois represent preliminary results of a comprehensive study that will address similar issues for the entire US renewable biofuel industry. In the next step of our analysis, the current coverage will be expanded to the Midwest region and later to a total of 28 states that are likely to supply corn and cellulosic biomass to the US ethanol industry. It would be ideal to solve the regional supply of bioenergy crops and optimal location of processing facilities simultaneously. However, the supply response model that provides input to the transshipment and facility location model used in the present study is already a very large-scale nonlinear mathematical programming model. The model used in this study is also a very large-scale mixed-integer programming model (including over 19,000 constraints and 150,000 variables, 3,000 of which are binary variables). Solving nonlinear mixed-integer programs of this size is generally very difficult (in this particular application, solving the transshipment model alone took nearly 3 hours of CPU time). Thus, solving the supply response and supply chain design problems simultaneously would be computationally intractable. A remedy to this deficiency is to incorporate the optimal locations of biorefineries (obtained from the MIP model) in the supply response model and solve the latter again with known refinery locations. This may lead to a near-optimal solution, if not exact. Another approach is to employ Lagrangian relaxation-based decomposition algorithms to handle the computational complexity. Also, instead of solving the model for each and every year of the planning horizon, the model could be solved only for a few benchmark years. This approach may result in an approximately optimal solution and may provide an equally valuable insight to policy makers, as well as the future investors in the bioenergy industry.

10.9 Appendix: Model Notation and Equations

10.9.1 Subscripts

$r \in R$: production regions
$j \in J$: candidate refinery locations
$k \in K$: blending terminals for ethanol
$l \in L$: livestock producing regions
$n \in N$: ethanol demand centers
$h \in \{$cattle for beef, cattle for dairy, hog$\}$: livestock type
$m \in \{$corn, cellulosic biomass$\}$: crop type.

10.9.2 Factors and Parameters

$S_r^{m,t}$: Amounts of energy crop m supplied by region r, year t
$S_r^{F,t}$: Amount of corn for feed produced in year t.
D_n^t: Demand for ethanol by demand center n and year t
b_j^m, v_j^m, o_j^m: annualized fixed investment cost, processing cost, and other operational costs, respectively, for refinery type m built at location j
β^m: Minimum utilization rate of refinery by type
α^m, γ: Conversion factors from biomass and corn to ethanol and from corn to DDGS
p^F, p^G: Prices of feed corn and DDGS
ε^h: Maximum blending of DDGS by livestock type h
q^h: Annual feed demand by livestock type h
$I_l^{h,t}$: Livestock inventory by type h, in region l, year t
t_{rj}^m: Transportation costs per unit amount and unit distance for feedstock m, region r and refinery j
t_{jk}^E, t_{kn}^E: Costs of delivering ethanol from refinery j to demand blending terminal k and from blending terminal k to demand centers n.
t_{rl}^F, t_{jl}^G: Transportation cost of feed corn and DDGS, respectively, to region l.

If a refinery is to be built at any candidate location, we impose a minimum capacity $C_{\min}^m$ and a maximum capacity $C_{\max}^m$ in order for the model not to choose very small facilities or extremely large capacities beyond practical values.

10.9.3 Model Variables and Variable Types

Binary Variables

$x_j^{m,t} = 1$ if a type-m ethanol refinery is built at location j in year t; $x_j^{m,t} = 0$ otherwise;

Nonnegative Variables

$y_{rj}^{m,t}$: Amount of corn for ethanol or biomass shipped from production region r to refinery j, in year t

$C_j^{m,t}$: Capacity of type-m ethanol refinery at built at location j, year t

$e_{jk}^{m,t}$: Amount of type-m ethanol shipped from refinery j to blending terminal k, year t

e_{kn}^t: Amount of type-m ethanol shipped from blending terminal k to demand center n, year t

f_{rl}^t: Amount of corn for feed shipped from production region r to livestock farm l, year t

g_{jl}^t: Amount of DDGS shipped from refinery j to livestock farm l, year t

Using the above notation, the problem is formulated as a linear mixed-integer mathematical program described algebraically below:

$$Min \sum_{r,j,m,t} t_{rj}^m \cdot y_{rj}^{m,t} + \sum_{j,m,t} (b_j^m \cdot x_j^{m,t} + o_j^m \cdot c_j^{m,t}) + \sum_{j,k,m,t} (v_j^m + t_{jk}^E) \cdot e_{jk}^{m,t} + \sum_{k,n,t} t_{kn}^E \cdot e_{kn}^t$$
$$+ \sum_{r,l,t} (p^F + t_{rl}^F) \cdot f_{rl}^t + \sum_{j,l,t} (p^G + t_{jl}^G) \cdot g_{jl}^t$$

The objective of the model is to minimize the total cost of all operations, including the transportation costs of raw materials and end products (ethanol, DDGS), costs of processing, and fixed investment costs associated with building refineries, and costs related to co-product deliveries. The following constraints are used to restrict the decision variables and link different components of the model:

Subject to:

$$\sum_j y_{rj}^{m,t} \leq S_r^{m,t} \quad \forall r,m,t \tag{10.1}$$

$$\sum_r y_{rj}^{m,t} \cdot \alpha^m \geq \sum_k e_{jk}^{m,t} \quad \forall j,m,t \tag{10.2}$$

$$\sum_k e_{jk}^{m,t} \leq c_j^{m,t} \quad \forall j,m,t \tag{10.3}$$

$$\sum_{j,m} e_{jk}^{m,t} \geq \sum_n e_{kn}^t \quad \forall k,t \tag{10.4}$$

$$\sum_k e_{kn}^t \geq D_n^t \quad \forall n,t \tag{10.5}$$

$$c_j^{m,t} \leq C_{max}^m \cdot x_j^{m,t} \quad \forall j,m,t \tag{10.6}$$

$$c_j^{m,t} \geq C_{\min}^m \cdot x_j^{m,t} \quad \forall j,m,t \tag{10.7}$$

$$\sum_k e_{jk}^{m,t} \geq \beta^m \cdot c_j^{m,t} \quad \forall j,m,t \tag{10.8}$$

$$C_j^{m,t+1} \geq C_j^{m,t} \quad \forall j,m,t \tag{10.9}$$

$$\sum_{r,m \in \{corn\}} y_{rj}^{m,t} \cdot \gamma \geq \sum_l g_{jl}^t \quad \forall j,t \tag{10.10}$$

$$\sum_l f_{rl}^t \leq S_r^{F,t} \quad \forall r,t \tag{10.11}$$

$$\sum_r f_{rl}^t + \sum_j g_{jl}^t \geq \sum_h q^h \cdot I_l^{h,t} \quad \forall l,t \tag{10.12}$$

$$\sum_j g_{jl}^t \leq \sum_h \varepsilon^h \cdot q^h \cdot I_l^{h,t} \quad \forall l,t \tag{10.13}$$

- Equation (10.1) is a supply constraint, defined for each production region and each year, which ensures that the amount of corn and biomass shipped from any given region to all biorefineries in the system cannot exceed the available supply in that region.
- Equation (10.2) is an in/out balance constraint, defined for each processing facility and year, which restricts the amount of ethanol produced by a biorefinery to the corresponding amount of raw inputs coming into that facility.
- In (10.3), the amount of ethanol produced by any facility is restricted to the processing capacity of that facility specified at the time of construction, which is also determined by the model as a decision variable.
- Equation (10.4) is an in/out balance constraint at the blending facilities. Equation (10.5) indicates that the ethanol transported to the demand centers should meet the input demand.
- By (10.6) and (10.7), the capacity of any plant at construction time cannot fall below a minimum and cannot exceed a maximum capacity, both of which are specified a priori (based on the sizes of existing processing plants and capacities of the plants currently being built). We also restrict the capacity utilization in any processing facility to be above a specified minimum percentage of the full capacity (if that facility is included in the system) (10.8).
- The model allows expanding the capacity of a previously built biorefinery over time (10.9), but this may occur at additional investment costs. Once a biorefinery is built at a given location and in a given year, the model assumes that it remains operational in the subsequent years throughout the planning horizon (i.e., closing and reopening a plant in a later year is not allowed).
- Equation (10.10) restricts the maximum amount of DDGS produced at the refinery. Equation (10.11) indicates the amount of available corn for feed. Equation (10.12) is a constraint of livestock feed demand. Equation (10.13) represents the maximum portion of DDGS for each livestock type.

References

API (2008) "*Shipping Ethanol Through Pipelines*." American Petroleum Institute. Available at http://www.api.org/aboutoilgas/sectors/pipeline/upload/pipelineethanolshipment-2.doc. Accessed on May 15, 2009.

Babcock BA, Hayes DJ and Lawrence JD (eds) (2008) "Using Distillers Grains in the U.S. and International Livestock and Poultry Industries," Midwest Agribusiness Trade Research and Information Center, Iowa State University, Ames, IA

Brat I and Machalaba D (2007) "Can Ethanol Get a Ticket to Ride?," *The Wall Street Journal*, Feb. 1, p. B1.

Brown R, Orwig E, Nemeth J and Subietta Rocha C (2007) "Economic potential for ethanol expansion in Illinois", Illinois Institute for Rural Affairs at Western Illinois University, Macomb, IL.

CRS (2007) "Ethanol and Other Biofuels-Potential for U.S.-Brazil Energy Cooperation." Congressional Research Service, Washington, DC.

Dantzig GB and Thapa MN (2003) Linear Programming: Theory and Extensions. New York: Springer.

Daskin MS (1995) Network and Discrete Location: Models, Algorithms, and Applications. New York: John Wiley and Sons.

De La Torre Ugarte DG and Ray DE (2000) "Biomass and bioenergy applications of the POLYSYS modeling framework." Biomass and Bioenergy 18: 291–308.

Dijkstra EW (1959) "A Note on Two Problems in Connexion with Graphs." Numerische Mathematik 1: 269–271.

Dooley FJ, Cox M and Cox L (2008) "Distillers Grain Handbook: A Guide for Indiana Producers to Using DDGS for Animal Feed", Department of Agricultural Economics, Purdue University, http://incorn.org/images/stories/IndianaDDGSHandbook.pdf. Accessed on May 15, 2009.

Drezner Z (1995) Facility Location. New York: Springer.

Eathington L and Swenson DA (2007) "Dude, Where's My Corn? Constraints on the Location of Ethanol Production in the Corn Belt." Department of Economics, Iowa State University.

Ellinger P (2008) Ethanol Plant Simulator. Department of Agricultural and Consumer Economics, University of Illinois at Urbana-Champaign, Urbana, IL.

English B, Short C and Heady EO (1981) "The Economic Feasibility of Crop Residues as Auxiliary Fuel in Coal-Fired Power Plants." Am J Agri Econ 63(4): 636–644.

Fuller SW, Randolph P and Klingman D (1976) "Optimizing Subindustry Marketing Organizations: A Network Analysis Approach." Am J Agri Econ 58(3): 425–436.

GAO (2007) "Biofuels: DOE Lacks a Strategic Approach to Coordinate Increasing Production with Infrastructure Development and Vehicle Needs", U.S. Government Accountability Office, GAO-07-713, Washington, DC.

Hess JR, Wright CT and Kenney KLK (2007) "Cellulosic biomass feedstocks and logistics for ethanol production." Biofuels, Bioproducts Biorefining 1(3): 181–190.

Hilger DA, McCarl BA and Uhrig JW (1977) "Facilities Location: The Case of Grain Subterminals." Am J Agri Econ 59(4): 674–682.

Kamm B, Gruber PR and Kamm M, eds. (2005) Biorefineries. Biobased Industrial Processes and Products, Wiley-VCH Verlag, Weinheim.

Kaylen M, Van Dyne DL, Choi YS and Blasé M (2000) "Economic feasibility of producing ethanol from lignocellulosic feedstocks." Bioresource Tech 72: 19–32.

Khanna M, Dhungana B and Clifton-Brown J (2008) "Costs of producing *Miscanthus* and switchgrass for bioenergy in Illinois." Biomass and Bioenergy 32(6): 482–493.

Khanna M, Önal H, Chen X and Huang H (2009) "Meeting Biofuels Targets: Implications for Land Use, Greenhouse Gas Emissions and Nitrogen Use in Illinois." See Chapter 17 in this book.

Kim Y, Mosier NS, Hendrickson R, et al. (2008) "Composition of corn dry-grind ethanol by-products: DDGS, wet cake, and thin stillage". Bioresource Tech 99(12):5165–5176.

Kumar A, Sokhansanj S and Flynn PC (2006) "Development of a Multicriteria Assessment Model for Ranking Biomass Feedstock Collection and Transportation Systems." Proceedings of 27th Symposium on Biotechnology for Fuels and Chemicals, 71–87.

Mahmudi H, Flynn PC and Checkel MD (2005) "Life Cycle Analysis of Biomass Transportation: Trains vs. Trucks." SAE Technical Papers, www.sae.org/technical/papers/2005-01-1551. Accessed on May 15, 2009.

Mahmudi H and Flynn P (2006) "Rail vs Truck Transport of Biomass." Appl Biochem Biotechnol 129(1): 88–103.

Mapemba LD (2005) "Cost to Deliver Lignocellulosic Biomass to a Biorefinery." Ph.D Dissertation, Oklahoma State University.

Mapemba LD, Epplin M, Taliaferro CM and Huhnke RL (2007) "Biorefinery Feedstock Production on Conservation Reserve Program Land." Rev Agri Econ 29(2): 227–246.

Morrow WR, Griffin WM and Matthews HS (2006) "Modeling Switchgrass Derived Cellulosic Ethanol Distribution in the United States." Environ Sci Technol 40(9): 2877–2886.

Nienow S, McNamara KT and Gillespie AR (2000) "Assessing plantation biomass for co-firing with coal in northern Indiana: A linear programming approach." Biomass Bioenergy 18: 125–135.

NREL (2007) "A National Laboratory Market and Technology Assessment of the 30x30 Scenario", National Renewable Energy Lab, Technical Report/TP-510-40942, January.

Peluso T, Baker L and Thomassin PJ (1998) "The Siting of Ethanol Plants in Quebec." Can J Region Sci 21(1): 73–86.

RFA (2008) "Industry Statistics", Renewable Fuel Association, accessed 28 November 2008, http://www.ethanolrfa.org/industry/locations. Accessed on May 15, 2009.

Scheffran J and Bendor T (2009) "Bioenergy and Land Use – A Spatial-Agent Dynamic Model of Energy Crop Production in Illinois." Int J Environ Pollution: in press.

Singh V, Johnston D, Naidu K, Rausch KD, Belyea RL and Tumbleson ME (2004) Effect of modified dry grind corn processes on fermentation characteristics and ddgs composition. Proceedings of the Corn Utilization & Technology Conference, Indianapolis, IN. June 7–9.

Sokhansanj S, Kumar A and Turhollow AF (2006) "Development and implementation of integrated biomass supply analysis and logistics model (IBSAL)." Biomass Bioenergy 30(10): 838–847.

U.S.EPA (2007) "Renewable Fuel Standard Implementation." Available at: http://www.epa.gov/OTAQ/renewablefuels/index.htm. Accessed on May 15, 2009.

Tembo G, Epplin FM and Huhnke RL (2003) "Integrative investment appraisal of a lignocellulosic biomass-to-ethanol industry." J Agri Res Econ 28(3): 611–633.

Tursun D, Kang S, Onal H, Ouyang Y and Scheffran J (2009) "Optimum Biorefinery Locations and Transportation Network for the Future Biofuels Industry in Illinois.": In: M. Khanna (ed.), Transition to a Bioeconomy: Environmental and Rural Development Impacts, Proceedings of Farm Foundation/USDA Conference, St. Louis, Missouri, October 15–16, 2008, Farm Foundation, Oak Brook.

Wallace R, Ibsen K, McAloon A and Yee W (2005) "Feasibility Study for Co-Locating and Integrating Ethanol Production Plants from Corn Starch and Lignocellulosic Feedstocks." NREL/TP-510-37092 Revised January Edition: USDA/USDOE/NREL.

Worldwatch (2006) "Biofuels for Transportation, Global Potential and Implications for Sustainable Agriculture and Energy in the 21st Century", Washington, D.C.: Worldwatch Institute.

Wooley R, Ruth M, Sheehan J, Ibsen K, Majdeski H and Galvez A (1999) "Lignocellulosic biomass to ethanol – Process design and economics utilizing co-current dilute acid prehydrolysis and enzymatic hydrolysis – Current and futuristic scenarios". Report No. TP-580-26157, National Renewable Energy Laboratory, Golden, CO.

Wright M and Brown RC (2007) "Establishing the optimal sizes of different kinds of biorefineries." Biofuels, Bioproducts and Biorefining 1(3): 191–200.

Chapter 11
The Capital Efficiency Challenge of Bioenergy Models: The Case of Flex Mills in Brazil

Peter Goldsmith, Renato Rasmussen, Guilherme Signorini, Joao Martines, and Carolina Guimaraes

Abstract Bio-based energy sources have received increasing interest in recent years as petroleum prices have risen, geo-political instability has increased, and climate change has been in evidence. Extensive farming systems producing bio-based feedstocks, such as maize and sugarcane, are the models most widely used. Similar models are planned for dedicated cellulose crops such as miscanthus and eucalyptus. Bioenergy feedstock production that follows the current commercial agricultural model may inefficiently employ capital as the spatial density of the system, and the relative gravimetric density of the feedstock and volumetric density of the fuel products are low. The example of ethanol production in Mato Grosso, Brazil demonstrates the key concepts of density and capital intensity that are so critical to the efficient use of capital.

11.1 Introduction

Researchers are discussing and analyzing the performance of liquid fuel bioenergy models (LFBM) as oil supply forecasts have turned bearish, energy demand has expanded dramatically with economic growth among developing countries, and correspondingly, petroleum prices have risen. LFBM models, such as the Midwest maize model, the Sao Paulo sugarcane model, and the Northwest cellulosic waste model have varying degrees of system efficiencies. Second- and third-generation alternatives involving cellulosic and other higher density feedstocks are expected to significantly improve system efficiency (Hamelinck and Faaij, 2006).

P. Goldsmith (✉)
Department of Agricultural and Consumer Economics, University of Illinois, Urbana-Champaign, IL, USA
e-mail: pgoldsmi@illinois.edu

An earlier version of this paper was presented at "The Socio-Economic Impacts of Energy in the Past, Present, and Future: A Comparison of Brazil and the United States," November 21-24, 2008, Ilhabela, Sao Paulo, Brazil

M. Khanna et al. (eds.), *Handbook of Bioenergy Economics and Policy*,
Natural Resource Management and Policy 33, DOI 10.1007/978-1-4419-0369-3_11,
© Springer Science+Business Media, LLC 2010

Liquid fuel bioenergy models involve production systems that are spatially disperse and involve types of fuels and feedstocks that have relatively low energy
densities. Ceteris paribus this leads to inefficient use of assets. At the same time,
their use of renewable and low-cost feedstocks may help compensate for weak density attributes. We introduce in this manuscript the analytical concept of asset turns
as an alternative, and possibly better way, to understand the competitiveness and the
long-term viability implications of low density LFBM. Asset turns is the relationship between energy and the underlying assets to produce or move the energy from
production through to utilization. Formally, it is the ratio of energy in the form of
megajoules to dollar of underlying assets.

Consider the Midwest maize liquid fuel bioenergy model (LFBM), the dominant commercial system in place in the United States. In a stylized form, maize is
grown in a 75-km radius of the dry mill. The grain is harvested and transported
or stored. The mill operates 350 days a year. The ethanol, 1/3rd of the mill's output and 100% of the energy output, is transported by truck or rail to refineries for
blending with gasoline and entrance into the transportation fuel supply channel.
The maize yields about 400 l per ton of maize. Mill locations are generally distant from refiners and high population centers. Dried distillers grains and soluables,
a second third of the mill's output, is shipped to local feed mills or livestock producers, often close to ethanol plants, as a medium protein feed product. The final
third is the carbon dioxide, which is either vented or sold into industrial marketing
channels.

Sugarcane-to-ethanol models involve harvesting a wet feedstock and transporting the material directly to a processing facility where it is converted into either
sugar or alcohol. In general, seventy percent of the farms are owned or controlled
by the mill to assure supply, and the decision whether to produce sugar or alcohol
is simply a function of the relative wholesale prices (Martines-Filho et al., 2006). In
Mato Grosso, the subject of this manuscript, sugarcane is produced, harvested, and
transported 15 km directly to the mill. Sugarcane is perishable so there is no storage.
Plants operate for only about 200 days a year, the length of the harvest. A metric
ton of sugarcane will yield about 85 l of ethanol. Additional by-products are 135 kg
of dry cellulose-laden bagasse, which is currently burned for electricity, and 605 kg
of water. There is no refining capacity nearby so most alcohol is exported 1,500 km
from the State by truck to refiners in the East, with the remainder used locally in the
form of pure ethanol fuel.

This research uses the Mato Grosso LFBM to address three objectives heretofore
unaddressed in the bioenergy literature: (1) measure LBFM asset turns (utilization
efficiency) for the sugarcane mills of Mato Grosso; (2) show how asset utilization
improves by including maize as a complementary feedstock; and (3) demonstrate
how the industrial metric, asset turns, helps managers, policymakers, and investors
assess the long-term viability of LBFM and trouble shoot for system improvements.
In other related work with our model we analyze: the efficiency of asset utilization
from production through to utilization of other LFBMs such as the Midwest maize
and Sao Paulo sugarcane models, and other energy systems such as wind, solar,
nuclear, and hydro.

11.2 Literature Review

Biofuels are inherently less dense than fossil fuels in terms of megajoules (MJs) per unit of feedstock (Hamelinck and Faaij, 2006) as measured by volumetric (per liter) or gravimetric (per kilogram) density. The lack of feedstock energy density raises costs per unit of energy harvested and produced (Hamelinck and Faaij, 2006).

Spatial density, MJs per square kilometer, too is critical, especially for bioenergy models. All energy models attempt to minimize losses and costs where feedstock production, energy production, and consumption are spatially distant from each other. For example, ceteris paribus, individuals producing and consuming their own energy increases spatial density, while transmitting power over power lines lowers density. Agricultural-based bioenergy models differ from other alternative fuel models, such as wind, solar, hydro, and nuclear in that feedstock production and fuel manufacturing are separate activities and geographically disperse. The lack of spatial density raises the postharvest cost of preparation, assembly, and preprocessing (Overend, 1982; Searcy et al., 2007).

There would appear to be an important interplay between the density and the cost of the feedstock. The lower the cost of the feedstock, the less concern there would be about the presence of low energy (either spatial or gravimetric) density values (Fig. 11.1). The issue of energy density and efficient use of assets becomes very interesting when comparing extensive models such as a wheat straw (see Kerstetter and Lyons, 2001) with nuclear, hydro, solar, or wind models. Low spatial and volumetric dense models such as wheat straw will have low asset turns if significant aggregation and transport assets are required to bring the feedstock to the fuel-manufacturing facility. Alternatively, wind, solar, and hydro combine the feedstock assemblage and fuel-manufacturing stages, which dramatically increases asset turns.

The extensiveness of LFBM challenges system efficiency and competitiveness with fossil fuel-based models because of the cost of assembly and preprocessing, as well as the complexities of balancing and matching assets and associated product flows (Arjona et al., 2001; Gallagher et al., 2005; Carolan et al., 2007). Moving a heavy and wet product, the case of sugarcane, forces plants to locate close to supply and limits the scale economy because of associated aggregation and storage inefficiencies (Nguyen and Prince, 1996). Balancing labor, transport vehicles, and transport trucks reduces the preprocessing costs (Arjona et al., 2001). Optimizing transport and adding a second feedstock, such as sorghum, to follow sugarcane can reduce capital costs by 25% to produce 175 million liters of ethanol because a smaller more fully operational plant can be utilized (Nguyen and Prince, 1996).

Feedstock production research highlights poor economies of scale and associated high costs of moving product from the farm to the processing plant. Costs of inputs including feedstock and transport costs increase with plant capacity as the draw area widened for 18 maize dry mills in the Midwest United States (Gallagher et al., 2005). As a result, maximum plant size was relatively small at 250 million liters per year.

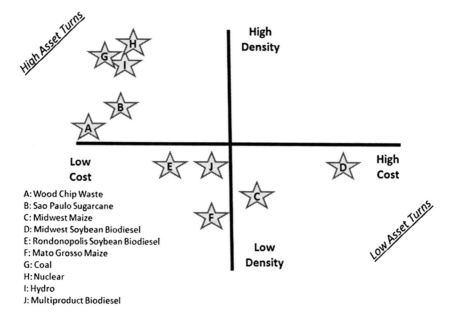

Fig. 11.1 Stylized model of key liquid biofuel competitiveness drivers: Pre-processing asset turns

11.2.1 Measures of Capital Utilization

Efficiently using capital use drives firm performance and is a central tenet of managerial accounting analysis (Altman, 1968; Nissim and Penman, 2001; Coelli et al., 2002). For example, capital use intensity or asset turns is one of the three most stable ratios associated with firm performance (Gombola and Ketz, 1983). Net returns to the firm can be generated by either increasing margins or increasing asset turns (Nissim and Penman, 2001). That is, overall firm performance can be enhanced (reduced) by increasing (decreasing) the quantity of margin from each sale or increasing (decreasing) the flow of sales from each asset. A common example of this concept is in retail where shelf facings for retailers are considered small pieces of real estate assets. Low-margin retail products need high turns to justify the usage of the shelf space. Similarly, high-margin products are not as dependent on turns to provide high returns on assets. Applied to comparative energy analysis, businesses employing energy-dense feedstock are not as dependent on asset turns as those where energy density is low. Put another way, energy models utilizing low-density feedstocks require lower valued assets.

The interplay of the cost of feedstock and asset turns is critical like the interrelationship between retail product margin and real estate assets. Ceteris paribus, dedicated feedstocks are more costly and have higher asset specificity than by-products. Higher asset specificity leads to higher levels of risk the more dedicated and nonfungible a feedstock. By-products have the advantage over dedicated energy crops in that their lower costs allow for systems to remain competitive at low energy turns, even if energy density is lacking. Energy densities vary significantly across by-product markets. For example, urban municipal waste has the benefit of high spatial and gravimetric densities when compared to rural waste; see Graf and Koehler (2000).

11.3 Methodology

Understanding the energy flow occurring in Mato Grosso's ethanol industry is complex. The size of the processing mills affects the size of its feedstock draw area, the number of its associated sugarcane suppliers, and the overall level of technology adopted to produce and transport the sugarcane. The relationship between the mill, the transportation capital, and feedstock production are related in a nonlinear fashion within a temporal dimension of seasonality. A methodology capable of capturing the feedback processes, stocks and flows, time delays, and other sources of dynamic complexity facilitates the interpretation of the state of the LFBM system given a set of assumptions.

System dynamics modeling methods and the software tool STELLA^{TM}, from IEEE Systems, are used to address the complexities of the Mato Grosso ethanol complex. The model allows the creation of various spatial, size, and dynamic configurations involving the cane and maize supply and its movement to and processing by the 11 operating mills in the State. Data are provided by extensive field work at the mills in Mato Grosso in 2007, plus supplementation with publically available secondary data.

Four stages and three exchange nodes provide a stylized overview of the bioenergy systems model (Fig. 11.2). In this paper, we focus on node two, *fuel manufacturing*. The creation of the fuel occurs in the manufacturing stage either from a farmed feedstock, sugarcane or maize.

Mato Grosso produces about 850 (2007) million liters of ethanol, which is about 5% of Brazil's output. The local market consumes 20% of ethanol produced in the State, with the remaining exported 2,000 km to markets in the heavily populated Eastern part of the country. The mills own approximately 70% of land used for the sugarcane production. The sugarcane harvest occurs when rains are not excessive so equipment can function, and when plant growth warrants cutting. Eleven mills operated on an average between 126 and 223 days a year in Mato Grosso in 2007 in the Southwest and South central regions of the State (Fig. 11.3). Thus, the mills, and the associated system-wide assets, are idle a significant part of the year.

The idle period reduces the asset turns of the mills resulting in two key effects, reduced productivity levels and a disincentive to employ new technologies and a more modern capital plant. Plants close when feedstock availability and quality decline. At the same time, ethanol prices begin to rise. The capital plant, the capacity and level of technology, match the level of available feedstock supply.

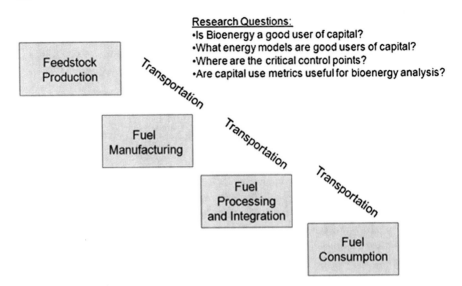

Research Questions:
- Is Bioenergy a good user of capital?
- What energy models are good users of capital?
- Where are the critical control points?
- Are capital use metrics useful for bioenergy analysis?

Fig. 11.2 Stylized model: Systems approach

If the season could be lengthened, then the capital costs would be spread over increased income, capital turns would increase, and greater capital investment could be warranted, hence the idea of the flex mill.

The flex-mill concept (see Signorini et al., 2007) extends the annual operating period of a sugarcane mill by introducing maize as a feedstock once the sugarcane season has concluded. Increasing the harvest window with maize allows plants to operate closer to their full capacity and thereby increase their asset turns. The plants on average operate 181 days per year (2005–2007) and have an opportunity to increase the operating period an additional 149 days (Table 11.1). One plant (D) operating 233 days on sugarcane as the feedstock would only utilize 107 days of maize supply. Another plant (K) operating only 126 days on sugarcane, could introduce 204 days of maize to its plant.

Sugarcane yields about 85 l of ethanol per metric ton compared with about 400 for maize. Sugarcane though moves through the capital plant relatively quickly, in about 7 hours, compared with about 50 h for maize. Cane is less dense than maize because of its high water content but has significantly higher throughput because of its readily available sugars. For example, assume sugarcane ferments five times faster than maize, thus the fermentation time ratio would be 5. If the yield ratio of maize to cane is 4, then the rate of ethanol production for the mill when maize is the feedstock would be 80% of the output when cane is the feedstock. Sugarcane mill assets turn faster than a dry maize mill because of the longer time for maize fermentation. Conceptually complementing sugarcane with maize as a feedstock raises the asset turns for a mill because asset turns during the idle period are zero.

Mills in Mato Grosso currently process about 9.5 million metric tons of sugarcane. Individual plants range in capacity from 14 to less than 1,000 tons per day

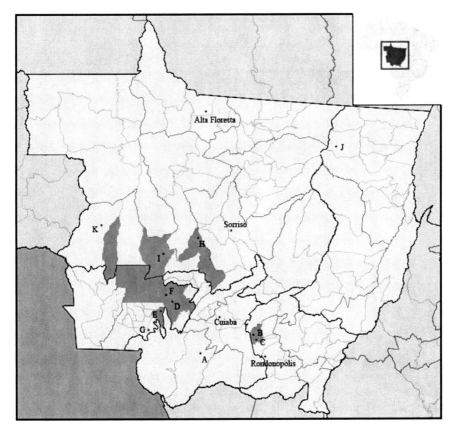

Fig. 11.3   Producer region of sugarcane (main cities), 2006
Source: Agriculture Secretary of Mato Grosso.

(Table 11.2). The smallest plant would demand less than 100 metric tons of maize, while the largest would require over 1,700 metric tons. The range in annual demand would be between 14 and 200 thousand metric tons of maize. Mills in Mato Grosso could utilize over 750 thousand metric tons of maize, or 12% of the State's 2007 crop, if all plants maximize the usage of maize as a complementary feedstock.

Currently, mills in Mato Grosso maintain no feedstock storage. Systems when maize is utilized would require additional storage and transport assets postproduction. Maize harvest occurs in June while the sugarcane harvest ends in December, causing the system to store maize for 6 months.

While cost data are plentiful, sugarcane mill capital cost data in Brazil are not readily available. We chose to follow Nguyen and Prince (1996) using regression analysis to estimate asset costs for the 11 Mato Grosso plants ranging from 13 to 270 million liters. These estimates were validated against single plant estimates of McAloon et al. (2000) and Hassuani et al. (2005).

Table 11.1 Days of operation for ethanol mills in Mato Grosso

| Mill | Days of processing | | Days of maintenance |
	Sugarcane	Maize	
A	128	202	35
B	187	143	35
C	198	132	35
D	223	107	35
E	212	118	35
F	214	116	35
G	128	202	35
H	197	133	35
I	199	131	35
J	181	149	35
K	126	204	35
Average	181	149	35
St. Dev.	35	35	0
Max	223	204	35
Min	126	107	35

Source: Signorini et al. (2007).

Maize is currently not used as a feedstock in sugarcane facilities. Introducing maize into a sugarcane facility requires two new sets of assets. First would be the grain handling and hammer mill (preprocessing) equipment, which would add 10% to the capital cost (Singh, 2008). The second would be a doubling of the fermentation equipment capacity due to the longer steeping times of maize vs. sugarcane (Singh, 2008). This would add 16% to the capital cost (Nguyen and Prince, 1996). Maize fermentation requires different and more expensive fermentation bacteria. So we assume as a starting point that introducing maize would raise capital and operating costs by 25%.

11.4 Results

Ethanol production increases 36% in Mato Grosso by the introduction of maize during the idle season. This results in an increase of an extra 305 million liters of ethanol (Table 11.3). The average mill would increase ethanol production by 28 million liters. The leading plant could increase ethanol production by 89% and produce 47% of its ethanol from maize under a flex-mill configuration. The average plant in Mato Grosso only stores 16 million liters of ethanol or 21% of annual production, so additional storage or transport is necessary to manage the increased output. Though not addressed here, the assets underlying the shipment portion of the feedstock from the farm to the plant will see higher asset turns when they are able to be used for both sugarcane and maize transport.

Table 11.2 Flex-mill feedstock usage

Mill	Sugarcane utilization Yearly	Maize utilization	Sugarcane utilization Daily	Maize utilization
	(– mt– – – – – – – – – – – – – – – – – – – –)			
A	159,091	30,379	1,243	150
B	151,705	14,037	811	98
C	295,455	23,833	1,492	181
D	1,863,636	108,200	8,357	1,011
E	965,909	65,053	4,556	551
F	3,083,523	202,245	14,409	1,743
G	397,727	75,947	3,107	376
H	984,659	80,437	4,998	605
I	1,132,727	90,225	5,692	689
J	272,727	27,166	1,507	182
K	238,636	46,750	1,894	229
Average	867,800	69,479	4,370	529
St. Dev.	869,072	51,107	3,879	469
Max	3,083,523	202,245	14,409	1,743
Min	151,705	14,037	811	98
Total	9,545,795	764,272	48,067	5,816

Source: Sindalcool, Mato Grosso, and authors' calculations.

Table 11.3 Ethanol production from the flex-mill simulation model

Mill	Ethanol production per year			Ethanol from maize		Ethanol production per day		Storage	% of cane ethanol
	From cane	From maize	Total	%Increase	%from	From cane	From maize		
A	14,000,000	12,151,563	26,151,563	87%	46%	109,375	60,156	864,000	6%
B	13,350,000	5,614,853	18,964,853	42%	30%	71,390	39,265	NR	NR
C	26,000,000	9,533,333	35,533,333	37%	27%	131,313	72,222	10,338,000	40%
D	164,000,000	43,279,821	207,279,821	26%	21%	735,426	404,484	42,137,000	26%
E	85,000,000	26,021,226	111,021,226	31%	23%	400,943	220,519	7,913,000	9%
F	271,350,000	80,897,804	352,247,804	30%	23%	1,267,991	697,395	41,592,000	15%
G	35,000,000	30,378,906	65,378,906	87%	46%	273,438	150,391	596000	2%
H	86,650,000	32,174,860	118,824,860	37%	27%	439,848	241,916	17,389,000	20%
I	99,680,000	36,090,171	135,770,171	36%	27%	500,905	275,497	28,284,000	28%
J	24,000,000	10,866,298	34,866,298	45%	31%	132,597	72,928	1,050,000	4%
K	21,000,000	18,700,000	39,700,000	89%	47%	166,667	91,667	6,507,000	31%
Average	76,366,364	27,791,712	104,158,076	50%	27%	384,536	211,495	15,667,000	21%
St. Dev.	76,478,359	20,442,753	96,921,112	24%	21%	341,343	187,739	15,385,933	20%
Max	271,350,000	80,897,804	352,247,804	89%	23%	1,267,991	697,395	42,137,000	16%
Min	13,350,000	5,614,853	18,964,853	26%	30%	71,390	39,265	596,000	4%
Total	840,030,000	305,708,835	1,145,738,83	36%				156,670,000	

Source: Sindalcool, Mato Grosso, and authors' calculations.

Also, Mato Grosso already has an excess supply of ethanol, thus all additional production would be exported. This would require greater levels of transportation assets, or as has been discussed, might provide sufficient quantity to make a pipeline feasible. Also discussed are new uses for ethanol such as E-Diesel, where ethanol substitutes for 15% of the diesel fuel. Diesel fuel is currently imported into the State of Mato Grosso.

Without the addition of maize, the sugarcane plants will generate between 282 thousand and 5.7 million gigajoules per year on assets that range from $51 to $213 million of plant capital (Table 11.4). The asset turns vary significantly as ethanol yield per ton of sugarcane and days of operation per year vary significantly across the 11 plants. The smallest plant (A) produces only 5.7 gigajoules per $ of mill capital, while the most efficient user of capital, plant F, is about 5 times as efficient in its use of its assets (Fig. 11.4). It is important to remember that there is a deterministic element in our estimated relationship between plant size and asset turns. A regression parameter was used to estimate the capital costs associated with each size plant in Mato Grosso.

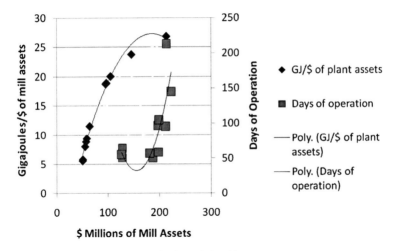

Fig. 11.4 Key sugarcane mill asset utilization relationships

Following Signorini (2007), we use a maize processing efficiency level of.55 (Table 11.5). That is, per ton of raw feedstock the plant will yield 55% of the ethanol output compared with sugarcane as a feedstock. The plants in Mato Grosso are able to raise the level of ethanol production from 11 to 80 million liters of ethanol per year or between 21 and 47% to their energy output.

Not surprisingly, maize-related capital costs benefit those plants that have the shortest operating seasons. Plants A, G, and K could augment their asset base up to 80% when introducing maize and still improve their overall asset use efficiency (Table 11.6). If capital costs were 40% or greater it generally would not pay for the plants to go "Flex." Increasing maize usage efficiency though would allow for greater investment in maize-related capital.

Table 11.4 Typical Mato Grosso sugarcane-processing mill energy asset turn

	A	B	C	D	E	F	G	H	I	J	K
Season ethanol production (mi L)	14	13.35	26	164	85	271.35	35	86.65	99.68	24	21
Daily ethanol production (L)	109,375	71,390	131,313	735,426	400,493	1,267,991	273,438	439,848	500,905	132,597	166,667
Energy production (GJ) per year	295,400	281,685	548,600	3,460,400	1,793,500	5,725,485	738,500	1,828,315	2,103,248	506,400	443,100
# of days operating	128	187	198	223	212	214	128	197	199	181	126
Days in maintenance	35	35	35	35	35	35	35	35	35	35	35
Sugarcane mill assets (mi USD)	50.9	50.5	58.5	145.5	95.7	213.2	64.1	96.7	104.9	57.2	55.3
Sugarcane energy	5.8	5.6	9.4	23.8	18.7	26.9	11.5	18.9	20	8.9	8.0

Table 11.5 Flex-mill ethanol production from maize

	A	B	C	D	E	F	G	H	I	J	K
Days possible to adopt corn	202	143	132	107	118	116	202	133	131	149	204
Corn process Relative efficiency	0.55	0.55	0.55	0.55	0.55	0.55	0.55	0.55	0.55	0.55	0.55
Potential corn ethanol production (L millions)	12.1	5.6	9.5	43.2	26	80.9	30.3	32.1	36	10.8	18.7
Potential energy production from corn (GJ)	256,398	118,473	201,153	913,204	549,049	1,706,944	640,995	678,890	761,503	229,279	394,570
Energy (GJ)/Year (Corn + Sugarcane)	551,798	400,158	749,753	4,373,604	2,342,548	7,432,429	1,379,495	2,507,205	2,864,751	735,679	837,670

Table 11.6 Varying maize expansion costs and the impact on asset turns*

%of original cost	A	B	C	D	E	F	G	H	I	J	K
120%	9.0	6.6	10.7	25.0	20.4	29.0	17.9	21.6	22.8	10.7	12.6
130%	8.3	6.1	9.9	23.1	18.8	26.8	16.5	19.9	21.0	9.9	11.7
140%	7.7	5.7	9.2	21.5	17.5	24.9	15.4	18.5	19.5	9.2	10.8
150%	7.2	5.3	8.6	20.0	16.3	23.2	14.3	17.3	18.2	8.6	10.1
160%	6.8	5.0	8.0	18.8	15.3	21.8	13.4	16.2	17.1	8.0	9.5
170%	6.4	4.7	7.5	17.7	14.4	20.5	12.7	15.3	16.1	7.6	8.9
180%	6.0	4.4	7.1	16.7	13.6	19.4	12.0	14.4	15.2	7.1	8.4
190%	5.7	4.2	6.8	15.8	12.9	18.3	11.3	13.6	14.4	6.8	8.0
200%	5.4	4.0	6.4	15.0	12.2	17.4	10.8	13.0	13.7	6.4	7.6

*Highlighted areas reflect capital cost levels that reduce asset use efficiency.

Table 11.7 Average asset turns for a flex-mill configuration in Mato Grosso

		Cane portion	Maize portion	Flex mill* (cane and maize)	Difference
Total ethanol produced	Liters	840,030,000	305,708,835	1,145,738,835	36%
Total energy produced	Gigajoules	17,724,633	6,450,456	24,175,089	36%
Total installed capital	$millions	1,113		1,336	20%
Asset turns		15.92		18.10	14%

*Assuming 20% additional capital cost for maize.

Table 11.8 Asset turn comparisons for selected energy production plants

Type of energy model	Capital cost		Asset turns		Capital turn rate
	per KW	Rank	MJ/$ K	Rank	(compared to photovoltaic)
Advanced gas/oil	594	2	53	1	8x
Advanced gas turbine	398	1	79	2	11x
Wind	1208	3	26	3	4x
Coal-fired plant	1290	4	24	4	3x
Hydro	1500	5	21	5	3x
Flex-mill ethanol (Mato Grosso)............................			19		
Geothermal	1880	7	17	6	2x
Sugarcane ethanol (Mato Grosso)..........			16		
Nuclear	2081	8	15	8	2x
Fuel cell	4520	9	7	9	
Photovoltaic	4751	10	7	9	

Source: Nuclear Energy Agency, 2005; and authors' calculations.

The total capital stock for ethanol production in Mato Grosso is US $1.113 billion (Table 11.7. Asset turns for Mato Grosso increase from 15.9 megajoules per dollar of capital in the sugarcane only configuration to 18.1 (+14%) when maize is introduced into the mill, assuming a 20% increase in capital to handle the maize. Capital use efficiency actually falls 9% below sugarcane only levels to 14.5 megajoules per asset dollar if 50% more capital is needed for maize processing.

Finally, in a separate paper we analyze the asset utilization for other energy models. The asset turn values of 16–19 megajoules per dollar of ethanol production capital in Mato Grosso is on the low end when compared to other bioenergy processing facilities (Table 11.8). The mills though are about twice as efficient when compared to photovoltaic cells.

Here, we present an analysis of just the manufacturing plant. LFBMs are much more complex systems. Feedstock, fuel manufacturing, and fuel utilization not only may be spatially dispersed but also may involve low energy density feedstocks and fuels resulting in poor energy asset turns. Other energy systems involve much higher spatial, volumetric, and/or gravimetric densities resulting in superior energy asset turns to the Mato Grosso flex-mill model. A full systems analysis from feedstock production through to utilization would provide a more accurate comparison of energy models.

References

Altman E (1968) "Financial Ratios, Discriminant Analysis and the Prediction of Corporate Bankruptcy," J Financ **23**, 589–609.

Arjona E, Bueno G, and Salazar L (2001) "An Activity Simulation Model for the Analysis of the Harvesting and Transportation Systems of a Sugarcane Plantation," Compu Electron Agr **32**, 247–264.

Carolan J, Joshi S, and Dale B (2007) "Technical and Financial Feasibility Analysis of Distributed Bioprocessing Using Regional Biomass Pre-Processing Centers," J Agric Food Industrial Org **5**, 10.

Coelli T, Grifell-Tatje E, and Perelman S. (2002) "Capacity utilisation and profitability: A decomposition of short-run profit efficiency," Int. J. Production Economics, **79**, 261–278.

Gallagher P, Brubaker H, and Shapouri H (2005) "Plant Size: Capital Cost Relationships in the Dry Mill Ethanol Industry," Biomass Bioenerg **28**, 565–571.

Gombola M and Ketz J (1983) "Financial Ratio Patterns in Retail and Manufacturing Organizations," Financ Manage **12**, 45–56.

Graf A and Koehler T (2000) "Oregon Cellulose-Ethanol Study an Evaluation of the Potential for Ethanol Production in Oregon Using Cellulose-Based Feedstocks," *Prepared by Bryan & Bryan Inc Colorado for submission to the Oregon Office of Energy. June.*

Hamelinck C and Faaij A (2006) "Production of Advanced Biofuels," Int Sugar J **108**, 168–175.

Hassuani SJ, Leal MRLV, de Carvalho Macedo I, Centro de Tecnologia Canavieira, and Programa das Nações Unidas para o Desenvolvimento (2005) Biomass power generation sugar cane bagasse and trash. CTC PNUD.

Kerstetter J, Lyons J, W.S. University, and C.E.E. Program (2001) *Logging and Agricultural Residue Supply Curves for the Pacific Northwest*, Washington State University, Cooperative Extension Energy Program.

Martines-Filho J, Burnquist H, and Vian C (2006) "Bioenergy and the Rise of Sugarcane-Based Ethanol in Brazil," Choices, **21**, 91–96.

McAloon A, Taylor F, Yee W, Ibsen K and Wooley R (2000) "Determining the Cost of Producing Ethanol from Corn Starch and Lignocellulosic Feedstocks," National Renewable Energy Laboratory, Colorado.

Nguyen M and Prince R (1996) "A Simple Rule for Bioenergy Conversion Plant Size Optimisation: Bioethanol from Sugar Cane and Sweet Sorghum," Biomass Bioenerg, **10**, 361–365.

Nissim D and Penman S (2001) "Ratio Analysis and Equity Valuation: From Research to Practice," Rev Acc Stud **6**, 109–154.

Overend R (1982) "The Average Haul Distance and Transportation Work Factors for Biomass Delivered to a Central Plant," Biomass, **2**, 75–79.

Searcy E, Flynn P, Ghafoori E and Kumar A (2007) "The Relative Cost of Biomass Energy Transport," Appl Biochem Biotech **137**, 639–652.

Singh, V. (2008) Professor, the Department of Agricultural and Biosystems Engineering, the University of Illinois. Personal Communication.

Signorini G, Goldsmith PD, Martines JG, Guimaraes CP and Rasmussen R (2007) "Flex-mill: Integrating corn and sugarcane produced in Mato Grosso," Working Paper. Department of Agricultural and Consumer Economics, The University of Illinois and the Departamento de Economia, Adminstração e Sociologia, the University of São Paulo.

Part IV
Environmental Effects of Biofuels and Biofuel Policies

Chapter 12
Could Bioenergy Be Used to Harvest the Greenhouse: An Economic Investigation of Bioenergy and Climate Change?

Bruce A. McCarl, Thein Maung, and Kenneth R. Szulczyk

12.1 Introduction

Bioenergy interest has been greatly stimulated by the fuel price rises in the late 2000s. Bioenergy is seen as a way to protect against the rising fossil fuel prices and the political insecurity of importing petroleum from the Middle East. Furthermore, growing evidence suggests that combustion of fossil fuels is precipitating climate change (Intergovernmental Panel on Climate Change 2007). Thus, at present three factors may influence the prospects for bioenergy: (1) increases in crude oil prices, (2) concerns for national energy security matters, and (3) concerns for climate change and global warming.

All three of these factors have an impact on the production and consumption of liquid biofuels such as ethanol and biodiesel. However, increasing petroleum prices do not matter as much for biopower – biomass used in electricity production. This is because electricity generation uses little petroleum and instead relies on coal and natural gas, which are abundant in the United States (Table 12.1). Also, because of the possibility of substitution among various fuel sources in electricity production, any increase in oil and gas prices will urge power producers to switch to other fuel sources, especially coal (Sweeney 1984). However, concern for climate change and the introduction of a cap and trade permit system for greenhouse gases (GHG) could stimulate interests in biopower.

EPA data show that electric power generation is the biggest source of US greenhouse gas emissions, followed by the transportation sector (Fig. 12.1). Burning coal produces more carbon dioxide (CO_2) than any other method of generating energy, with coal used to generate more than half of US electricity (Table 12.1). Agriculture may offer a way to reduce net GHG emissions, and thereby help mitigate the risks of climate change (McCarl and Schneider 2000). Agricultural products, crop residues,

B.A. McCarl (✉)
Department of Agricultural Economics, Texas A&M University, College Station, TX, USA
e-mail: mccarl@tamu.edu

M. Khanna et al. (eds.), *Handbook of Bioenergy Economics and Policy*,
Natural Resource Management and Policy 33, DOI 10.1007/978-1-4419-0369-3_12,
© Springer Science+Business Media, LLC 2010

Table 12.1 Percent of net electricity generation by different fuel sources, 1990 and 2005

Fuel type/year	1990 (%)	2005 (%)
Coal	52.65	50.04
Natural gas	12.31	18.67
Nuclear	19.05	19.39
Petroleum	4.18	3.03
Hydro	9.67	6.59
Biomass	1.51	1.54
Geothermal	0.51	0.38
Solar	0.01	0.01
Wind	0.09	0.36

Source: Energy Information Administration.

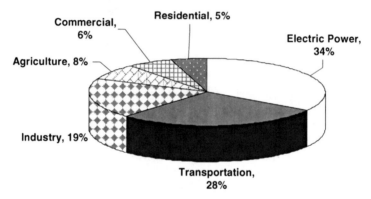

Fig. 12.1 US CO_2 emissions by sector in 2005
(Source: Inventory of U.S. Greenhouse Gas Emissions and Sinks (EPA))

and wastes may be used as substitutes for fossil fuel products to fuel electric power plants or as inputs into processes making liquid biofuels.

Because plant growth absorbs CO_2 while combustion releases it, using agricultural products to generate energy generally involves recycling of CO_2. This suggests that bioenergy producers or consumers would not need to buy GHG or carbon emission permits (assuming a GHG trading market exists) when generating biopower or consuming liquid biofuels. As long as bioenergy does not require acquisition of potentially costly emission permits, carbon permit prices could raise the market value of agricultural products. As a result, agricultural producers can gain income by supplying biofeedstocks, while energy producers can effectively reduce GHG emissions and carbon permit expenditures.

Before embracing bioenergy as a GHG mitigation mechanism, one must fully consider the GHGs emitted when growing, harvesting, and hauling feedstocks, and then converting them into bioenergy. In addition, one must also consider the market effects and possible offsetting effects of production induced elsewhere. Two issues arise from this: (1) What are the GHG offsets obtained when using a particular form of bioenergy and what does this imply for comparative economics of feedstocks? (2) When bioenergy production reduces traditional commodity production does the indirect market effect reduce net GHG effects?

In this chapter, we examined the technology and economics of bioenergy. First, an agricultural model, FASOMGHG, is introduced. The agricultural model allows the researchers to simulate the agricultural markets, because bioenergy competes for feedstocks with the agricultural industries. Second, an overview of bioenergy possibilities is examined and how the bioenergy industry relates to the agricultural markets. Finally, the agricultural model is used to predict the future production levels of bioenergy given various fossil fuel and GHG prices, and whether bioenergy could mitigate climate change.

12.2 Modeling Background

The Forest and Agricultural Sector Optimization Model-Greenhouse Gas version, herein referred to as FASOMGHG, is used to predict production levels over time for biofuels and biopower. FASOMGHG is a mathematical programming model that contains markets for bioenergy, biofeedstocks, and byproducts plus conventional agricultural and forestry commodities, accounting for market interactions, hauling and processing costs, and GHG emissions. Each activity may release or sequester GHG. The US agriculture is divided into 63 production regions. Each region has unique climate, soil fertility, and water resources. Producers can produce the primary crops and livestock shown in Table 12.2. FASOMGHG also includes constraints on land usage and production activities for each region. Producers take the primary crops and livestock and can produce the secondary products shown in Table 12.3. Processing activities are modeled in 11 aggregate market regions for the United States. Furthermore, FASOMGHG allows primary and secondary products to be imported or exported. Currently, the bioenergy commodities are not internationally traded (Lee 2002, Adams et al. 2005).

Table 12.2 Primary crops and livestock in FASOMGHG

Category	Activity
Primary crops	Barley, citrus, corn, cotton, hay, oats, potatoes, rice, silage, sorghum, soybeans, sugar beets, sugarcane, tomatoes, and wheat
Energy crops	Hydrid poplar, switchgrass, and willow
Livestock	Beef cattle, dairy cattle, hogs, horses and mules, poultry, and sheep
Miscellaneous	Eggs

Source: Adams et al. (2005).

Table 12.3 Major secondary products in FASOMGHG

Category	Activity
Animal products	Beef, chicken, edible tallow, nonedible tallow, pork, turkey, and wool
Bio-energy	Biodiesel, ethanol, and electricity
Corn wet mill	Corn oil, corn starch, corn syrup, dextrose, high fructose corn syrup, and gluten feed
Dairy products	American cheese, butter, cream, cottage cheese, ice cream, and milk
Potato products	Dried potatoes, frozen potatoes, and potato chips
Processed citrus products	Grapefruit and orange juice
Refined sugar items	Refined cane sugar and refined sugar
Soybeans	Soybean meal and soybean oil
Sweetened products	Baking, beverages, confection, and canning

Source: Adams et al. (2005).

12.2.1 Lifecycle Accounting

GHGs are emitted during the entire production lifecycle arising from fossil fuels and other inputs used to produce biofeedstocks and transform them into bioenergy. Generally, GHG emissions are generated when production inputs are manufactured, and when the bioenergy feedstock is grown, harvested, hauled, and processed into fuel or electricity. FASOMGHG contains net GHG emissions and sequestration for a large variety of agricultural activities.

The net GHG contribution of bioenergy production depends on the GHG emissions encountered during the biofeedstock to fuel to byproduct lifecycle. This contribution varies by feedstock, bioenergy product type, and region. Employing data from regions which have high potential for feedstock production, Table 12.4 shows estimates across a number of possibilities for use of crop or cellulosic ethanol in place of gasoline, biodiesel in place of diesel, and biopower produced with the biofeedstock used to both cofire 5% and fire 100%.

When corn-based ethanol is used, the percentage reduction in net GHG emissions is about 31% relative to the use of gasoline as shown in Table 12.4. The table also shows that the emission-offset rates are higher for electricity mainly because the feedstock requires less processing than coal, and thus little transformative energy once it is at the power generation site. In addition, cofiring power plants generally have a higher degree of emission-offset rates. This is because they have higher efficiency in terms of feedstock heat recovery and require only a small amount of feedstock.

Table 12.4 Percentage offset of net GHG emissions from the usage of a biofeedstock

Feedstock commodity being used	Form of bioenergy being produced				
	Liquid fuels			Electricity	
	Crop ethanol	Cellulosic ethanol	Biodiesel	Cofire at 5 %	Fire with 100% biomass
Corn	30.5				
Hard Red Winter Wheat	31.5				
Sorghum	38.5				
Softwood residue		80.0		99.2	97.4
Hardwood residue		79.9		99.0	96.5
Corn residue		75.1		93.7	88.1
Wheat residue		73.8		95.6	91.4
Cattle manure				99.6	96.5
Switch grass		68.6		94.3	89.5
Hybrid poplar		61.9		94.1	89.1
Willow		67.7		96.6	93.7
Soybean oil			70.9		
Sugarcane	64.8				
Corn oil			55.0		
Sugarcane bagasse		90.1		100.0	100.0
Lignin				91%	86%

Liquid fuels, as compared to electricity, have relatively lower offset rates. Grain-based ethanol has the lowest offset rate, while cellulosic ethanol and biodiesel from soybean oil have relatively higher offset values. The differential offset rates are due to the use of emission-intensive inputs at varying rates when producing feedstocks (for example, growing corn requires a large amount of fertilizers) and emission-intensive transformation processes in producing ethanol. Consequently, FASOMGHG also includes the standard byproducts and their markets; for example, DDGS is produced as a byproduct from the ethanol dry grind process and can raise the corn GHG offset from 17 to 31%. However, the DDGS is blended with cattle feeds and cattle produce methane gas from enteric fermentation. Thus, FASOMGHG is able to account for these complex interactions. If GHG prices were to rise in the future, then there would be a shift in production away from grain-based ethanol and toward cellulosic ethanol and electricity.

12.2.2 Leakage

Due to market forces and other factors, net GHG emission reductions within a region can be offset by increased emissions in other regions. For example, rising corn prices can help reduce net GHG emissions in the United States but stimulate increased emissions in other parts of the world due to expanded corn production. The net results are increased GHG emissions (Murray et al. 2004; Searchinger et al. 2008).

Many forms of this leakage phenomenon which are being discussed in many circles today include:

- Forests converted to crop land
- Reversion of Conservation Reserve Program (CRP) lands in the United States into crop land
- Expansions of crop acres in Brazil and Argentina at the expense of grasslands and rainforest

Leakage consideration suggests that bioenergy project GHG offsets need to be evaluated under broad national and international accounting schemes so that both the direct and indirect implications of project implementation are examined including offsite-induced leakage. McCarl (2006) shows that international leakage can easily offset approximately 50% of the domestic diverted production, when GHG offsets per acre are equal and an even higher share of the net GHG gains if acres with higher emissions are involved. Searchinger et al. (2008) also show that net GHG emissions would increase when acres are directly replaced by rainforest reductions. FASOMGHG includes three types of leakages; it allows changes in land use, land taken out of CRP and changes in foreign production although the latter is not associated with carbon prices.

12.3 Bioenergy Production Possibilities

Producing bioenergy involves different feedstocks, opportunity costs, byproducts, and GHG emissions. Consequently, an overview of the production possibilities for ethanol, biodiesel, and biopower is provided in this section and how they are represented in FASOMGHG.

12.3.1 Ethanol

Gasoline has two potential substitutes: butanol and ethanol. Both can be produced from sugar/starch crops, agricultural residues, and wood byproducts via fermentation. Butanol, in general, has better fuel properties than ethanol; however, butanol can dissolve in water and can be toxic from long-term exposure (Product Safety Assessment n-Butanol, 2006). Thus, this toxicity could prevent wide-scale adoption. Further, the EPA recently banned the use of MTBE as an oxygenate. The ban is fueling a strong demand for ethanol, because one gallon of ethanol offsets two gallons of MTBE (Reynolds 2000). Thus, this research focuses on ethanol as an additive to gasoline.

   Biorefineries have three available technologies to produce ethanol. The first is the dry grind, and producers convert the sugar and starch feedstocks to ethanol. The feedstock is ground, steeped in water, then yeast ferments the sugars into ethanol, and the ethanol is separated from the mixture. Starch crops have one additional

Table 12.5 Ethanol and DDGS yields from sugar and starch crops

Feedstock	Sugar content (%)	Starch content (%)	Ethanol chemical yield (gal/ton dry feed stock)	DDGS (lbs/ethanol gallon)
Barley	–	50–55	67.6–74.8	
Corn (dry grind)	–	72	96.1–96.7	5.9–6.4
Grain sorghum	–	67–73.8	82.7–106.1	7.9
Oats	–	64.0	86.5–87.0	9.9
Potato	–	15.0	20.3–20.4	6.7
Rice grain	–	74.5	100.7–101.2	5.3
Sugar beet	16–17.3	–	20.0–21.8	14.2
Sugarcane		–	15.9–16.7	14.9
Sweet sorghum	13.0	–	16.1–16.2	7.9
Sweet potato	–	26.7	36.1–36.3	6.7
Wheat	–	57.9	87.0–87.4	7.3–9.2

Source: Szulczyk et al. (2009).

processing stage called hydrolysis, where the starch is converted to sugar. The ethanol yields are shown in Table 12.5. Many of the sugar and starch crops are exported or used in human food and animal feeds. Thus, a large ethanol industry could increase food prices because of the residual demand for feedstocks. The dry grind produces distiller's dried grain with solubles (DDGS), which is blended with animal feeds. However, to transport DDGS, it has to be dried to less than 10% moisture content, increasing a biorefinery's energy costs.

FASOMGHG contains the sugar and starch fermentation shown in Table 12.5 along with DDGS. FASOMGHG does not contain wet distiller's grains (WDG) or modified distiller's grains (MDG). WGS is dried to a 65% moisture content and MDG is created by mixing WGS with grains until the mixture contains 50% moisture. Although, WGS and MDG require less energy to dry, WGS and MDG have a short shelf life, restricting their use to locations near the ethanol plant. Further, MDG and WGS would freeze in winter and spoil quickly in summer. Finally, the dry grind produces CO_2 as a byproduct. The food industry uses liquefied CO_2 to freeze, chill, and preserve food while the drink industry uses CO_2 to carbonate beverages. This CO_2 is included in the GHG emissions in FASOMGHG, because the CO_2 is eventually released into the atmosphere.

The second technology for ethanol is the corn wet mill and exclusively utilizes yellow dent corn as a feedstock. The wet mill separates corn kernels into a variety of products, making corn wet milling more capital intensive than the dry grind. However, it produces a variety of valuable products. The products are shown in Table 12.6. Two products, corn gluten meal and feed, are used in animal feeds. The other products arise from starch production and are the opportunity cost of ethanol production. A corn wet mill could sell starch directly to the markets or process the starch into corn syrup, dextrose, ethanol, or high fructose corn syrup. High fructose corn syrup is used as a sweetener by the beverage and

Table 12. 6 Corn wet mill possibilities

Input	Output
1 bushel corn	31.5 lbs of starch or 2.8 gallons of ethanol
	and 1.5 lbs of corn oil
	and 2.6 lbs of corn gluten meal
	and 13.5 lbs of corn gluten feed
1 pound of starch	1.3 lbs of corn syrup
	or 1.19 lbs of dextrose
	or 1.41–1.54 lbs high fructose corn syrup (HFCS)

Sources: National Corn Growers Association 2007; Rausch and Belyea (2006);
Light (2006).
Note: HFCS comes in two types. The sweeter HFCS is used in carbonated drinks
while the less sweet HFCS is used in everything else.

confection industries. FASOMGHG contains markets for all products from the corn
wet mill.

The third technology is lignocellulosic fermentation and is still in the experimental stage. Producers manufacture ethanol from crop and wood residues, and
energy crops, like hybrid poplar, switchgrass, and willow. Lignocellulosic fermentation breaks down the cellulose and hemicellulose from these feedstocks into five
types of sugars. Thus, it requires more processing and is the most expensive process,
but the feedstocks are the cheapest. The likely ethanol yields from lignocellulosic
feedstocks are shown in Table 12.7 along with their energy content. FASOMGHG

Table 12. 7 Lignocellulosic ethanol yields and energy content

Feedstock	Ethanol yield (gal/ton of feedstock)	Energy content (BTU/ton of feedstock)
Crop residues		
Bagasse	49.8–90.7	16,376,000
Barley straw	60.00–74.4	14,894,000
Corn stover	51.25–92.78	15,186,000
Oat straw	60.00–62.4	NA
Rice straw	45.67–83.81	14,008,000
Sorghum straw	39.70–71.79	NA
Wheat straw	49.47–89.54	15,066,000
Lignin	–	18,222,000
Wood residues		
Softwoods	55.96–87.33	NA
Hardwoods	48.03–84.71	16,092,000
Energy crops		
Hybrid popular	46.90–82.39	15,444,000
Switchgrass	43.67–78.78	NA
Willow	NA	14,336,000
Miscellaneous		
Manure	–	15,408,000

Sources: Domalski et al. (1986); Szulczyk et al. (2009).

contains production possibilities for crop residues, energy crops like hybrid poplar, switchgrass, and willow, and wood residues.

FASOMGHG accounts for the CO_2 gas created from the fermentation and the byproduct of lignin. Lignin is a fiber that is extracted from the mixture before fermentation; it could be cofired with coal to produce electricity. However, FASOMGHG does not include the byproducts of furfural and methane gas. Furfural could be used to make carpet fibers and methane gas could be collected and burned for heat and energy from the anaerobic fermentation of waste water.

FASOMGHG limits the amount of crop residues that can be harvested. Farmers leave some crop residues on the fields, because they reduce soil erosion and increase organic matter in the soil. Further, FASOMGHG allows ethanol to be produced from hard and soft wood residues. Thus, the ethanol industry competes with the lumber industry, because these wood residues could be processed into paper, particleboard, and mulches. Finally, FASOMGHG allows producers to switch land use. Producers could grow perennial energy crops like hybrid poplar, willow, and switchgrass, but they switch land use away from crops, pastures, or forests.

12.3.2 Biodiesel

Biodiesel is produced from vegetable oils and tallow. The main sources for the United States are soybean oil, corn oil, tallow, and yellow grease. All these oils can be blended with animal feeds and sold to cattle, poultry, and swine producers. Further, soybean oil and corn oil could be exported or used as human foods. A large biodiesel industry would thus cause higher food prices because of the demand for biodiesel feedstocks. Biodiesel production is quite efficient and is approximately a one-to-one gallon conversion of oil into biodiesel (Szulczyk and McCarl 2009). Additionally, biodiesel refineries could be small and may not require large amounts of capital. Thus, the vegetable oil and animal rendering plants can easily append a biodiesel production line.

FASOMGHG contains biodiesel production from four industries. First, soybean oil is produced from soybean crushing facilities. One pound of soybeans yields on average 0.19 pounds of oil and 0.41 pounds of soybean meal. Soybean meal is high in protein and is blended with animal feeds. Second, the corn wet mill supplies corn oil. Corn is the only feedstock in the United States that can be used to produce both ethanol and biodiesel. Third, the beef cattle industry supplies both edible and nonedible tallow. From 100 lbs of meat one can obtain on average 5.4 pounds of edible tallow and about 11.0 pounds of nonedible tallow (Swisher 2004). Finally, biodiesel could be manufactured from waste cooking oil which comes in two types: yellow grease and brown grease. Yellow grease comprises less than 15% of free fatty acids, while brown grease exceeds this. The biodiesel industry would more than likely use yellow grease, because it involves less processing and cleaning. Yellow grease is one of the cheapest feedstocks for biodiesel but requires higher processing and conversion costs. Approximately, one pound of oil produced from a corn wet mill or soybean crushing facility yields 0.13 pounds of yellow grease (Canakci 2007).

This ratio is expected to increase over time as the biodiesel industry expands and the infrastructure improves for collecting, storing, and transporting yellow grease.

FASOMGHG does not contain glycerol production or a glycerol offset. Glycerol is a byproduct of the biodiesel industry and is used in pharmaceutical, cosmetic, and chemical industries. However, glycerol is a relatively small market and a large biodiesel industry could saturate the market, causing price to drop (Bender 1999, Ortiz-Canavate 1994). Therefore, the glycerol price may not be high enough to cover glycerol purification and higher capital costs.

12.3.3 Biopower

FASOMGHG allows both the lignocellulosic ethanol and bioenergy producers to compete for feedstocks from crop residues, wood residues, or energy crops. Electric power plants can cofire biomass with coal up to 100% biomass. However, a power plant has to invest in more capital to handle higher cofire rates. The energy contents of different feedstocks are shown in Table 12.7 and 32% of the heat energy can be converted to electricity (Spath et al. 1999).

12.4 Economics of Biofeedstock

FASOMGHG contains two types of costs: endogenous and exogenous. The capital, production, and storage costs are exogenous, because bioenergy producers are assumed to be small relative to the market. Moreover, the feedstock and hauling costs are assumed to be endogenous and determined within FASOMGHG. The bioenergy producers compete with other industries for feedstocks. For example, higher feedstock demands lead to higher feedstock prices; thus producers switch to other feedstocks to reduce their costs. Further, transporting and hauling costs could comprise a significant portion of the costs because of the low bulk density of biomass. As the distance traveled increases between farmers and bioenergy producers, the hauling cost increases exponentially (see French 1960). FASOMGHG uses French's approximation for hauling costs. Hauling costs depend on the producers' production capacity, crop yields, and cost of hauling and harvesting the feedstocks, and technology. FASOMGHG incorporates the following technology parameters:

- Crop yields increase over time at rates forecasted by USDA (Interagency Agricultural Projections Committee 2008).
- Ethanol yields increasing over time where ethanol producers are assumed to attain 90% of theoretical chemical yield in 30 years, when total efficiency attains 90% of theoretical (Szulczyk et al. 2009).
- The efficiency for lignin-electricity generation increases to 42% (Spath et al. 1999), increasing 1.1% annually. The energy efficiency occurs as producers upgrade or build new electric generation facilities.

The biodiesel industry does not have any technological improvement, because biodiesel production is already quite efficient at 97% of theoretical (Szulczyk and McCarl 2009).

12.5 Predicted Bioenergy Production

For the analysis in this chapter, FASOMGHG is used to solve several scenarios such as varying fossil fuel and carbon prices and then predict future biofuel and biopower production levels. Thus, FASOMGHG allows researchers to predict the impact on the US agricultural markets as if the United States incorporated a cap and trade system for greenhouse gases. The energy prices are exogenous in FASOMGHG and are defined as

- The gasoline and diesel fuel prices are wholesale prices expressed in dollars per gallon. The prices range from $1 to $4 per gallon, agreeing with the 25-year energy price forecasts from Office of Integrated Analysis and Forecasting (2006). Further, ethanol and biodiesel prices are adjusted for the lower energy content using the lower heating value.
- The coal price is expressed in dollars per ton and its base price is $24 per ton.
- The carbon equivalent price is expressed in dollars per equivalent metric ton. The agricultural industries emit or sequester carbon dioxide, methane, and nitrous oxide. These greenhouse gases are put in equivalent terms by using the 100-year global warming potential; carbon dioxide is defined as 1, methane as 21, and nitrous oxide as 310 (Adams et al. 2005, Cole et al. 1996).

The results are reported by bioenergy type.

12.5.1 The Case of Ethanol

Ethanol is used currently as a fuel additive and as a substitute for gasoline. Current gasoline engines with no engine modifications can operate up to 15% ethanol by volume for gasoline–ethanol blends while flexible fuel vehicles can use up to 85% ethanol by volume. Furthermore, 1.6 gallons of ethanol displaces one gallon of gasoline, because ethanol's lifecycle GHG emissions are adjusted to reflect ethanol's lower energy content. The predicted ethanol production level in millions of gallons of ethanol is shown in Fig. 12.2 for various wholesale gasoline prices and Fig. 12.3 for various carbon equivalent prices. Data are also shown in Table 12.8 and the gasoline price is fixed at $2 per gallon for the carbon equivalent prices. All time paths are identical for ethanol. Ethanol is restricted to its known production levels for 2000 and 2005, which are 1.7 and 6.0 billion gallons of ethanol. Then new production capacity is constrained to grow at a maximum of 1.2 billion gallons per year, because only a handful of companies like Fagen International LCC build ethanol facilities.

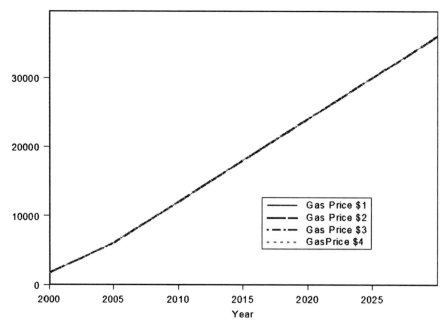

Fig. 12.2 Predicted aggregate US ethanol production (million gallons)

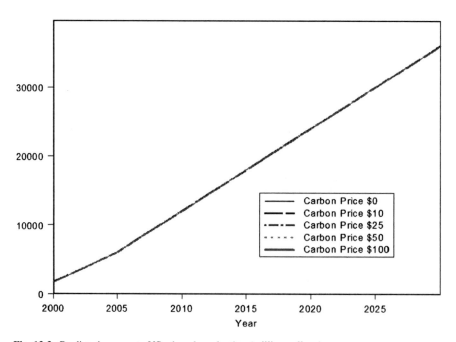

Fig. 12.3 Predicted aggregate US ethanol production (million gallons)

Table 12. 8 Results from FASOMGHG

	2000	2005	2010	2015	2020	2025	2030
Ethanol (millions of gallons)							
Gas price $1, CO_2 price $0	1,701.7	6,006.0	12,012.0	18,018.0	24,024.0	30,030.0	36,036.0
Gas price $2, CO_2 price $0	1,701.7	6,006.0	12,012.0	18,018.0	24,024.0	30,030.0	36,036.0
Gas price $3, CO_2 price $0	1,701.7	6,006.0	12,012.0	18,018.0	24,024.0	30,030.0	36,036.0
Gas price $4, CO_2 price $0	1,701.7	6,006.0	12,012.0	18,018.0	24,024.0	30,030.0	36,036.0
Ethanol (millions of gallons)							
Gas price $2, CO_2 price $0	1,701.7	6,006.0	12,012.0	18,018.0	24,024.0	30,030.0	36,036.0
Gas price $2, CO_2 price $10	1,701.7	6,006.0	12,012.0	18,018.0	24,024.0	30,030.0	36,036.0
Gas price $2, CO_2 price $25	1,701.7	6,006.0	12,012.0	18,018.0	24,024.0	30,030.0	36,036.0
Gas price $2, CO_2 price $50	1,701.7	6,006.0	12,012.0	18,018.0	24,024.0	30,030.0	36,036.0
Gas price $2, CO_2 price $100	1,701.7	6,006.0	12,012.0	18,018.0	24,024.0	30,030.0	36,036.0
Ethanol (millions of gallons)							
Corn wet mill, CO_2 price $0	1,531.5	4,359.9	9,825.0	12,101.9	12,101.9	12,101.9	12,101.9
Lignocellulosic, CO_2 price $0	0.0	0.0	0.0	2,802.8	10,095.7	16,147.0	22,325.6
Corn wet mill, CO_2 price $100	1,531.5	4,875.2	9,788.6	9,788.6	9,788.6	9,788.6	9,788.6
Lignocellulosic, CO_2 price $100	0.0	0.0	174.5	2,977.3	13,626.2	19,329.8	24,794.7
Biodiesel (millions of gallons)							
Gas price $1, CO_2 price $0	5.3	250.1	2,739.9	2,126.5	2,550.1	2,592.7	2,693.4
Gas price $2, CO_2 price $0	5.3	250.1	3,102.7	3,100.6	3,222.2	3,265.6	3,312.8
Gas price $3, CO_2 Price $0	5.3	250.1	3,610.7	3,716.1	3,816.4	3,863.1	3,912.7
Gas price $4, CO_2 Price $0	5.3	250.1	3,765.6	3,985.3	4,019.6	4,126.9	4,429.5
Biodiesel (millions of gallons)							
Gas price $2, CO_2 price $0	5.3	250.1	3,102.7	3,100.6	3,222.2	3,265.6	3,312.8
Gas price $2, CO_2 price $10	5.3	250.1	3,146.6	3,135.4	3,245.8	3,291.8	3,573.9
Gas price $2, CO_2 price $25	5.3	250.1	3,183.3	3,158.7	3,232.8	3,367.8	3,549.4
Gas price $2, CO_2 price $50	5.3	250.1	3,225.5	3,048.1	3,207.7	3,406.2	3,394.7
Gas price $2, CO_2 price $100	5.3	250.1	3,254.9	3,153.3	3,398.3	3,470.0	3,083.6

Table 12.8 (continued)

	2000	2005	2010	2015	2020	2025	2030
Biopower (100 MW)							
Gas price $2, CO_2 price $0	11.9	12.2	15.5	0.0	0.0	0.0	0.0
Gas price $2, CO_2 price $10	14.9	16.3	18.6	5.8	21.1	28.9	38.8
Gas price $2, CO_2 price $25	30.1	28.7	40.2	36.8	54.9	83.4	89.0
Gas price $2, CO_2 price $50	36.0	38.0	52.7	49.8	69.4	87.5	107.3
Gas price $2, CO_2 price $100	113.7	119.8	171.9	221.2	259.8	312.4	334.3
Corn price ($ per bushel)							
Gas price $2, CO_2 price $0	2.55	3.27	3.44	3.42	3.22	3.05	2.93
Gas price $2, CO_2 price $10	2.59	3.30	3.42	3.45	3.21	3.04	2.98
Gas price $2, CO_2 price $25	2.67	3.39	3.42	3.52	3.26	3.05	3.08
Gas price $2, CO_2 price $50	2.78	3.47	3.51	3.55	3.34	3.11	3.19
Gas price $2, CO_2 price $100	3.07	3.80	3.89	3.87	3.60	3.37	3.43
Soybean Price ($ per bushel)							
Gas price $2, CO_2 price $0	6.33	11.41	11.69	11.31	11.65	11.70	11.90
Gas price $2, CO_2 price $10	6.43	11.50	11.79	11.40	11.75	11.79	11.98
Gas price $2, CO_2 price $25	6.57	11.65	11.93	11.55	11.66	11.92	11.97
Gas price $2, CO_2 price $50	6.89	11.89	12.02	11.79	11.91	11.96	12.16
Gas price $2, CO_2 price $100	7.73	12.35	12.28	12.28	12.39	12.44	12.87
Beef slaughtered price ($ per CWT)							
Gas price $2, CO_2 price $0	81.71	78.54	80.41	80.87	85.09	87.57	92.14
Gas price $2, CO_2 price $10	82.89	79.82	81.58	81.53	86.34	88.93	93.07
Gas price $2, CO_2 price $25	85.29	81.71	83.75	81.60	88.25	90.34	96.08
Gas price $2, CO_2 price $50	89.13	86.46	86.55	84.05	91.82	95.17	103.01
Gas price $2, CO_2 price $100	100.44	94.95	94.94	95.17	99.91	103.21	107.64

Alterations in the carbon equivalent price have a minor impact on the total size of the ethanol industry, as shown in Fig. 12.3. However, under a carbon equivalent price there is a composition shift within the technologies used for ethanol production due to differences in their lifecycle emissions. A carbon equivalent price slightly contracts the ethanol production from the corn wet mills, but boosts production from lignocellulosic feedstocks with corn stover as the dominant feedstock. The results are shown in Fig. 12.4 and data are in Table 12.8. This is because lignocellulosic fermentation is more GHG efficient than the other technologies. Crop residues also do not stimulate a great deal of leakage, because they provide a joint product.

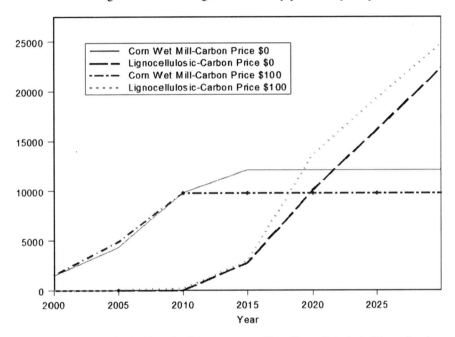

Fig. 12.4 Predicted US ethanol production – corn wet mill vs. lignocellulosic (million gallons)

12.5.2 The Case of Biodiesel

Biodiesel substitutes for #2 diesel fuel, and diesel engines with no modifications could use up to 100% of biodiesel by volume. Furthermore, we assume 1.05 gallons of biodiesel displaces one gallon of diesel fuel, adjusting the lifecycle GHG emissions for the different energy content between the two fuels. As shown in Fig. 12.5 and in Table 12.8, the aggregate US biodiesel production is in millions of gallons and a higher wholesale diesel fuel price boosts biodiesel production. For years 2000 and 2005, US biodiesel production is constrained at its known values, which are 5 and 250 million gallons, respectively. The predicted biodiesel production is shown in Fig. 12.6 for carbon equivalent prices; the diesel fuel price is fixed

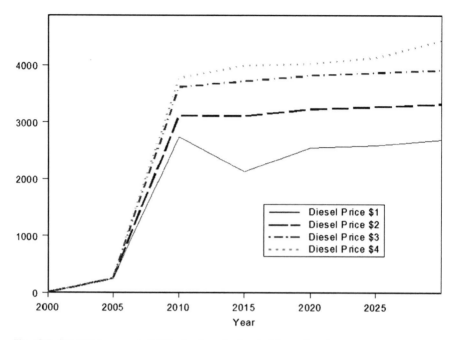

Fig. 12.5 Predicted aggregate US biodiesel production (million gallons)

at $2 per gallon. Even though the GHG efficiency for biodiesel is extremely high, carbon equivalent prices have a small impact on biodiesel production. The reason is carbon equivalent prices boost the biopower. Moreover, biodiesel industry mainly relies on soybeans and corn as feedstocks, but also uses some tallow and yellow grease.

12.5.3 The Case of Biopower

Gasoline and diesel fuel prices have a minimal impact on the US biopower production. As shown in Fig. 12.7 and in Table 12.8, the aggregate US biopower production is in 100 megawatts of electricity and the predicted biopower production greatly expands from higher carbon equivalent prices. However, when the carbon equivalent price is $0, biopower production drops to zero for all gasoline prices. Thus, the future market production of biofeedstocks for power generation may depend on the following factors: (1) the price of coal, (2) the price of GHG emissions, (3) the heat content of biofeedstocks, and (4) the costs of biofeedstock production. Maung and McCarl (2008) show that for crop residues to have economic potential in electricity generation, mostly in the form of cofiring, either the current price of coal or the price of GHG has to increase significantly.

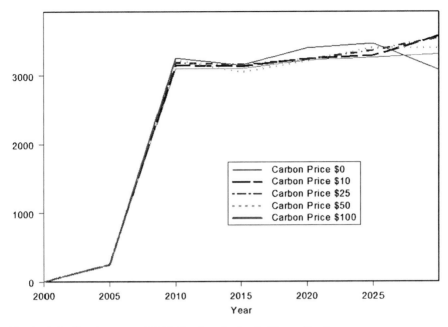

Fig. 12.6 Predicted aggregate US biodiesel production (million gallons)

This result not only applies to crop residues but also to switchgrass, willow, poplar, and other feedstocks. The finding also indicates that generally feedstocks with higher heat content will have better potential in generating electricity than those with lower heat content. This implies that if farmers decide to invest in feedstocks, they should invest in feedstocks with relatively higher heat content such as bagasse. Moreover, results from our study suggest that if we are to induce electric power producers to consume feedstocks such as crop residues without any reliance on coal or GHG price increase, the production costs must be reduced by about 50%. But given the current market conditions of feedstocks, cost reductions of 50% will not be easy to achieve without drastic improvements in production technologies and integration of biofeedstock markets at the farm and industry levels.

Because coal is abundant in the United States, its current and historical price have been relatively low and stable compared with the prices of natural gas and petroleum (see Fig. 12.8). Without GHG price increases, electricity producers may be encouraged to choose coal for power generation as oil and gas prices increase.[1] The future competitiveness of biofeedstock and bio-based electricity would likely depend on how pricing GHG evolves over time and on the advancement in feedstock production and biopower generation technologies.

---

[1] This had happened in the past. During the energy crisis in the 70s, as oil and gas prices increased, electricity producers switched to coal. As a result, the demand for coal increased along with its price (Fig. 12.4).

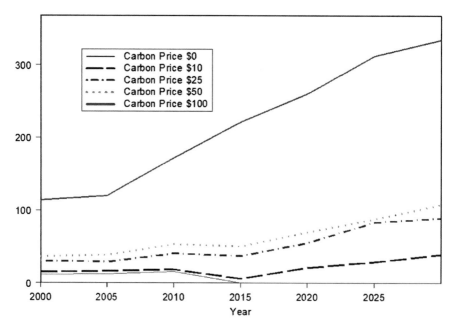

Fig. 12.7 Predicted aggregate US biopower (100 MW) (Source: Energy Information Administration)

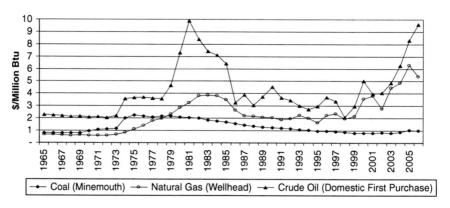

Fig. 12.8 Average annual real fossil fuel prices, 1965–2006

12.5.4 GHG Mitigation Strategy

The GHG mitigation role of agriculture changes as the prices of CO_2 and gasoline change. The national GHG summary (see Fig. 12.9) as a function of the CO_2 and gasoline prices shows that under the scenario of low gasoline and low CO_2 prices, the predominant strategy involves agricultural soil sequestration. With low

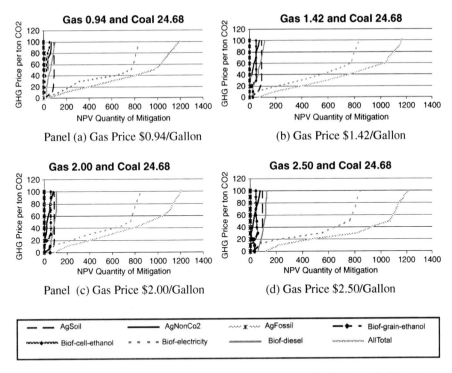

Fig. 12.9   GHG mitigation strategy use for alternative gasoline and GHG/carbon dioxide prices

gasoline prices but higher CO_2 prices, the strategy is dominated by biofeedstock-fired electricity because of its higher GHG offset rate. Higher gasoline prices may induce more liquid bioenergy production. However, GHG contributions of ethanol and biodiesel are limited because of their lower offset rates.

Figure 12.9 also shows that even at zero CO_2 price, a reduction in CO_2 emissions can take place only as a consequence of increased gasoline prices, which is a complementary policy. The figure suggests that if one were really after GHG mitigation, one would depend mainly on bio-based electricity. An important finding across all these scenarios involves the portfolio composition between agricultural soil sequestration and bioenergy. Specifically, when the prices of fossil fuel and CO_2 are low, agricultural soil sequestration is the predominant strategy as sequestration can be achieved more economically by changes in tillage practices that are largely complementary with existing production. But, if the CO_2 price gets higher, then a land-use shift occurs from traditional production into bioenergy strategies. Consequently, the gains in carbon sequestration effectively discontinue, topping out the potential for agricultural soil carbon sequestration (McCarl and Schneider 2001). This shift occurs because of higher fossil fuel or CO_2 prices, any of which induces a shift of land to biofeedstocks.

Another important finding involves the relative shares of cellulosic and grain-based ethanol. When the gasoline price is high and CO_2 prices are low, grain-based ethanol dominates the production, but as the CO_2 price gets higher, the production shifts to cellulosic ethanol. This is largely due to the efficiency of GHG, which induces the shift from grain-based ethanol to cellulosic ethanol as the CO_2 price increases.

12.5.5 Food Prices

Biofuels have several criticisms. First, not only could a large biofuel industry divert feedstocks away from human food and animal feeds, but it also could increase prices on agricultural products from the stronger demand for feedstocks. Second, a carbon equivalent price can penalize cattle producers, because cattle emit methane gas from enteric fermentation.

FASOMGHG predicts corn prices in real terms in Fig. 12.10 and Table 12.8. The corn prices are in dollars per bushel. Higher carbon equivalent prices cause corn prices to be higher. However, corn prices peak in 2015 and begin to taper off. As already mentioned, ethanol producers begin to boost their production using agricultural residues and energy crops as feedstocks, lowering their demand for corn. Similarly, soybean prices are shown in Fig. 12.11 and the price is in dollars per bushel. However, the rapid expansion of the biodiesel industry causes a large

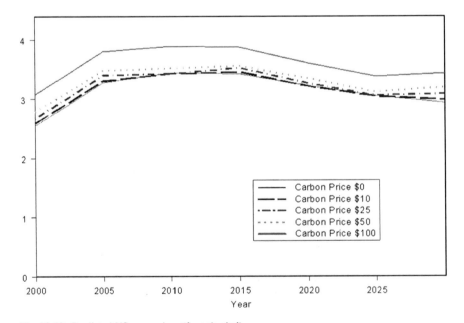

Fig. 12.10 Predicted US corn prices ($ per bushel)

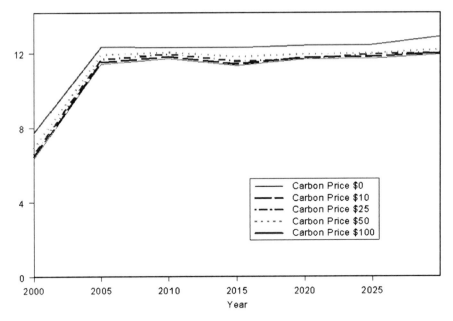

Fig. 12.11 Predicted US soybean prices ($ per bushel)

demand for soybeans, causing their price to increase over time. Further, higher carbon equivalent prices cause soybean prices to be higher. Finally, Fig. 12.12 shows prices for slaughtered cattle in dollars per hundred pounds of meat. As expected, higher carbon equivalent prices cause higher meat prices, even though the biodiesel industry creates more soybean meal. Consequently, the higher agricultural prices from a GHG cap and trade system can put US agricultural exports at a disadvantage.

12.6 Concluding Remarks

In this chapter, we address several major issues related to bioenergy and GHG offsets. Generally, GHG offset effects are different among different forms of bioenergy. Grain-based ethanol has the lowest offset rates, while cellulosic ethanol and biodiesel have relatively higher offset rates, followed by biopower which has the highest offset values. In the commodity markets, leakage created by induced replacement production overseas may offset the gains in domestic GHG emission reduction (Murray et al. 2004).

Economically, GHG prices could play an important role in inducing the production and consumption of bioenergy. As GHG prices increase, the production would likely shift from grain-based ethanol toward cellulosic ethanol, biodiesel, and bioelectricity. Other factors which can influence the market penetration of bioenergy

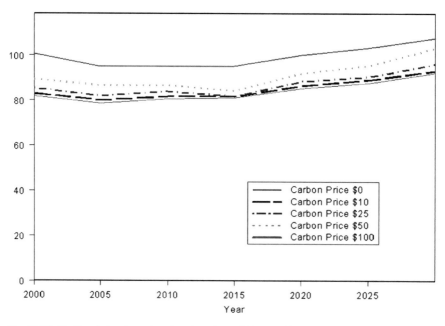

Fig. 12.12 Predicted slaughtered cattle prices ($ per 100 pounds)

include prices of fossil fuels, heat content of biofuels, costs of production especially transport/hauling cost, and improvements in production technologies.

Bioenergy may help the United States reduce petroleum imports, thus improving energy security. However, bioenergy feedstocks produced at large scale may have substantial opportunity costs, because they replace traditional food/feed crops or compete with them for limited agricultural lands. The feedstocks for biodiesel and corn-based ethanol compete with the cattle feed and human food supplies, thus potentially increasing food prices. However, lignocellulosic feedstocks are less competitive relying in part on crop residues plus on energy crops that have higher ethanol yields per acre than corn. Additionally, the impact on the US trade deficit would be ambiguous, because a large ethanol and biodiesel industry could reduce the demand for petroleum imports, but at the same time the United States would be exporting less agricultural products either because of their reduced supply or their increased use by the ethanol industry.

Bioenergy and GHGs are complexly intertwined. In terms of policy implications, the arguments in this chapter suggest that current promotion of commodities like corn ethanol may not in fact be contributing much to GHG reductions particularly after considering leakage. Thus, policies on GHG emission reduction need to be carefully designed to avoid leakage issues. Results in this chapter also suggest that the severity of leakage can be lessened with reliance on residue and waste products, with emphasis on cellulosic ethanol and bio-based electricity.

References

Adams DM, Alig RD, McCarl BA, et al. (February 2005). "FASOMGHG Conceptual Structure, and Specification: Documentation" Unpublished, Texas A&M University. Available at http://agecon2.tamu.edu/people/faculty/mccarl-bruce/papers/1212FASOMGHG_doc.pdf (access date: 6/31/08).

Bender M (1999) "Economic Feasibility Review for Community-Scale Farmer Cooperatives for Biodiesel." Bioresource Technol 70:81–87.

Canakci M (2007) "The Potential of Restaurant Waste Lipids as Biodiesel Feedstocks." Bioresource Technol 98:183–190.

Cole CV, Cerri C, Minami K, et al. (1996) "Agricultural Options for the Mitigation of Greenhouse Gas Emissions." In Climate Change 1995: Impacts, Adaptation, and Mitigation of Climate Change: Scientific-Technical Analysis. Cambridge, England: Cambridge University Press.

Domalski ES, Jobe Jr TL, and Milne TA (1986) Thermodynamic Data for Biomass Conversion and Waste Incineration. Golden, CO: Solar Energy Research Institute. Available at http://www.nrel.gov/docs/legosti/old/2839.pdf (access date: 8/5/07).

French BC (1960) "Some Considerations in Estimating Assembly Cost Functions for Agricultural Processing Operations." J Farm Econ 42:767–778.

Interagency Agricultural Projections Committee. (February 2008). USDA Agricultural Projections to 2017. Washington, DC: U.S. Department of Agriculture, Report OCE-2008-1. Available at www.usda.gov/oce/commodity/archive_projections/USDAAgriculturalProjections2017.pdf (access date: 12/12/08).

Intergovernmental Panel on Climate Change (IPCC) (2007) Climate Change 2007 – The Physical Science Basis: Contribution of Working Group I to the Fourth Assessment Report of the IPCC. New York, NY: Cambridge University Press.

Lee HC (December 2002). An Economic Investigation of the Dynamic Role for Greenhouse Gas Emission Mitigation by the U.S. Agricultural and Forest Sectors. PhD Dissertation, Texas A&M University.

Light RH Received email on (July 1, 2006) from Light@admworld.com.

Maung TA, McCarl BA (2008) "Economics of Biomass fuels for Electricity Production: A Case Study with Crop Residues." Available at http://ageconsearch.umn.edu/bitstream/6417/2/467464.pdf.

McCarl BA, Schneider UA (2000) "U.S. Agricultural's Role in a Greenhouse Gas Mitigation World: An Economic Perspective." Rev Agric Econ 22:134–159.

McCarl BA, Schneider UA (2001), "The Cost of Greenhouse Gas Mitigation in US Agriculture and Forestry", Science, Volume 294 (21 Dec), 2481–2482.

McCarl BA (2006) "Permanence, Leakage, Uncertainty and Additionality in GHG Projects." In Quantifying Greenhouse Gas Emission Offsets Generated by Changing Land Management, Editor G.A. Smith, A book developed by Environmental Defense.

Murray BC, McCarl BA, and Lee H-C (2004) "Estimating Leakage From Forest Carbon Sequestration Programs", Land Econ 80(1), 109–124.

National Corn Growers Association (2007) "Energized 2007 World of Corn." Washington, DC: National Corn Growers Association. Available at http://www.ncga.com/WorldOfCorn/main/production1.asp (access date: 8/5/07).

Office of Integrated Analysis and Forecasting. (February 2006). Annual Energy Outlook 2006 with Projections to 2030. Washington, DC: U.S. Department of Energy, Energy Information Administration. Available at www.eia.doe.gov/oiaf/aeo/(access date: 7/19/06).

Ortiz-Canavate J (1994) "Characteristics of Different Types of Gaseous and Liquid Biofuels and Their Energy Balance." J Agr Eng Res 59:231–238.

Product Safety Assessment n-Butanol (2006) Dow Chemical Company. Available at http://www.dow.com/productsafety/finder/nbut.htm (access date: 01/25/09).

Rausch KD, Belyea RL (2006) "The Future of Coproducts from Corn Processing." Appl Biochem Biotech 128:47–86.

Reynolds RE (May 15, 2000). *The Current Fuel Ethanol Industry Transportation, Marketing, Distribution, andbTechnical Considerations.* Bremen, IN: Downstream Alternatives Inc. Available at http://www.ethanolrfa.org/objects/documents/111/4788.pdf (access date: 4/17/06).

Searchinger T, Heimlich R, Houghton RA, et al. (2008) "Use of U.S. Croplands for Biofuels Increases Greenhouse Gases through Emissions from Land-Use Change." Science 319:1238–1240

Spath PL, Mann MK, Kerr DR (1999) *Life Cycle Assessment of Coal-fired Power Production.* Golden, CO: National Renewable Energy Laboratory, Report NREL/TP-570-25119. Available at http://www.nrel.gov/docs/fy99osti/25119.pdf (access date: 8/6/07).

Sweeney JL (1984) "The Response of Energy Demand to Higher Prices: What Have We Learned?" Am Econ Rev 74:31–37.

Swisher K (2004) "Market Report 2003: One of the Best Years – Then Came December 23rd." *Render* Available at http://rendermagazine.com/April2004/MarketReport2003.pdf (access date: 11/1/06).

Szulczyk KR, McCarl BA (2009) "Market Penetration of Biodiesel." Working Paper, Department of Agricultural Economics, Texas A&M University.

Szulczyk KR, McCarl BA, Cornforth GC (2009) "Market Penetration of Ethanol." Working Paper, Department of Agricultural Economics, Texas A&M University.

Chapter 13
A Simple Framework for Regulation of Biofuels

Deepak Rajagopal, Gal Hochman, and David Zilberman

Abstract In this chapter, we develop a framework to regulate producers of biofuel-blended fuels using GHG emissions standards, while accounting for heterogeneity and uncertainty. To this end, we categorize the net GHG emissions caused by a regulated site into two parts: direct and indirect emissions. Direct emissions arise both at and away from the final regulated site but are directly attributable to the final output. Indirect emissions, on the other hand, are emissions not attributable to any single regulated entity, but are caused by the interaction of aggregate supply and demand. An example is emissions due to indirect land use change (ILUC). The two parts, i.e., direct and indirect emissions, are computed using the best available data. The sum of the site-specific direct emissions per unit of output and average indirect emissions per unit of output is, then, compared to the emission threshold or standard.

13.1 Introduction

Economic forces, as well as demand for energy security, no doubt are providing incentives for producing and blending biofuels as substitute fuel. At the same time, in order to tackle global warming, governments are beginning to regulate GHG emissions from electricity and transportation sectors (LCFS[1], RGGI[2], ETS[3]).

D. Rajagopal (✉)
Energy and Resources Group, University of California, Berkeley, CA, USA
e-mail: deepak@berkeley.edu

[1] Low Carbon Fuel Standard (LCFS) for transportation fuels. State of California Executive Order S-01-07 http://gov.ca.gov/index.php?/executive-order/5172/

[2] The Regional Greenhouse Gas Initiative (RGGI) comprising 10 states in the US North East. A mandatory market-based program for reducing GHG emissions from electricity sector http://www.rggi.org/home

[3] Emission Trading Scheme (ETS) in the European Union. A market-based mechanism or reducing GHG emissions from electricity sector http://ec.europa.eu/environment/climat/emission/index_en.htm

M. Khanna et al. (eds.), *Handbook of Bioenergy Economics and Policy,*
Natural Resource Management and Policy 33, DOI 10.1007/978-1-4419-0369-3_13,
© Springer Science+Business Media, LLC 2010

First-best carbon taxes are for a variety of reasons not likely to be a part of the reality in the near future. Second-best instruments under consideration (or under implementation) include standards (such as California's LCFS for transportation) and cap and trade programs (such as RGGI in the northeast USA and ETS in the European Union both of which apply to electricity power generation). Producers of energy from fossil-based fuels are considering blending biofuels (or biomass) as an option for complying with GHG regulation. In a world with biofuels in the mix, such policies require adopting a lifecycle view of emissions, i.e., an approach that takes into account not just the emissions from combustion of the final fuel but also emissions associated with production of the fuel. Although carbon emitted during combustion of biofuel is sequestered during photosynthesis, the emissions related to the inputs used in production of the biofuel could either reduce or increase GHG emissions relative to its fossil substitute (Farrell et al. 2006, Rajagopal and Zilberman 2008, Tilman et al. 2006).

Moreover, the GHG intensity of biofuels is heterogeneous. Biofuels can be produced from a diverse set of feedstock (such as corn, sugarcane, oilseeds, etc.) using a diverse set of production technologies. The cultivation and processing of each type of feedstock can be carried out in a variety of ways with widely varying carbon intensities (Rajagopal and Zilberman 2008b). Another complicating factor is indirect effects, an example of which is indirect land use change (ILUC) emissions. Biofuels increase demand for agricultural land, which induces land use changes in regions that substantially affect global carbon sequestration (regions that are also efficient in producing biofuel crops). Furthermore, trade causes land use changes to occur in regions different from the place of production and/or consumption of biofuels. Therefore, regulating biofuels should account for such indirect emissions, if indeed the regulators' goal is to lower, or at least mitigate, carbon emissions, which are a global pollutant. The challenge of regulating biofuel is then augmented by uncertainty.

The goal of this chapter is to derive a regulatory framework that can be used for regulation of biofuel-blended energy. It can be used to determine whether a producer of biofuel-blended energy meets a set GHG emission standards. We begin by categorizing GHG emissions into direct and indirect emissions (Section 2) and propose a way to calculate the site-specific emissions (Section 3). We then suggest a site-specific methodology for regulating GHG emissions, a methodology that extends current methods by introducing heterogeneity (Section 4), as well as accounting for uncertainty (Section 5) and market forces. Before concluding, we briefly describe lifecycle-based regulation within the broader context of regulation of GHG externalities (Section 6).

13.2 Categorizing Lifecycle Emissions

We classify emissions into two categories: direct and indirect.

13.2.1 Direct Emissions

Direct emissions comprise all emissions directly related to production of final output (e.g., gasoline or biofuel or a blend). Direct emissions are classified into two subcategories, namely

- *Direct on-site emissions*: These are emissions at the regulated site, which directly relate to the production of the final product. For example, if the regulated site is an ethanol biorefinery, then these are emissions from combustion of coal or natural gas used in converting corn or sugarcane to ethanol. Suppose, for instance, that the regulated site is a biorefinery. For US ethanol corn production, direct on-site emissions comprise 55% of total direct emissions (see Fig. 13.1).
- *Direct off-site emissions*: These are emissions emanating off-site that are directly attributed to intermediate inputs used to produce the final good. For instance, ethanol producers use crops. Crops use fertilizers, which are a large source of emissions both at the farm site and the fertilizer-production site. From Fig. 13.1, we can see that 45% of the direct emissions are off-site, with fertilizer production and use accounting a large share of total emissions.

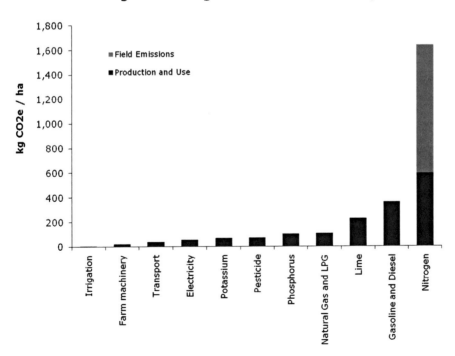

Direct greenhouse gas emissions from corn production

Fig. 13.1 Life cycle GHG emissions for US corn ethanol based on EBAMM model

13.2.2 Indirect Emissions

When food or cropland is diverted to biofuel production, it will have two types of effects, namely, extensive and intensive effects. GHG emissions that accompany such changes are referred to as indirect emissions. For instance, demand for biofuel raises the price of agricultural commodities, which raises the rent to land, thereby allowing marginal land to enter production, i.e., the extensive effect. Emissions due to the extensive effects arise from conversion of nonagricultural land into farmland (for example, emissions from clearing trees and pastures), and cultivation on converted land (for example, emissions from use of inputs like fertilizer). The former is what is currently known in the literature as indirect land use change (ILUC) emissions. An illustration of this is shown in Fig. 13.2. On the other hand, higher output prices result in more intensive use of inputs like fertilizers and irrigation on existing farmland, which is the intensive effect.[4]

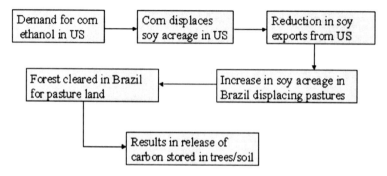

Fig. 13.2 An example of ILUC pathway

13.3 Calculating Emissions

13.3.1 Calculation of Direct Emissions

Several studies calculate direct emissions for major biofuel pathways such as corn ethanol and soy biodiesel in the United States, sugarcane ethanol in Brazil, and rapeseed biodiesel in Europe. A detailed review of this literature can be found in Rajagopal and Zilberman (2008b). Most of these studies use a process-LCA

---

[4]Theoretically speaking, this is an indirect effect, which results in agriculture becoming more input intensive and perhaps emission intensive. This can be modeled without much difficulty and incorporated into our framework provided reliable data are available. However, we believe this to be insignificant and second-order effect compared to the magnitude of ILUC emissions and hence do not discuss this further.

approach[5] and report a single number for the average carbon intensity. For example, Farrell et al. (2006) using the EBAMM model calculated that average emission intensity for corn ethanol produced in the United States equals 77 gCO_2e/MJ[6] and the average intensity of gasoline to be 94 gCO_2e/MJ. But it should be pointed out that such LCA numbers represent the emission intensity for a particular combination of inputs and processes (usually assumed to represent the industry's average although no single production facility may use this particular process). For instance, Farrell et al. (2006) assume that the average ethanol refinery uses a mix of 40% coal and 60% natural gas to produce the energy required for production of ethanol from corn; the direct on-site emissions are, therefore, appropriately weighted by the average carbon intensity of coal and natural gas. However, no given refinery may use both fuels. Depending on whether a biorefinery uses coal or natural gas for processing corn to ethanol, the total direct emissions for corn ethanol will equal 91 or 58% of net GHG emissions from gasoline, respectively (Rajagopal and Zilberman 2008a). Figure 13.3 shows emissions from ethanol produced under various scenarios of direct emissions relative to emissions from gasoline (indirect emissions are held constant[7]). We see that only production facilities of type C through F would be able to be in compliance with the standard that is set to equal the emissions from gasoline (Section 4 describes other thresholds that regulators could choose). The net GHG emissions are sensitive also to assumptions about the characteristics of cultivation of feedstock such as the type of tilling, quantity of inputs, and productivity. In essence, there exists considerable heterogeneity in production functions and hence in direct emissions.

Building on Caswell et al. (1990) and Khanna and Zilberman (1997), we develop a mechanism to measure direct emissions. To this end, let $y = f_i(x,e)$ denote the quantity of biofuel produced by a biorefinery that uses on-site the effective input e (e.g., natural gas or coal) to convert the feedstock x to biofuel[8]. The function $f_i(x,e)$ has the regular properties of a neoclassical production function and i denotes the technology used. For simplicity, the amount of input e used by the biorefinery (effective units) equals the amount of input e actually applied toward the production process. In addition, the effective amount of feedstock consumed by the biorefinery, i.e., x, equals the amount produced off-site at the farm level.

---

[5]A detailed description of the technique of LCA can be found in Rajagopal and Zilberman (2008a).

[6]gCO_2e/MJ - grams of carbon dioxide equivalent emissions per megajoule of energy.

[7]In all these scenarios, ILUC emissions are constant and take a value equal to one-third that estimated by Searchinger et al. (2008). This value was chosen because we expect their estimate of ILUC to be more than three times larger compared to ours. For more details see Hochman et al. (2008).

[8]The idea of using effective input as opposed to applied input is to account for differences in efficiency of in the use of a given quantity of input depending on the technology. For example, natural gas may be used more efficiently by a gas turbine of new vintage compared to that of an old vintage. Another example is from irrigation where a given quantity of water applied to crops may be utilized more effectively with drip irrigation than furrow irrigation.

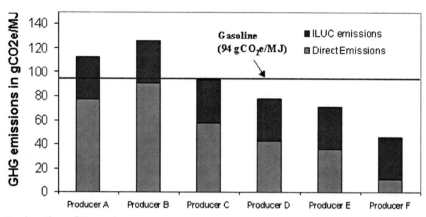

Explanation of Scenarios

Producer A: Average corn producer in US (based on Farrell et. al.)

Producer B: Coal based conversion of corn to ethanol all else equal to A

Producer C: Gas based conversion of corn to ethanol all else equal to A

Producer D: Gas based conversion, 39% improvement in corn yield, 25% reduction
              in energy for processing of corn to ethanol all else equal to A

Producer E: Sugarcane ethanol from Brazil on average

Producer F: Cellulosic ethanol from switchgrass on average (based on Farrell et. al.)

Fig. 13.3 Heterogeneity in direct emissions

The feedstock is a function of the amount of effective units c (e.g., fertilizers and/or other chemicals used), i.e., $x = f_j(c)$. The function $f_j(c)$ has the regular properties of a neoclassical production function and j denotes the current technology used. The effectiveness of input c is the ratio of effective to applied input units and assumed to depend on the technology j and the environment. Let α denote the applied amount of input c under technology j and let α be the land quality index which is an indicator of the environment, which assumes values between 0 and 1. The effective ratio, then, is formally defined as $h_j(\alpha) = e_j(\alpha)/a$. The measure of quality used here is the feedstock's ability to retain the input (e.g., fertilizer). The amount of input not retained by the feedstock is a source of emissions. If everything is certain, then the direct emission from biofuel production equals the sum of emissions arising at the farm due to use of polluting farm chemicals and emissions arising at the biorefinery during conversion of feedstock to biofuel using polluting energy sources. Mathematically this is expressed as,

$$Z = g_c(a.\gamma_j(\alpha)) + g_e(e) \tag{13.1}$$

where, $\gamma_j(\alpha) \geq 1 - h_j(\alpha)$ and $g_k()$ denote the amount of emissions produced.

We are now ready to define the average total direct emissions, which now address heterogeneity as

$$f_D = \frac{Z}{y} = \frac{g_c\left(a \cdot \gamma_j(\alpha)\right) + g_e(e)}{y} \tag{13.2}$$

13.3.2 Calculation of Indirect Emissions

Different from direct emissions, indirect emissions arise from the interaction of aggregate supply and demand, and therefore, are not site-specific. In other words, each individual farmer or for that matter even an individual biofuel refinery can be assumed to have a negligible impact on the aggregate supply and hence the extensive (or the intensive) margin of agriculture. However, taken collectively at the level of a region or a nation, the aggregate quantity of biofuel consumed or produced can be sizeable enough to induce an effect on prices and hence an effect on land use change.

The total amount of land use change and the indirect emissions needs to be calculated using a multimarket or general equilibrium model[9]. An illustration is shown in Fig. 13.4. The average indirect emissions per unit of biofuel are then obtained by dividing the total indirect emissions by the total quantity of biofuel produced or consumed within the regulated region. Therefore, analogous to the idea of a price-taking producer in a competitive market, we propose that each regulated facility be assigned a number equal to average amount of indirect emissions per unit of biofuel produced (or consumed) within the region. This number can also be adjusted using a scaling factor, which may reflect the differences in efficiency between different feedstocks and different conversion technologies or a combination of the two.

13.3.3 Ex post Direct Emissions and Ex ante Indirect Emissions

This section explains why the direct emissions are computed ex-post while the indirect emissions are computed ex-ante. Producers need to be aware of the threshold value of emissions they will be required to comply with. In our framework, since indirect emissions are taken by each individual regulated site as given, regulated

---

[9]Searchinger et al. (2008), using the FAPRI (partial equilibrium) model, compute that producing 56 billion liters of corn ethanol (requiring 140 million tons of corn at a corn-to-ethanol conversion rate of 2.7 gallons of ethanol per bushel of corn) in the United States would cause global agricultural acreage to expand by 10.8 million hectares. By allocating this acreage across different types of land with differing stocks of carbon, they calculate indirect emission from land-use change as 106.4 gCO_2e per MJ of ethanol. However, we find this estimate to be high going by past historical trends (see Hochman et al. 2008). Others have similarly tried to calculate the effect of increase in gasoline prices on induced land-use change (Keeney and Hertel 2008, Thompson et al. 2008).

Fig. 13.4 An equilibrium
approach to calculate ILUC
and indirect emissions

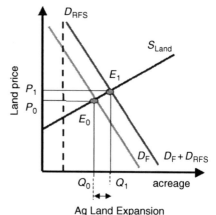

Ag Land Expansion

S_{Land} – supply of land
D_F – demand for land for food
D_{RFS} – demand for land for meeting biofuel mandate
E_0, E_1 – equilibrium before and after the mandate
P_0, Q_0 – price and quantity before mandate
P_1, Q_1 – price and quantity after mandate
Q_1-Q_0 – increase in agricultural land due to biofuels

producers are responsible for ensuring that their net direct emissions per unit of output satisfies the constraint given in Equation (13.3) below. Therefore, in order to be able to clearly specify such a target value to producers, regulators must ex-ante determine a value for average indirect effect that will be assigned to each producer and remain fixed over a given time frame. Ex-ante indirect emissions are of course uncertain (they are also ex-post uncertain). But using the equilibrium framework and with historical data, one can compute the indirect emissions to a good first-degree approximation. In any case, with the passage of time these estimates can be improved.

13.4 A Target Number and a Framework for Regulation Given Uncertainty

Suppose for instance, regulation aims to establish an upper bound for GHG emissions per unit of fuel. Let $\bar{f}$ be the upper bound, which is to be compared to the emission measure of each regulated site. For instance, this could be set equal to the average emission intensity of gasoline (94 gCO_2e/MJ according to Farrell et al. 2006). Alternatively, this number may be set equal to the emission intensity of marginal fuel, which may be more carbon intensive (such as gasoline from tar sands or coal) or less carbon intensive (such as compressed natural gas). See Table 13.1 for a list of the average GHG emission intensities per unit of energy contained in the fuel for different types of energy sources. In any case, if f_D denotes the ex-post direct site-specific emissions per unit output and f_I the ex-ante average indirect effect, the

Table 13.1 Average GHG emission intensity for different types of fossil energy sources (Based on Farrell and Brandt 2006)

Fuel	gC/MJ	gCO$_2$/MJ	Relative to coal or NG or conv. oil
Natural gas (NG)	15	56	
Conventional oil	25	92	67% more than gas
Coal	28	103	95% more than gas
Enhanced oil recovery	25–30	92–110	0–20% more than conventional oil
Heavy oil	30–35	110–128	20–40% more than conventional oil
Tar sands	30–35	110–128	20–40% more than conventional oil
Coal to liquids (CTL)	40–50	147–183	40–80% more than coal
Gas to liquids (GTL)	30	110	100 more than gas

sum $f_D + f_I$ represents the overall emissions per unit of biofuel from a given site. To reiterate, f_D is computed ex-post using an LCA style approach, whereas f_I is computed ex-ante using economic equilibrium models.

Depending on the regulatory framework, the regulated site may have to provide certification showing, $f_D + f_I \leq \bar{f}$ to get a permit. Alternatively, the regulator may need to inspect the site and show that $f_D + f_I > \bar{f}$ to close the site. Since the site takes indirect emissions f_I as given, this effectively requires ensuring that the direct emissions f_D satisfies the constraint,

$$f_D < \bar{f} - f_I \tag{13.3}$$

13.5 Uncertainty in Calculation of Emissions

In reality, both direct and indirect emissions are uncertain. With regard to direct emissions, even if we were able to ex-post observe all the intermediate inputs used in the production of biofuels, the total emissions are uncertain because of inherent randomness in the production (or the accompanying pollution generation) processes. For instance, one important source of uncertainty in direct emissions is related to the emission of nitrous oxide (N_2O)[10] through the processes of nitrification and denitrification of ammonium fertilizers applied in agricultural fields. The emissions of N_2O vary with climate and soil type. Another source of uncertainty is in ILUC emissions, which are calculated ex-ante.

13.5.1 Modeling Direct Emissions with Uncertainty

The value of the pollution function, which measures the direct emissions, is uncertain. For example, the amount of fertilizer applied that is not retained by the

---

[10]Nitrous oxide is a potent GHG with a global warming potential (GWP) 170 to 290 times that of carbon dioxide.

feedstock is a source of emissions, albeit an uncertain one. The amount of pollution in this context depends on factors such as soil quality, timing of application, weather conditions, etc., which are affected by random factors. To introduce uncertainty, we begin with Equation (13.1) but now assume that the parameter γ_j is a random variable distributed on the segment $[\beta, 1 - h_j(\alpha)]$ (see Section 3.1 for explanation of terms). The amount retained by the feedstock and the pollution cannot sum to more than 1. On the other hand, the lower bound β may be negative. For instance, biomethanation, a second-generation biofuels technology, contributes to the reduction of uncontrolled methane emissions from decomposition of plant and animal wastes. We are now ready to define the direct emission, which now addresses both heterogeneity and uncertainty, as

$$f_{\mathrm{D}} = \frac{E(Z)}{y} = \frac{E\left(g_c(a \cdot \gamma_j(\alpha)) + g_e(e)\right)}{y} = \frac{E\left(g_c(a \cdot \gamma_j(\alpha))\right) + g_e(e)}{y} \qquad (13.4)$$

Instead of using the expected value, a regulator could be conservative and choose a precautionary principle, in which case he or she may set f_{D} equal to the worst-case scenario, i.e.,

$$f_{\mathrm{D}} = \frac{\max_{\gamma_j}\left(g_c\left(a \cdot \gamma_j\right)\right) + g_e(e)}{y} \qquad (13.5)$$

The merits and demerits of each approach, i.e., precautionary vs. promotional, albeit important are beyond the scope of this chapter.

13.5.2 Uncertainty in Indirect Emissions

Unlike direct emissions, which are calculated ex-post, ILUC emissions are calculated ex-ante and this is the reason we consider them uncertain (see Section 3.3). We also assume that this is uncertainty that is inherent when it comes to prediction of the future and is not an artifact of poor representation of the model and its parameters used in prediction of ILUC. We believe that a good first-degree approximation can be obtained using a well-developed multimarket or general equilibrium model in conjunction with reliable historical data on land use change. In the absence of a better model, a point estimate of ILUC emissions produced by the model can serve as a proxy for the expected value of the random variable.

13.6 Implementing This Framework

We now briefly discuss the steps required to implement this regulatory framework. One option, with regard to the setting of an upper bound on emissions from biofuel, is to set this relative to the emissions from gasoline; for example, no higher than net

GHG emissions from gasoline and reliable estimates of this exist today (see table for data on average emission intensity of various fossil fuels – Table for 13.1. Two, a variety of models based on process LCA methodology such as GREET[11] EBAMM (Farrell et al. 2006), BESS[12] and BEACCON (Plevin and Mueller 2008) can calculate GHG emission intensities of biofuels. But they do not model uncertainty. Models such as these can be improved and then adopted as the standard tool for measurement and certification of GHG emissions of a given biofuel facility. Three, calculation of indirect emissions requires data on the quantity and type of lands that were converted from nonfarm use to farm use world-wide and the net change in carbon stored on those parcels of land due to such conversion. The former can be calculated using economic equilibrium models and the latter can be calculated using existing GIS-based databases of soil and above-ground carbon in different agro-ecological environments. It is worth emphasizing that what one can calculate from equilibrium models of ILUC is total change in land use between two points in time. The challenging calculation is then in ascribing a share of this total change in land use to biofuels after controlling for changes in land use induced by other factors that influenced agricultural demand and supply such as economic growth, weather, inflation, and exchange rate shocks between the same two points in time.

13.7 Policy

If the goal is to produce biofuel efficiently and minimize carbon emissions and damage to the environment, then the first best policy is a carbon tax and payment for environmental services. Levying a carbon tax shifts production from fossil fuel to biofuel and induces greater supply of clean fuel. It, however, brings on land conversion and a loss of biodiversity. Therefore, a policy to price clean air should be paired with a policy to price environmental services (Hochman et al. 2008). Politically, a carbon tax may not be a viable option. It may not be feasible to levy a tax on a global public bad. A second best policy is the next possibility.

A fuel tax based on LCA is currently proposed by some state and national governments. They are easy to impose because fuel consumption is observable. Different from existing fuel taxes, a second best fuel tax should vary according to fuel types – with dirtier fuels taxed more heavily. LCA could then be used to classify fuels according to their carbon emissions. Such a tax may also account for other local externalities such as traffic congestion. This policy also has a problem of double counting.

An alternative second best solution, which bans biofuel production if it has limited environmental benefit, is LCA thresholds or certification standards. Only biofuels that have sufficient small life-cycle emissions can be used. Governments

---

[11] The Greenhouse Gases, Regulated Emissions, and Energy Use in Transportation (GREET) Model. Available online at http://www.transportation.anl.gov/modeling_simulation/GREET/

[12] Biofuel Energy System Simulator (BESS) Model. Available online at http://www.bess.unl.edu/

may account toward mandate or offer subsidies only to those biofuels that are certified to meet the desired standards. Note that standards might be different between countries, because local environmental amenities are different. Standards are currently being proposed in the United States.

Because carbon emissions are a global public bad, policy ought to be coordinated between all countries. More specifically, international environmental agreements should account for the cost of deforestation (e.g., destruction of rain forests in Brazil). Landowners do not capture all the benefit from their efforts to preserve the environment. The benefit, in terms of biodiversity and carbon sequestration, accrues to people around the world. Therefore, landowners should be paid for the environmental services their land provides. To this end, an international agreement, which will internalize the negative externalities from fuel production and consumption, needs to be established.

13.8 Conclusion

Even if a first best GHG tax is imposed on all GHG-emitting fuels, so long as there is no tax on GHG emissions from land conversion and land use, biofuels can result in leakage, i.e., effective GHG emissions due to a blend may be above the level accepted by the regulator. In the absence of carbon tax, the implementation of second best mechanisms such as carbon standards or cap and trading requires calculation of all associated and induced emissions upstream of final combustion. With this in mind, we have outlined a framework that can be applied to the regulation of GHG emissions from biofuel-blended energy sources. Ours is a hybrid approach that combines detailed LCA style calculation of all direct emissions and a multimarket or general equilibrium type calculation of market-induced indirect effects. This approach can be implemented in practice given existing data and can account for heterogeneity and uncertainty. It can also be extended to the regulation of nongreenhouse gas externalities. But significant improvements in the methodology including standardization of the life cycle assessment model and computable equilibrium model to be used for calculating emissions are needed before this type of framework is adopted for regulation.

Acknowledgments Funding support for research leading to this publication was provided by the Energy Biosciences Institute and the Farm Foundation.

References

Caswell M and Lichtenberg E, and Zilberman D (1990) " The Effects of Pricing Policies on Water Conservation and Drainage" Am J Agri Econ 72(4) 883–890.
Farrell AE, Plevin RJ, Turner BT, Jones AD, O'Hare M, and Kammen DM (2006) "Ethanol Can Contribute to Energy and Environmental Goals," Science 311 5760: 506–508.
Farrell AE and Brandt AR (2006) Risks of the oil transition, Environ Res Lett 1(1).
Hochman G, Rajagopal D and Zilberman D (2008) "Regulation of GHG emissions from biofuel blended energy", *Farm Foundation conference on Transition to a Bioecomy, Environmental and Rural Development Impacts*, St. Louis, Missouri, October 15th–16th.

Keeney R and Hertel TW (2008) The Indirect Land Use Impacts of US Biofuel Policies: The Importance of Acreage, Yield, and Bilateral Trade Responses, GTAP Working Paper (forthcoming), Center for Global Trade Analysis, Purdue University, West Lafayette.

Khanna M and Zilberman D (1997) "Incentives, precision technology and environmental protection" Ecol Econ 23(1) October 25–43.

Plevin RJ and Mueller S (2008) The effect of CO_2 regulations on the cost of corn ethanol production, Environ Res Lett 3(2).

Rajagopal D and Zilberman D (2008a) Prices, Policies and Environmental Life Cycle Analysis of Energy, *Farm Foundation conference on Lifecycle Footprint of Biofuels*, Washington DC, January.

Rajagopal D and Zilberman D (2008b) "Environmental, Economic and Policy Aspects of Biofuels", Foundations and Trends in Microeconomics,Vol. 4, No. 5.

Searchinger T, Heimlich R, Houghton RA et al. (2008) "Use of U.S. Croplands for Biofuels Increases Greenhouse Gases Through Emissions from Land Use Change," Science 319, 5867: 1238–1240.

Tilman D, Hill J, and Lehman C (2006) "Carbon-Negative Biofuels from Low-Input High Diversity Grassland Biomass," Science 314, 5805:1598–1600.

Thompson W, Meyer S and Westhoff P (2008) Potential for Uncertainty about Indirect Effects of Ethanol and Land Use in the Case of Brazil, Working paper, Food and Agricultural Policy Research Institute, University of Missouri.

Chapter 14
Market and Social Welfare Effects of the Renewable Fuels Standard

Amy W. Ando, Madhu Khanna, and Farzad Taheripour

Abstract This chapter evaluates the welfare and greenhouse-gas effects of the Renewable Fuel Standard (RFS) in the presence of biofuel subsidies. In our numerical model, demand for gasoline and ethanol stems from consumer demand for driving miles, but all fuels have congestion and environmental external costs. Our estimates of the effects of ethanol mandates on greenhouse gases and social welfare (relative to the status quo) are sensitive to assumptions about the gasoline supply elasticity. The impact of the mandate, by itself, on greenhouse gas emissions ranges from -0.5 to -5% relative to the status quo and is reduced when the mandate is accompanied by a tax credit. The welfare costs of the mandate relative to the socially optimal policy range from \$60 B to \$115 B depending on the elasticity of gasoline supply. The provision of a tax credit in addition to the mandate leads to additional deadweight losses that range from \$1.1 to \$12 billion. An ethanol mandate policy provides assured demand for ethanol and therefore supports the domestic ethanol industry, particularly the cellulosic biofuel industry. However, such policy may harm the well-being of the country as a whole, even relative to the ethanol support policy that was in place before the current mandate was passed.

14.1 Introduction

The United States has supported domestic biofuel production through tax credits and import tariffs as a means to reduce reliance on foreign oil. Concern about climate change and rising oil prices has provided further impetus to legislation in the form of mandates and new subsidies for biofuel production. Early regulation sought to promote production and use of first-generation biofuels, such as corn ethanol, that have the potential to lower greenhouse gas emissions by about 20% relative to baseline levels. However, the recently enacted Energy Information and Security Act

A.W. Ando (✉)
Department of Agricultural and Consumer Economics, University of Illinois, Urbana-Champaign, IL, USA
e-mail: amyando@illinois.edu

M. Khanna et al. (eds.), *Handbook of Bioenergy Economics and Policy*, Natural Resource Management and Policy 33, DOI 10.1007/978-1-4419-0369-3_14, © Springer Science+Business Media, LLC 2010

of 2007 (EISA) and the Food Conservation and Energy Act of 2008 (FCEA) contain ethanol quantity mandates and subsidies designed to encourage production of the next generation of cellulosic biofuels. Controversy exists over the likely effects of these policies on the environment, energy use, and food markets (Martin 2008). This paper evaluates the likely impacts of these policies on fuel prices, fuel quantities, and social welfare including negative externalities from driving and consumption of motor fuel.

Binding mandates for biofuel consumption are, by definition, expected to raise the level of biofuels blended with gasoline above levels that would have been supported by a free market. While mandates can be expected to lower greenhouse gas emissions per mile traveled due to a substitution effect (by displacing gasoline), they also affect total consumption of biofuels and gasoline by changing their prices. With upward sloping and inelastic supply functions for biofuels and gasoline, mandates are likely to raise the cost of biofuels and lower the domestic price of gasoline; the latter change may offset some of the substitution effect.[1] Research has yet to determine the extent to which biofuel mandates will reduce emissions, the implementation costs of any emission reductions, and the implications of mandates for social welfare. Recent debate over biofuel policy has paid substantial attention to greenhouse gas emissions, but in evaluating the impact of biofuel mandates on social welfare, it is important to recognize other environmental effects of fuel consumption as well. These include the externalities associated with miles traveled in the form of congestion, accidents, and conventional air pollution.

In analyzing the effects of mandates, one must also consider the biofuel subsidies that accompany the mandate. While these subsidies (particularly at differential rates for biofuels produced from corn vs. cellulosic feedstocks) would enhance the substitution effect mentioned above by lowering the price of biofuels relative to gasoline and creating incentives for substituting cellulosic biofuels for corn ethanol, they lower the costs of miles traveled and create perverse incentives to increase miles traveled and total fuel consumption. Khanna et al. (2008) analyze the effect of the corn-ethanol subsidy by itself and show that it increased miles traveled and caused a slight increase in greenhouse gas emissions (we will henceforth refer to that paper as KAT).

This chapter extends the framework developed in KAT to examine the welfare and greenhouse-gas effects of the RFS in the presence of biofuel subsidies. We consider a stylized model of an economy with homogeneous consumers that benefit from vehicle miles traveled (VMT) which are produced by blending gasoline and biofuels but suffer disutility from congestion and greenhouse gas emissions. Gasoline and biofuels are modeled as being imperfect substitutes in the production of VMT while corn ethanol and cellulosic ethanol are considered to be perfect substitutes. This approach permits sensitivity analysis over the assumption about substitutability between gasoline and ethanol. Those two fuels have previously been

---

[1] Rajagopal et al. (2007) estimate that the provision of ethanol in the United States lowered the domestic price of gasoline by 3%.

modeled both as perfect complements (Vedenov and Wetzstein 2008) and as perfect substitutes (de Gorter and Just (2009), while the truth (given constraints on the stock of motor vehicles and their capacity for using ethanol) probably lies somewhere in between.

We use this model to first analyze whether the mandated mix of biofuels is close to the socially optimal levels that internalize the externalities caused by motor vehicles. We compare the social welfare and environmental implications of the RFS with and without biofuels subsidies. Finally, we explore the effects of having an overall binding RFS but with the flexibility to meet it using either corn ethanol or cellulosic ethanol. Given the uncertainty about the costs of cellulosic ethanol production and the responsiveness of its production to fuel prices, as well as the degree of substitutability between gasoline and biofuels at mandated levels, we examine the robustness of our model results to various parametric assumptions.

14.2 Related Literature

Several studies have examined the effects of biofuel subsidy policies on fuel prices, consumption, and social welfare. Rajagopal et al. (2007) estimate that the ethanol tax credit lowered the price of gasoline by 3% and is likely to have led to a net gain in social welfare for the United States. Much of this gain ($11 billion) is due to the increase in surplus of gasoline consumers and corn producers ($ 6.4 billion) in the United States. On the other hand, de Gorter and Just (2009) find that the ethanol tax credit leads to a loss in social welfare in the presence of farm subsidies. Their numerical analysis assumes a perfectly elastic supply curve for gasoline with no effect of ethanol policy on oil prices. While the tax credit to ethanol lowers the need for farm subsidies and improves the terms of trade for corn exports, the revenue costs of the tax credit and the loss in surplus of corn consumers due to higher prices result in a net loss in social welfare of $1.3 billion. Taheripour and Tyner (2008) show that the benefits of the subsidy are more likely to be passed on to the corn producers as the share of corn used for ethanol increases and the supply elasticity of corn production decreases. Gardner (2007) estimates the deadweight losses due to the ethanol subsidy to be $91 M in the short run and $6.65 B in the long run. He also finds that these subsidies are many times larger than those that would be generated with a deficiency payment policy that directly subsidizes corn. The environmental benefits of ethanol would need to be valued at least $0.23 per gallon of ethanol to offset the deadweight losses of the subsidy. In examining the welfare effects of the ethanol subsidy, these studies do not include the environmental damages associated with fuel consumption.

Unlike these studies, Vedenov and Wetzstein (2007) incorporate the environmental and fuel security externalities associated with biofuels and consider the multiplier effects of government subsidies. They find that the optimal subsidy for ethanol should be positive at 0.22 per gallon, primarily because it improves fuel security valued at $0.17 per gallon fuel and has a multiplier effect on incomes, and thus tax revenues for the government (a 1% increase in the ethanol subsidy increases government revenue by 0.52%). Considering only environmental (and not security) externalities, KAT show that an optimal policy would tax ethanol ($0.04 per gallon) but give a tax credit relative to the tax on gasoline, which should be $0.08 per gallon. The ethanol subsidy of $0.51 per gallon has a marginally negative impact on aggregate carbon emissions relative to the baseline and leads to a loss in social welfare (relative to the optimum) of $19 billion.

Only a few studies have examined the economic effects of an RFS and of other biofuel policies in the presence of an RFS (Gallagher et al. 2003; de Gorter and Just 2008; Tyner and Taheripour 2008). Gallagher et al. (2003) show that an RFS of 5 billion gallons of ethanol together with a ban on the use of MTBE as a fuel additive lowers gasoline prices by 2% while raising the price of ethanol by 10% in 2015 relative to baseline levels. Gasoline consumption thus falls by much more (0.13 billion barrels) than the increase in ethanol consumption to meet the mandate (0.01 billion barrels), due to the high cost of the blended fuel relative to baseline levels. As a result, social welfare (without including environmental benefits) decreases by 6% but emissions of air pollutants also decline (in areas using reformulated gasoline). de Gorter and Just (2008) examine the effects of a tax credit and an import tariff in the presence of a blend mandate assuming that the price of oil is exogenous and that gasoline and ethanol are perfect substitutes. They show that in the presence of a mandate, an import tariff raises the domestic ethanol price by reducing imports and requiring an increase in domestic supply. A tax credit with a mandate results in a subsidy to fuel consumers and higher fuel consumption. de Gorter and Just (2009) examine the effects of a tax credit in the presence of a price-contingent farm subsidy and show that while the tax credit reduces the costs of the loan rate program, the latter increases the tax costs of the tax credit. Comparing the effects of various biofuel policies, Tyner and Taheripour (2008) show that the costs of the RFS to fuel consumers decrease as price of oil increases (and the gap between the cost of ethanol and gasoline decreases) while the cost of the ethanol subsidy to the government increases as the price of oil increases (and the production of ethanol increases).

One key feature of motor-fuel markets is treated with wildly different assumptions in the energy economics literature: the own-price elasticity of gasoline and/or petroleum supply. de Gorter and Just (2008, 209) assume petroleum supply is perfectly elastic. However, Gallagher et al. (2003) assume an elasticity of 10, Austin and Dinan (2005) estimate a value of 2, and Rajagopal et al. (2007) use 0.25. The appropriate value depends heavily on length of "run," since short-run gasoline production is heavily capacity constrained. Given the disparate treatment of this parameter in the literature, we explore sensitivity of policy outcomes to different assumptions about gasoline supply. Existing studies also differ in their assumptions about the elasticity of substitution between gasoline and ethanol. While de Gorter and Just (2009) assume the two are perfectly substitutable, Vedenov and Wetzstein (2008) assume they are perfect complements. Our analysis considers them to be imperfect substitutes and we explore the sensitivity of our results to assumptions about the degree of substitutability between the two.

14.3 Background: Motor-Fuel Technology and Policy

Federal motor-fuel excise-tax credits for ethanol have been present for decades. The tax credit peaked at $.60/gallon in 1984, fell to $.54/gallon in 1990 and was phased down to $.51/gallon between 1998 and 2005. The Energy Policy Act of 2005 also instituted the first Renewable Fuels Standard (RFS) program, which required modest levels of renewable fuels to be blended into US motor fuel: 4 billion gallons in 2006, increasing to 7.5 billion gallons in 2012. New generations of policies expand these supports. The FCEA of 2008 provides various new incentives in the form of cost-sharing and subsidies to encourage the production and use of cellulosic feedstocks for fuel production. These include a differential subsidy for corn ethanol ($0.45 per gallon) and cellulosic ethanol ($1.01 per gallon). That cellulosic ethanol tax credit is set to expire at the end of 2012, but the history of ethanol tax credits in the United States is one of regular extensions of ethanol tax credit policy. More controversially, the EISA passed in 2007 expanded the RFS program to demand 36 billion gallons of ethanol per year by 2022, of which 21 billion gallons must be produced from second-generation feedstocks that have the potential to reduce greenhouse gas emissions by at least 50% relative to baseline levels (Yacobucci and Schnepf 2007).[2] Life cycle analysis shows that the carbon emissions associated with an energy equivalent gallon of ethanol are 88% lower than those of gasoline while those from corn ethanol are 18% lower (Farrell et al. 2006). These analyses disregard the carbon emissions from the land-use changes due to biofuel production. Searchinger et al. (2008) show that the land use emissions caused by biofuel production have the potential to eliminate reduction in carbon emissions from biofuel consumption.

The goals of the new RFS are stated in the enabling legislation as total volumes of different types of renewable fuels; it is therefore a quantity mandate. The Environmental Protection Agency (EPA) implements this policy by establishing an annual percent blend mandate.[3] The EPA calculates the blend standard for a given year by dividing the total desired volume of renewable fuel (for example, nine billion gallons in 2008) by the total volume of gasoline that is forecast to be sold in that year and multiplying by 100. An obligated party (refiners, blenders, and some importers of gasoline) calculates its total renewable-fuel volume obligation for the year by multiplying its actual gasoline production in that year by the blend standard established for that year. There is a market for renewable-fuel volume credits under this program that can be traded among obligated parties and banked for 1 year (though an obligated party must meet at least 80% of its obligation with current-year

---

[2]The administering agency, EPA, has the authority to alter the quantity mandates if there is inadequate domestic ethanol supply or if the mandates would yield serious environmental damage; however, even under heavy political pressure from the livestock industry in 2008, the EPA declined to alter the mandate.

[3]This discussion is based on background information from http://www.epa.gov/otaq/renewablefuels/index.htm, U.S. EPA (2007), and Beveridge and Diamond, *New RFS Program Requirements*, accessed at http://www.bdlaw.com/news-202.html [5/2008].

credits). Thus, the market is forced to produce such that the fraction of renewable fuel in the total quantity of fuel sold meets a set percentage target. The actual volume of renewable fuel could accidentally be greater or less than the total volume stated in the law if gasoline sales turn out to be different than forecasted by the EPA. For example, an unexpected recession could yield VMT below the forecast, in which case the blend target chosen by EPA would yield less renewable fuel than desired. However, there is no reason to expect EPA to chronically over or underestimate fuel use and thereby choose blend requirements that fail to achieve the total quantity goals stated in EISA. Thus, we assume that EPA will choose the blend mandate each year to reach the legally established total quantities, and we analyze the policy as a quantity mandate.

14.4 Model

We take the basic model presented in KAT as our starting point. However, that paper analyzed a model in which there were only two types of fuel (gasoline and ethanol) and the only policies allowed were taxes and subsidies on miles and fuels. In this paper, we distinguish between corn ethanol (e_c) and biomass ethanol (e_b) and we analyze the impact of different approaches to implementing the RFS.

The key features of the model are summarized as follows. We assume n homogeneous consumers obtain utility from individual miles driven, m, and disutility from congestion (including lost time, accidents, and noncarbon pollution), which depends on aggregate miles driven $M = nm$ and from aggregate carbon emissions C. The marginal external damages due to emissions and congestion are assumed to be constant and equal to ϕ and θ, respectively. The fuel for one consumer's vehicle consists of a flexible mix of g gallons of gasoline, e_c gallons of corn ethanol, and e_b gallons of biomass ethanol. We assume aggregate miles M are produced according to a constant elasticity of substitution production function with aggregate inputs $G = ng$, $E_c = ne_c$, and $E_b = ne_b$ and constant returns to scale:

$$M = [\alpha(\gamma G^{-r} + (1 - \gamma)(E_c + E_b)^{-r})^{-1/r}] \qquad (14.1)$$

Cellulosic and corn ethanol are perfect substitutes in production of miles, but the function in Equation (14.1) allows for flexibility in the degree of substitutability between gasoline and ethanol and in the shares of G, E_c, and E_b in the ultimate fuel blend used in vehicles.

The total cost of producing each type of fuel is a function of the quantity of fuel produced and is represented by $C_g = C_g(G)$, $C_{ec} = C_{ec}(E_c)$, and $C_{eb} = C_{eb}(E_b)$. The carbon emissions generated epend on the emissions intensity of a gallon of each fuel $\delta_g, \delta_{ec}, \delta_{eb}$, with $\delta_g > \delta_{ec} > \delta_{eb}$. Congestion externalities are assumed to be a function of the number of aggregate miles driven.

In the absence of any policy, the market will ignore the external costs. Miles will be produced with a combination of fuels G^0, E_c^0, and E_b^0 that minimizes total private production costs of a given level of mileage. An implicit supply curve for miles will be given by the resulting private marginal cost curve for miles, and total miles will be determined by the intersection of supply and demand.

The status quo before the passage of the RFS was a set of taxes and tax credits that yielded a positive net tax on gasoline t_g and a negative net tax on ethanol t_e that was not differentiated by feedstock. Under these circumstances, miles will be produced with a combination of fuels G^{sq}, $E_c{}^{sq}$, and $E_b{}^{sq}$ that minimizes total net production and tax costs of a given level of miles $\tilde{M}$:

$$\underset{G,E_c,E_b}{Min} \; C_g(G) - C_{ec}(E_c) - C_{eb}(E_b) + t_g G + t_e(E_c + E_b) \quad s.t. \quad M(G, E_c, E_b) = \tilde{M}$$
(14.2)

The solution to this problem yields private total and marginal cost functions $C_P^{sq}(M)$ and $MC_P^{sq}(M)$. It was established in KAT that status quo taxes do not approximate the first-best policy, so miles are not produced using social-cost-minimizing combinations of fuels.

To find the welfare maximizing choices of fuels, one must solve the social planner's problem:

$$\underset{G,E_c,E_b}{Max} \int_0^M D(M)dM - \emptyset(\delta_g G + \delta_{ec}E_c + \delta_{eb}E_b) - \theta M - C_g(G) - C_{ec}(E_c) - C_{eb}(E_b)$$
(14.3)

That solution will yield first-best values of G^*, $E_c{}^*$, and $E_b{}^*$ for given externality costs and parameterizations of demand and the production function. The characteristics of the solution to that problem are described in KAT. The optimal policy involves Pigouvian taxes on miles and each kind of fuel according to the externalities associated with them. KAT also demonstrates how an ethanol subsidy has two countervailing effects on welfare and carbon emissions: a subsidy causes a favorable substitution effect in which a given number of miles are produced with a higher fraction of the less-damaging ethanol fuel, but a subsidy also causes the number of miles driven to increase by reducing the cost of fuel. Simulations in KAT indicate that the current ethanol subsidy is welfare-reducing.

The RFS policy requires large quantities of different types of ethanol to be produced and consumed in total: $E_c = k_c$ and $E_b = k_b$. The RFS policy will be binding for a range of miles up to $\bar{M}$.[4] The range of vehicle miles traveled below $\bar{M}$ is of most interest, since for larger values of M the policy is not binding and has no effect on cost curves, market outcomes, or social welfare. The conceptual treatment below

---

[4]The level of miles at which the RFS is not binding may be extremely high or may not exist at all because the mandate fixes not only the overall quantity of ethanol but also the individual quantities of corn and cellulosic ethanol. However, there is likely to be some level of miles above which the policy is not binding, and $\bar{M}$ should be interpreted as such.

largely abstracts from the possibility that the mandate established in the RFS might not be binding.[5]

While the two types of ethanol still function as perfect substitutes in production of miles, blenders are not free to arbitrage between them under the RFS program. The marginal costs of producing those quantities of ethanol – which will equal the prices of the two kinds of ethanol unless ethanol producers have strong bargaining power with blenders – will be:

$$\bar{P}_{ec} = t_e + \frac{\partial C_{ec}}{\partial E_c}\bigg|_{k_c} \quad \bar{P}_{eb} = t_e + \frac{\partial C_{eb}}{\partial E_b}\bigg|_{k_b}. \tag{14.4}$$

Given these constraints, we use the production function in Equation (14.1) to determine the total quantity of gasoline G that will be used to produce any given quantity of aggregate miles M given the requirement that k_c gallons of corn ethanol and k_b gallons of biomass ethanol must go in the mix. If we define $K_1 = (1 - \gamma)(K_c + K_b)^{-r}$, then

$$M = [\alpha(\gamma G^{-r} + K_1)^{-1/r}] \rightarrow G(M|k_c,k_b) = \left(\frac{\left(\frac{M}{\alpha}\right)^{-r} - K_1}{\gamma}\right)^{-1/r} \tag{14.5}$$

The private total cost of producing miles will be

$$C_m(M|k_c,k_b) = \bar{P}_{ec}k_c + \bar{P}_{eb}k_b + P_g(G(M))G(M|k_c,k_b) \text{ where}$$

$$P_g(G(M)) = t_g + \frac{\partial C_g}{\partial G}\bigg|_{G(M|k_c,k_b)} \tag{14.6}$$

The presence of a binding constraint means that the private sector could have lowered the total cost of producing a given amount of fuel by choosing smaller quantities of ethanol.

The associated marginal cost function will be

$$\begin{aligned} MC_m(M|k_c,k_b) &= P_g(G(M|k_c,k_b))\frac{\partial G(M|k_c,k_b)}{\partial M} + G(M|k_c,k_b)\frac{\partial P_g}{\partial G}\frac{\partial G(M|k_c,k_b)}{\partial M} \\ &= \frac{\partial G(M|k_c,k_b)}{\partial M}\left[P_g(G(M|k_c,k_b)) + G(M|k_c,k_b)\frac{\partial P_g}{\partial G}\right] \end{aligned} \tag{14.7}$$

and the average cost function will be

$$AC_m(M|k_c, k_b) = \frac{\bar{P}_{ec}k_c + \bar{P}_{eb}k_b}{M} + \frac{P_g(G(M))G(M|k_c, k_b)}{M} \tag{14.8}$$

---

[5] However, we are careful in our numerical simulations to check that our scenarios have not strayed from a range where the mandate is binding; if that were to happen, we would permit the market to choose optimal quantities of ethanol that exceed the quantities in the mandate.

The first term in Equation (14.8) is the average fixed cost associated with the ethanol mandate; the second term is the average variable cost which depends only on gasoline.

To understand the impact of the RFS policy on the cost curves for miles (but not yet considering externality costs), we step back to consider the cost curves with no RFS policy. We are assuming a production function for miles that displays constant returns to scale, but the input prices (P_{eb}, P_{ec}, P_g) are endogenous, and will increase with the amounts of fuels used to produce miles such that the price of each fuel is equal to its marginal cost. Thus, the long-run average and marginal cost curves for miles (AC_{noRFS} and MC_{noRFS}) will be upward sloping, though not necessarily linear (see Fig. 14.1).

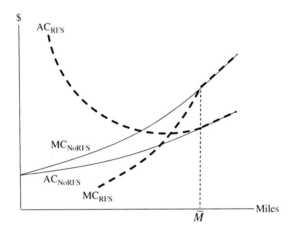

Fig. 14.1 Average and marginal costs with and without RFS mandate
Notes: (1) $\bar{M}$ is a level of miles above which the RFS policy does not bind. (2) The AC and MC of miles with no RFS mandate are both upward sloping because the prices of inputs (gasoline, cellulosic ethanol, corn ethanol) increase with the quantities of those inputs that are used. (3) The AC curve under the RFS mandate policy is U shaped because the minimum quantities of ethanol introduce a large initial fixed cost into the production of miles. (4) The MC curve under the RFS mandate policy intersects AC_{RFS} at its minimum point.

The RFS policy introduces a large initial fixed cost for the production of miles. This increases the average cost of producing miles and gives a U-shape to the average cost curve. The policy also changes the marginal cost to be a function only of the additional gasoline needed to produce more miles; that marginal cost curve intersects the minimum of the new average cost curve. The altered AC_{RFS} and MC_{RFS} are also shown in Fig. 14.1. If, as discussed earlier, some level of miles $\bar{M}$ exists beyond which the RFS policy is no longer binding, the average cost and marginal cost curves with and without the policy will coincide beyond that point.

As we see in Fig. 14.1, for some range of miles over which the policy is binding, the marginal cost of miles lies below the average cost of miles. If the demand curve for miles were to lie in this left-most part of the graph, marginal-cost pricing of miles would cause the blenders of gasoline and ethanol to earn negative profits; hence, price will not equal marginal cost in that range of miles. We assume the market price of fuel will be the weighted average of the prices of the gasoline and ethanol in the blended fuel. With average-cost pricing in the market for motor fuel, the consumer price of a mile will be the average cost of that mile. However, where the MC_{RFS} exceeds the AC_{RFS}, we assume that conventional marginal-cost pricing for miles will prevail.

Characteristics of market outcomes under this policy are highly sensitive to the exact circumstances of supply and demand curves for miles and motor fuels. We rely on numerical simulations to reveal the effects of the RFS policy under plausible representations of these markets. However, some general observations can be made. First, the RFS distorts production such that total private production costs are not minimized. This policy might reduce the carbon-related externality cost of fuel relative to status quo because it forces a larger fraction of fuel to be comprised of less carbon-intensive ethanol; however, the effect on total carbon emissions will also depend on whether the final market price of miles goes up or down as a result of this policy. The RFS does not address externality costs associated with miles and not carbon (e.g., congestion, accidents), and the RFS is not flexible enough to permit the socially optimal mix of fuels to emerge. Hence, it is unlikely that the outcome under the RFS policy will achieve the first-best level of welfare. However, it is difficult to evaluate (without numerical simulation) what the welfare effect of the RFS will be relative to the status quo. For example, AC pricing could yield a very high price of miles and cut driving in the United States, which would reduce deadweight loss under the status quo from having too many miles.

14.5 Numerical Analysis – Methods and Parameters

We seek to model market outcomes in 2015 when the new RFS has reached an important milestone: 15 billion gallons of corn ethanol must be used in that year, while the cellulosic ethanol mandate is 5.5 billion gallons. We assume constant elasticity supply curves for gasoline, corn ethanol, and cellulosic ethanol and a constant elasticity demand curve for miles. We assume that the price elasticity of corn ethanol supply is equal to 0.75 based on Rask (1998). We then assume that the supply curve for cellulosic ethanol is shifted upward by $1 relative to the supply curve for corn ethanol in 2015, and that price elasticity of supply of cellulosic ethanol is 0.75. We use a baseline price elasticity of gasoline supply equal to 0.25 (Rajagopal et al. 2007) and assume a fuel-price elasticity of miles driven of −0.4 (Johansson and Schipper 1997).

To parameterize supply and demand for that future year, we first parameterize and calibrate a base model for 2006, and then assume that demand for miles shifts out 2.5% each year after that.[6] We also assume that the supply curve for corn ethanol

---

[6]See http://www.fhwa.dot.gov/environment/vmt_grwt.htm#4foot

shifts out in 2015 relative to the curves that exist in 2006, though we assume no change in the supply curve for gasoline. To determine the size of the shift in corn ethanol supply we assumed that the capacity of corn ethanol production in 2015 will be about 13 billion gallons (existing capacity plus capacity of plants under construction). The Federal Highway Administration (FHWA) reports that total vehicle-miles traveled in 2006 were 3,014 billion. The Energy Information Administration (EIA) reports there were 141.8 billion gallons of gasoline, while the Renewable Fuels Association (RFA) reports there were 5.5 billion gallons of ethanol used in the United States in 2006. The EIA reports that average retail price of gasoline that year was $2.63 per gallon. We calculate the retail price of ethanol as the wholesale rack price plus taxes and a $0.30 per gallon markup, yielding $2.76 per gallon in 2006.[7] The FHWA shows that motor fuels are subject to a federal excise tax of $0.184 per gallon, with state taxes on both ethanol and gasoline at around $0.20 per gallon.

The largest and most stable policy to support ethanol production has been an excise tax credit of $0.51 per gallon, yielding a net tax rate on ethanol that is negative. We assume that this tax credit falls to $0.45 per gallon in 2015 as proposed in the recent Farm Bill. The external cost associated with congestion, accidents, and noncarbon pollution is assumed to be $0.085 per mile, and the externality cost of carbon emissions is $25 per metric ton (Parry and Small 2005). While studies vary in their estimates of the energy per gallon and CO_2 emissions per unit of energy, we use estimates by Farrell et al. (2006) to describe ethanol production today to estimate the emissions intensity of gasoline, corn ethanol, and cellulosic ethanol as 0.0032, 0.0017, and 0.0002 tons of carbon per gallon, respectively. Emissions of pollutants other than CO_2 are not uniformly lower for ethanol than for gasoline (Hill et al. 2006). Hence, we make the simplifying assumption that costs associated with non-CO_2 emissions are the same for ethanol and gasoline.

We assume that the elasticity of substitution between gasoline and ethanol in the production of miles is 2 and calibrate the model to find the other parameters of the production function, γ and α, that are consistent with baseline miles, gasoline, and ethanol use. Results using the parameters described above are in Table 14.1. Table 14.2 presents results from simulations that assume a much higher price elasticity of gasoline supply equal to 10. Table 14.3 demonstrates the sensitivity of our results to other key parameters – the elasticity of substitution between ethanol and gasoline and the vertical intercept of the cellulosic ethanol supply curve.

14.6 Results

We first simulate the production levels and prices of each of the fuels under the status quo policy (with a $0.387/gallon fuel tax and a $0.51/gallon tax credit for ethanol) if maintained in 2015. Results in Table 14.1, column A show that if a cellulosic

---

[7] www.neo.ne.gov/statshtml/66.html.

Table 14.1 Results of numerical analyses for gas supply elasticity = 0.25

	Status quo policy A	Optimal policy intervention B	Current mandate, no tax credit C	Current mandate with tax credit D	Optimal mix mandate with tax credit E	Tax credit to achieve aggregate mandate F
Policy interventions						
Gas tax ($/gal)	.387	0	.387	.387	.387	.387
Cellulosic ethanol tax ($/gal)	−.123	0	.387	−.623[a]	−.623[a]	−2.232[c]
Corn-ethanol Tax ($/gal)	−.123	0	.387	−.063[b]	−.063[b]	−1.672[c]
Miles tax ($/mile)	0	.025	0	0	0	0
Emissions tax ($/Kg)	0	.085	0	0	0	0
Prices						
Consumer price of miles ($/mile)	.219	.253	.218	.217	.217	.213
Producer price of gas ($/gal)	4.36	3.56	4.10	4.12	4.12	4.27
Producer price of corn ethanol ($/gal)	3.29	2.56	4.32	4.32	4.47	4.42
Producer price of cellulosic ethanol ($/gal)	3.29	2.60	5.31	5.31	4.80	4.98
Quantities consumed						
Miles (bill)	3,726	3,513	3,732	3,737	3,738	3,768
Gasoline (bill gals)	161	153	159	159	159	160
Corn ethanol (bill gals)	12.2	10.1	15.0	15.0	15.4	15.3
Cellulosic ethanol (bill galls)	3.8	3.2	5.5	5.5	5.1	5.2
Ethanol/gasoline (%)	9.08	8.02	11.5	11.4	11.4	11.4
Environmental and welfare effects[d]						
ΔConsumer surplus ($bill)	80.9	0	83.1	84.6	84.9	95.3
ΔProducer surplus – gasoline($ bill)	125.6	0	83.9	87.2	87.7	110.9
ΔProducer surplus – corn ethanol ($ bill)	8.2	0	22.2	22.2	24.5	23.7
ΔProducer surplus – cellulosic ethanol ($bill)	2.5	0	11.9	11.9	9.2	10.1
ΔTax revenue ($bill)	−251.0	0	−242.0	−247.4	−247.7	−286.6
Carbon emissions (mmt)	536	507	533	534	535	539
ΔExternal costs ($bill)	−18.8	0	−19.3	−19.7	−19.8	−22.4
ΔTotal welfare ($bill)	−52.7	0	−60.1	−61.2	−61.1	−69.0

[a]This tax is equal to $.387/gallon minus a $1.01/gallon tax credit.
[b]This tax is equal to $.387/gallon minus a $.45/gallon tax credit.
[c]These subsidies induce use of 20.5 billion gallons of ethanol using the welfare-maximizing mix.
[d]The benchmark for changes is the optimal policy scenario.

Table 14.2 Results of numerical analyses for gas supply elasticity $= 10$

	Status quo policy A	Optimal policy intervention B	Current mandate, no tax credit C	Current mandate with tax credit D
Policy interventions				
Gas tax ($/gal)	.387	0	.387	.387
Cellulosic ethanol tax ($/gal)	−.123	0	.387	−.623[a]
Corn ethanol tax ($/gal)	−.123	0	.387	−.063[b]
Miles tax ($/mile)	0	.085	0	0
Emissions tax ($/Kg of Carbon)	0	.025	0	0
Prices				
Consumer price of miles ($/mile)	.127	.196	.139	.134
Producer price of gas ($/gal)	2.34	2.30	2.32	2.33
Producer price of corn ethanol ($/gal)	2.43	1.95	4.32	4.32
Producer price of cellulosic ethanol ($/gal)	2.43	1.99	5.31	5.31
Quantities consumed				
Miles (bill)	4,682	3,894	4,463	4,527
Gas (bill gals)	205	173	192	195
Corn ethanol (bill gals)	9.74	8.28	15.0	15
Cellulosic ethanol (bill galls)	3.06	2.63	5.5	5.5
Ethanol/gasoline (%)	5.86	5.94	9.65	9.52
Environmental and welfare effects[d]				
ΔConsumer surplus ($bill)	187	0	152	166
ΔProducer surplus – gasoline($ bill)	7.6	0	4.5	5.2
ΔProducer surplus – corn ethanol ($ bill)	4.3	0	27.8	27.8
ΔProducer surplus – cellulosic ethanol ($bill)	1.2	0	13.7	13.7
ΔTax revenue ($bill)	−267	0	−263	−284
Carbon emissions (mmt)	674	567	641	650
ΔExternal costs ($bill)	−65.1	0	−50.2	−55.9
ΔTotal welfare ($bill)	−132	0	−115	−127

[a]This tax is equal to $.387/gallon minus a $1.01/gallon tax credit.
[b]This tax is equal to $.387/gallon minus a $.45/gallon tax credit.
[c]These subsidies induce use of 20.5 billion gallons of ethanol using the welfare-maximizing mix.
[d]The benchmark for changes is the optimal policy scenario.

ethanol technology were available in 2015 at a cost of $1/gallon higher than that of producing corn ethanol, then it would be profitable to produce 3.8 billion gallons of it and 12.2 billion gallons of corn ethanol. In the absence of a mandate, producers would produce the two types of ethanol, such that their marginal cost of production is equalized (given our assumption that the two types of ethanol are perfect substitutes for each other). The producer price of ethanol is thus $3.86 per gallon ethanol while that of gasoline is $4.36 per gallon of gasoline (Note that gasoline and ethanol are imperfect substitutes in our model; thus the price of ethanol and that of gasoline are not necessarily equal in energy equivalent terms). As compared to the socially

Table 14.3 Differences between socially optimal scenario and actual mandate policy

Outcomes	Baseline	Price elasticity of gas supply =.75	Elasticity of substitution = 5	Added cost of cellulosic = 2
ΔMiles (bill)	223.8	388.2	241.1	227.3
ΔGasoline (bill gals)	5.68	12.1	6.69	5.65
ΔCorn ethanol (bill gals)	4.87	5.85	3.47	4.77
ΔCellulosic ethanol (bill gals)	2.29	2.59	1.85	2.65
ΔCarbon emissions (mmt)	26.8	49.1	27.6	26.5
ΔWelfare ($ bill)	−61	−90.7	−60.6	−63.8

optimal policy, the status quo has higher consumer and producer surplus but greater external costs with the net loss in social welfare amounting to $52.7 billion.

In the socially optimal scenario, the carbon tax and miles tax induce the least cost mix of fuel conservation, fuel substitution, and miles reduction to internalize externalities. The socially optimal quantities of gasoline consumption and miles are about 6% lower than those found under the status quo while ethanol consumption is about 19% lower, despite its potential to reduce greenhouse gas emissions per mile. The reduction in fuel consumption results in considerable reduction in the producer prices of these fuels and thus a reduction in the producer surplus relative to the status quo. Consumers, however, do not benefit from the lower producer prices due to the wedge between the two introduced by the emissions and miles taxes. Both consumer and producer surplus is therefore lower in the socially optimal case. This is, however, more than offset by the higher tax revenues and lower externality costs which result in a net gain in social welfare relative to the status quo.

The ethanol mandate requires consumption of 15 billion gallons of corn ethanol and 5.5 billion gallons of cellulosic ethanol. We simulate that policy with no ethanol tax credits; those results are in column C of Table 14.1. As compared to the status quo policy, the ethanol mandate increases the consumption of ethanol from 16 billion gallons to 20.5 billion gallons. In the absence of any tax credits for ethanol, the mandate increases the per gallon producer price of corn ethanol and cellulosic ethanol relative to the status quo and the socially optimal scenario. The mandate results in displacement of 2 billion gallons of gasoline relative to the status quo. Given the fairly steep gasoline supply curve, this displacement lowers the price of gasoline by $0.26 per gallon (i.e., by 6%). As a result, miles consumed increase by 0.2%. However, the substitution of ethanol for gasoline more than offsets this miles effect and lowers carbon emissions by 6% relative to the status quo. External costs increase (after accounting for the miles externality effects) and total welfare is lower by $7 billion, relative to the status quo.

Column D shows that the provision of ethanol subsidies ($0.45/gallon and $1.01/gallon for corn and cellulosic ethanol, respectively) does not affect the cost of production of either type of ethanol (since the consumption mandate is binding). These subsidies do, however, lower the consumer cost of ethanol and thus marginally lower the cost per mile; that increases gasoline consumption and externality costs and lowers net social welfare relative to a mandate without the tax credit. As compared to the socially optimal case, the mandate with and without the tax credit results in excessive emphasis on substitution of ethanol for gasoline (despite its high costs) and insufficient emphasis on fuel conservation.

The optimal policy may be difficult to implement politically because it entails large taxes on emissions and miles, yielding a price of miles that is 16% higher than the status quo. Therefore, we explore two alternative approaches to achieving the overall mandate of 20.5 billion gallons of ethanol. First, we examine the impact of imposing an ethanol mandate in which the mix of corn and cellulosic ethanol is chosen to maximize welfare given their differential costs and effects on greenhouse gas emissions (see Column E). We find that the optimal quantities (15.4 billion gallons and 5.1 billion gallons of corn and cellulosic ethanol, respectively) are only slightly different from those mandated currently. Choosing an optimal mix of the two types of ethanol to meet the aggregate mandate would lead to a relatively small increase in social welfare (by $0.1 billion) primarily because it lowers the per unit cost of producing cellulosic ethanol which more than offsets the 1-MMT increase in carbon emissions.

Second, we analyze the use of a market-based tax-subsidy instrument (instead of a quantity instrument that mandates 20.5 billion gallons of ethanol) to achieve the same level of aggregate ethanol production (see Column F). With the status quo gasoline tax of $0.383 per gallon of gasoline, per gallon subsidies of $1.67 for corn ethanol and $2.23 for cellulosic ethanol yield the target total amount of ethanol in the welfare maximizing combination of 15.3 billion gallons of corn ethanol and 5.2 billion gallons of cellulosic ethanol. The high subsidy on ethanol lowers the cost of miles to $0.21 per mile instead of $0.25 under social optimality. This raises incentives to drive more and leads to an increase in greenhouse gas emissions and a deadweight loss of $69 billion. In general, Table 14.1 shows that the mandate policy creates a larger deadweight loss than the status quo, and that this loss increases when the mandate is accompanied or replaced by a tax credit for ethanol.

14.7 Sensitivity Analysis

We examine the sensitivity of some of the results in Table 14.1 to a change in the gasoline supply elasticity from 0.25 to 10; see the results presented in Table 14.2. We find three key differences in outcomes. First, when gasoline supply is more elastic, consumer surplus is much lower under the mandate as compared to the status quo than when the gasoline supply is inelastic. This is because the increase in per unit cost of a mile under the mandate policy now results in a very large reduction in the consumption of gasoline (but an insignificant reduction in the price of gasoline (less than 1 cent per gallon)). With an elastic gasoline supply curve, gasoline producer surplus is small under the status quo and somewhat lower under the mandate. However, the mandate also results in a significant reduction in greenhouse gas emissions (5%) relative to the status quo; hence, overall welfare is higher under the mandate policy than in the status quo if gasoline supply is highly elastic.

Second, we find that (even though the mandate is better than the status quo) the deadweight losses under the mandate (with or without the tax credit) are now two or more times higher than in the case with an inelastic gasoline supply curve. This is because the producer surplus of the gasoline suppliers does not increase as much when the gas price increases relative to the socially optimal case. On the other hand, the consumption of gasoline under the mandate is now considerably higher than in socially optimal and thus the externality costs are several times larger than with an inelastic gasoline supply curve.

Third, we find that tax credits with the mandate are even more distortionary with an elastic gasoline supply curve because they cause gasoline consumption to rise by 3 billion gallons. That increase raises externality costs. Social welfare is reduced by $12 billion just as a result of adding tax credits to the mandate policy.

We also examine the sensitivity of our results to the elasticity of substitution between gasoline and ethanol and to the cost of cellulosic ethanol production. Table 14.3 presents selective results comparing the socially optimal and the current mandate policy with tax credits. As compared to the results with our baseline parameters and elasticity of gas supply of 0.25, we find that changes in these two assumptions do not have significant impacts on carbon emissions or net social welfare. On the other hand, a change in the elasticity of gas supply from 0.25 to 0.75 does significantly increase emissions and deadweight losses under a mandate for reasons explained above.

14.8 Conclusions

Scholars and policymakers debate the likely effects of the new sweeping RFS policy that imposes large ethanol quantity mandates in the United States. This paper develops a framework to analyze the welfare and environmental effects of that RFS mandate policy. We compare outcomes under a mandate policy to the situation that results from premandate biofuel policies and the situation that would be produced by a socially optimal policy that internalizes the externalities associated with vehicle miles traveled and fuel use. Numerical simulations for the effect of the mandate in 2015 show that the mandate would lower gasoline price relative to the status quo – the extent of that decrease ranging from $0.02 per gallon to $0.26 per gallon depending on the elasticity of the gasoline supply curve for the parameter range considered here. The mandate raises the cost of ethanol by over a dollar a gallon. However, due to the relatively small share (9–12%) of ethanol in total fuel consumed, the mandate policy has a very small effect on the cost of a mile driven relative to the status quo. The mandate policy lowers greenhouse gas emissions only if the gasoline supply elasticity is large which leads to a slight increase in the cost of a mile; this is sufficient to reduce miles traveled and therefore greenhouse gas emissions. The impact of the mandate on social welfare relative to the status quo is positive (negative) if gasoline supply is highly elastic (inelastic) and the external benefits from a reduction in miles and emissions more than offset the increase in fuels costs due to the mandate.

Our analysis shows that a mandate is a second best policy to reduce environmental externalities. A carbon tax together with a tax on miles that internalizes the externalities associated with miles traveled raises social welfare by $60–$115 billion dollars relative to the mandate. The mandate overemphasizes one strategy to reduce emissions: substitution of ethanol for gasoline irrespective of the cost it imposes on consumers. It also fixes the quantities of the two types of ethanol. We find that the dominant causes of the deadweight losses from the mandate are the aggregate quantity mandated and the failure to force the prices of fuels to reflect their true social costs, rather than the incorrect mix of ethanol types mandated. A tax credit combined with the mandate benefits consumers and gasoline producers; the benefit for gasoline producers increases as the elasticity of the supply of gasoline decreases. The tax credit with the mandate increases miles driven relative to those with a mandate policy only; its negative impact on greenhouse gas emissions ranges from an additional 1 MMT to 9 MMT relative to a mandate, by itself, but at a cost to taxpayers that ranges from $5 to $21 billion depending on the elasticity of gasoline supply. The deadweight loss of the mandate-cum-tax credit relative to the socially optimal is higher with an elastic gasoline supply curve because the mandate-cum-tax credit then leads to a large reduction in gasoline consumption and a loss in fuel tax revenues.

This paper focuses on the implications of biofuel policies for energy markets, abstracting from possible implications of multisectoral interaction effects of biofuel policies. Large-scale production of biofuels may have implications for other markets such as food and feed markets and affect economic efficiency and welfare through these channels as well. In addition, biofuel subsidies could increase the costs of supporting biofuel production in a second-best world where subsidies are financed through distortionary taxes. Future research on ethanol policy would do well to pay attention to such general-equilibrium effects.

However, our partial-equilibrium model yields striking findings that highlight the importance of more careful debate about ethanol policy. We do find that exact estimates of the effects of ethanol mandates on greenhouse gases and social welfare (relative to the status quo) are highly dependent on the gasoline supply elasticity, but the qualitative story is robust. The mandate policy is far from optimal, with welfare costs (relative to the socially optimal policy) ranging from $60 B to $115 B depending on the elasticity of gasoline supply. The provision of a tax credit in addition to the mandate leads to larger deadweight losses that range from $1.1 B with a gasoline supply elasticity of 0.25 to $12 B with a gasoline elasticity of 10. An ethanol mandate policy provides assured demand for ethanol and, therefore, supports the domestic ethanol industry, particularly the cellulosic biofuel industry. However, such policy may harm the well-being of the country as a whole, even relative to the ethanol support policy that was in place before the current mandate was passed.

References

Austin D and Dinan T (2005) "Clearing the air: The Costs and Consequences of Higher CAFE Standards and Increased Gasoline Taxes." J Environ Econ Manage 50: 562−582.

de Gorter H and Just DR (2009) "The Welfare Economics of a Biofuel Tax Credit and the Interaction Effects with Price-Contingent Farm Subsidies." Am J Agr Econ 91(2), May.

de Gorter H and Just DR (2008) "The Economics of the U.S. Ethanol Import Tariff with a Blend Mandate and Tax Credit" J Agr Food Industrial Org 6(2), December: 1−21.

Farrell A E, Plevin RJ, Turner BR, Jones AD, O'Hare M, Kammen DM (2006) "Ethanol Can Contribute to Energy and Environmental Goals" Science 311:506−508.

Gallagher P, Shapouri H, Price J, Schamel G, Brubaker H (2003) "Some Long-Run Effects of Growing Markets and Renewable Fuel Standards on Additives Markets and the US Ethanol industry." J Pol Model 25 (6−7): 585−608.

Gardner B (2007) "Fuel Ethanol Subsidies and Farm Price Support." J Agr Food Industrial Org 5(2) 2007: 1−20.

Hill J, Nelson E, Tilman D, Polasky S, Tiffany D (2006) "Environmental, economic, and energetic costs and benefits of biodiesel and ethanol biofuels." Proc Natl Acad Sci 103(30): 11206−11210.

Johansson O and Schipper L (1997) "Measuring the Long-Run Fuel Demand of Cars: Separate Estimations of Vehicle Stock, Mean Fuel Intensity, and Mean Annual Driving Distance." J Transp Econ Pol 31(3): 277−292.

Khanna M, Ando A, Taheripour F (2008) "Welfare Effects and Unintended Consequences of Ethanol Subsidies." Rev Agr Econ Fall 30(3): 411−421.

Martin A (2008) "Fuel Choices, Food Crises and Finger-Pointing." New York Times, April 15, 2008.

Parry I and Small K (2005) "Does Britain or the United States Have the Right Gasoline Tax?" Amer Econ Rev 95(4): 1276−1289.

Rajagopal D, Sexton SE, Roland-Holst D, Zilberman D (2007) "Challenge of Biofuel: Filling the Tank Without Emptying the Stomach?" Environ Res Lett 2: 1−9.

Rask K (1998) "Clean air and renewable fuels: the market for fuel ethanol in the US from 1984 to 1993." Energy Econ 20(3): 325−345. ·

Searchinger TR, Heimlich R, Houghton RA, et al. (2008) "Use of U.S. Croplands for Biofuels Increases Greenhouse Gases through Emissions from Land-Use Change." Science 319 (5867): 1238−1240.

Taheripour F and Tyner W (2008) "Ethanol Subsidies, Who Gets the Benefits?," in Joe Outlaw, James Duffield, and Ernstes (eds), Biofuel, Food & Feed Tradeoffs, Proceeding of a conference held by the Farm Foundation/USDA, at St. Louis, Missouri, April 12−13 2007, Farm Foundation, Pak Brook, IL, 91−98.

Tyner W and Taheripour F (2008) "Policy Analysis for Integrated Energy and Agricultural Markets in a Partial Equilibrium Framework," Paper Presented at the Farm Foundation Conference, February 12−13, 2008, Atlanta, GA.

US EPA (2007) Renewable Fuel Standard under Section 211(o) of the Clean Air Act as Amended by the Energy Policy Act of 2005. Federal Register 72(227): 66171−66173.

Vedenov D and Wetzstein M (2008) "Toward an Optimal U.S. Ethanol Fuel Subsidy," Energy Econ 30 (5), September; 2073−2090.

Yacobucci B and Schnepf R (2007) Selected Issues Related to the Expansion of the Renewable Fuel Standard (RFS). CRS Report RL34265.

Chapter 15
US–Brazil Trade in Biofuels: Determinants, Constraints, and Implications for Trade Policy

Christine Lasco and Madhu Khanna

Abstract This chapter compares the cost and greenhouse gas (GHG) mitigation benefits of corn ethanol in the United States relative to sugarcane ethanol produced in Brazil and develops a stylized model to analyze its implications for the impact of US biofuel policies on social welfare and GHG emissions. The policies considered here include the $0.51 per gallon blender's subsidy for ethanol and the import tariff of $0.54 per gallon on sugarcane ethanol. Our analysis shows that the combined subsidy and tariff policy decreases welfare by about $3 B depending on assumptions about the extent of market power the United States has in the world ethanol market. These policies also provide negligible (in some cases negative) benefits in the form of GHG reduction. The results indicate that the United States would gain from removing domestic and trade distortions in the ethanol market. Increasing ethanol demand in the world market will entail expansion of Brazil's ethanol industry. We briefly discuss concerns about the environmental impacts of this expansion.

15.1 US–Brazil Trade in Biofuels: Determinants, Constraints, and Implications for Trade Policy

With growing global concern about energy security and climate change, biofuels have been viewed in many countries as a solution for reducing dependence on foreign oil, enhancing energy security, supporting rural economic development, and mitigating GHG emissions associated with fossil fuels. Biofuels production has grown rapidly since 2003 in the OECD countries and Brazil when oil prices began

C. Lasco (✉)
Department of Agricultural and Consumer Economics, University of Illinois, Urbana-Champaign, IL, USA
e-mail: mlasco2@illinois.edu

M. Khanna et al. (eds.), *Handbook of Bioenergy Economics and Policy*,
Natural Resource Management and Policy 33, DOI 10.1007/978-1-4419-0369-3_15,
© Springer Science+Business Media, LLC 2010

to rise above $25/barrel in 2003 to over $100 per barrel in early 2008. World production of ethanol tripled from 5 billion gallons to 16 billion gallons while biodiesel production grew 10-fold from 0.260 billion gallons to 2.37 billion gallons between 2001 and 2007 (FAO, 2008). The US and Brazil produce almost 90% of the global production of ethanol, with the main feedstock being corn in the United States and sugarcane in Brazil. Figure 15.1shows trends in ethanol production of the United States and Brazil.

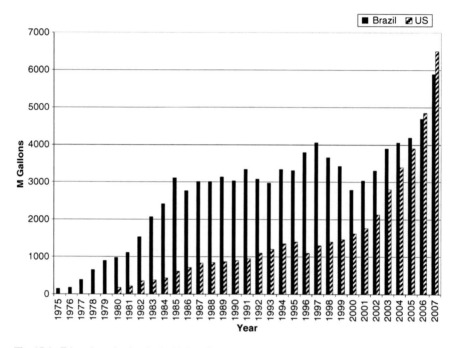

Fig. 15.1 Ethanol production in the United States and Brazil (Source: RFA, FNP)

Trade in biofuels is currently quite small; only 10% of global fuel production is traded with half of these sales consisting of exports from Brazil (de Vera, 2008). In 2007, 30% of Brazilian exports went to the United States, with Japan and the Netherlands as the other two main buyers of ethanol (UNICA, 2008). Demand for biofuels in the OECD countries is expected to expand significantly in the coming years primarily due to government mandates on the consumption of biofuels. This would most likely necessitate more ethanol trade due to the fact that most OECD countries do not have the capacity to produce enough biofuels to support their mandates. In addition, feedstocks like sugarcane, which can be converted to fuels with greater energy efficiency and at lower costs, are grown in tropical and subtropical countries like Brazil and not in OECD countries where biofuel consumption is currently being promoted (see de Vera, 2008).

Current US biofuels policy has, however, focused on promoting domestic production of biofuels to meet the mandated levels of ethanol consumption. To support the ethanol industry, the government has provided a tax credit of $0.51 per gallon of ethanol blended with gasoline. The tax credit together with the $0.38 per gallon fuel tax on both gasoline and ethanol implies that there is a net subsidy of $0.23 per gallon on ethanol on a volumetric basis. Imports are restricted through an ad valorem tariff of 2.5% and a specific tariff of $0.54 per gallon. The $0.54 tariff is meant to offset the $0.51 subsidy that would otherwise benefit the ethanol exporters as well. From an economic efficiency and a greenhouse gas (GHG) mitigation perspective this import policy is questionable. Brazil has relatively low costs of production of biofuels and considerable available land to expand production. Sugarcane ethanol also has a more favorable energy balance and lower GHG emissions than corn-based ethanol (Macedo et al., 2008). On the other hand, an import tariff could improve the terms of trade for the United States if it has market power in the world ethanol market.

The objective of this chapter is to analyze the implications of current US biofuels policy for fuel consumption, prices, GHG emissions, and social welfare[1] in the United States. We undertake this analysis by extending the framework developed in Khanna et al. (2008) to an open economy and allowing for the possibility that the United States could be a price taker or a price setter in the world ethanol market. In Section I, we briefly describe the development of the ethanol industry in Brazil and compare per gallon production costs and GHG emissions of sugarcane ethanol with those of corn ethanol in the United States. We also discuss the related literature analyzing the economic effects of US biofuel policies. Section II graphically illustrates the welfare effects of a biofuel tariff and subsidy policy and describes the data and assumptions underlying our numerical analysis. Section III presents results of the numerical simulation of the welfare and GHG impacts of US biofuel policy. Section IV concludes with a discussion of the expansion potential of Brazil's ethanol industry and concerns about its environmental implications.

15.2 Background

Brazil has been using ethanol mixed with gasoline as fuel since the 1930s when ethanol was mostly a byproduct of sugar production. Production of sugar and ethanol was controlled by the government to stabilize sugarcane prices. In 1975, Brazil instituted the Brazilian National Alcohol Program (PROÁLCOOL) in response to rising oil prices due to instability in the Middle East, low sugar prices, and surplus production of sugar. The goal of PROÁLCOOL was to decrease dependence on foreign imports of oil by encouraging the production and use of alcohol as fuel. The government encouraged the use of ethanol through a mix of incentives for producers and consumers. First, it mandated a 20% blend of ethanol in motor fuels. Second, it provided cheap capital for the establishment of autonomous ethanol

---

[1] Welfare refers to social surplus defined as the sum of consumer and producer surplus, and government revenue.

mills (mills not connected to sugar mills) that are dedicated to producing only ethanol. Funding was also made available for the development of pure ethanol cars, for infrastructure to make ethanol available at the pump, and for expansion of the sugarcane industry. Third, the consumption of ethanol was encouraged by keeping its prices competitive with gasoline. The government sets gasoline prices at levels where the ethanol price is about 59% of the price of gasoline. This pricing mechanism makes ethanol competitive even if its energy content is one-third lower than that of gasoline[2] (Rask, 1995; Papageorgiou, 2005). As a result of PROÁLCOOL, ethanol became widely available, and because it was cheaper than gasoline, demand for it increased dramatically. In 1999, the government started to liberalize sugar and ethanol markets by phasing out subsidies and allowing the price of ethanol to fluctuate with the market. Presently, government support to the ethanol industry is limited to a blend mandate of 20–25% ethanol in motor fuels. Unlike the US, there is a net tax on ethanol although the per gallon tax on ethanol is about 30% lower than the per gallon tax on gasoline. In 2003, flex fuel vehicles (FFVs) that can run on any blend of ethanol and gasoline were introduced. The availability of FFVs, together with rising gasoline prices greatly expanded the demand for ethanol in the past several years. In 2007, 90% of vehicle sales in Brazil were FFVs (ANFAVEA, 2008).

Brazil has a comparative advantage in the production of ethanol relative to the US. OECD reports that the cost of ethanol production[3] in Brazil in 2004 prices was $0.83 per gallon, while in the US it was $1.09 per gallon (assuming $2.1 per bushel corn price) (von Lampe, 2006). This implies that production cost in Brazil was 24% lower than in the US. After including transportation cost of $0.16 to US ports, ethanol from Brazil is about 9% cheaper than US ethanol (Tokgoz and Elobeid, 2006).

However, the extent to which the cost of sugarcane ethanol (from Brazil) at US ports is lower than that of corn ethanol depends critically on several factors, such as the cost of corn and sugarcane, the cost of transportation and the exchange rate. Recent data collected by us show that in 2007 prices, the production cost for a gallon of ethanol in Brazil was about $1.64, while the cost of transporting ethanol from Brazil to the US was about $0.42, which brings total cost at US ports to $2.06 per gallon (assuming an exchange rate of US $1 = 1.95 Reals). Compared to the US wholesale ethanol price of $2.24, Brazilian ethanol is about 8% less costly. The lower cost of ethanol production in Brazil could be attributed to higher ethanol yields per hectare and lower feedstock cost per gallon of ethanol. Ethanol yield per hectare (ha) in Brazil is about 1,725 gallons of ethanol while in the United States, the yield of corn ethanol per ha is about 1000 gallons. Higher ethanol yield follows from a higher feedstock yield of 75 tons of sugarcane per ha based on annualized values for a 6-year cycle, compared to 10 tons of corn per ha in the United States. Additionally, sugarcane plantations in Brazil are usually owned by the ethanol mill,

---

[2]Note that the alternative gasoline fuel also contains 20–25% ethanol so that pure ethanol contains 71–73% of the energy content in the alternative gasoline-ethanol blend.

[3]This estimate includes cost of feedstock and processing, without capital expenses.

which means that the feedstock is supplied to the mill at cost. This is unlike the case in the United States where many mills have to buy corn at market price.

GHG emissions from sugarcane ethanol are less than those of corn ethanol. An early study, by Oliveria et al. (2005), reports that GHG emissions from the manufacture of sugarcane ethanol are about 1.22 kg CO_2-eq per gallon.[4] Smeets et al. (2008) review several studies on GHG emission estimates including the study by Oliveria et al. (2005) and argue that Oliveria et al. (2005) underestimated diesel use in sugarcane production. Using data from Macedo et al. (2004), Smeets et al. (2008) provide updated estimates of GHG emissions in Sao Paulo, Brazil. Their estimates for life cycle GHG emissions are 1.42–1.5 kg CO_2-eq per gallon,[5] with the variation coming from assumed values of sugarcane yield, ethanol yield, and the amount of electricity generated by ethanol mills from bagasse. Depending on the technology, emissions can range from 1.88 kg CO_2-eq per gallon (worst case, no electricity cogeneration) to 1.07 kg CO_2-eq per gallon (best case, electricity cogeneration). Macedo et al. (2008) estimate GHG emissions in selected mills in Sao Paulo, Brazil. They report total life cycle emissions of 1.65 kg CO_2-eq per gallon of anhydrous ethanol.[6] The studies above include emissions from agricultural production of sugarcane, as well as ethanol processing in the mill. GHG emissions reported above do not include emissions from land use change and soil carbon sequestration. According to Macedo et al. (2008), more than 60% of the emissions from production of sugarcane come from soil emissions (33%), trash burning (19%), and fertilizer production (11%). The range of estimates for life cycle GHG emissions of corn ethanol in the United States is 3.6–6.02 kg CO_2-eq per gallon of corn ethanol (Farrell et al., 2006; Wu et al., 2007; Liska et al., 2009). Based on the most recent estimate by Macedo et al. (2008) and the range of estimates provided by the above studies on GHG emissions from corn ethanol, GHG emissions from sugarcane ethanol in Brazil are about 54–72% lower than those from corn ethanol in the United States.

Sugarcane production (hence, sugar and ethanol) in Brazil is primarily located in the center-south region of the country. The north-northeast areas were once the center of sugarcane production is Brazil, but in the mid 1950s, the fertile flat lands of the center-south attracted sugarcane production; production in this region has continued growing while production in the north-northeast has remained flat. In 2007, 7.8 M ha of sugarcane was harvested in Brazil, with over 50% of sugarcane going into ethanol production. The center-south accounted for 87% of total production, most of which (71%) was from the state of Sao Paulo (UNICA, 2008).

Future expansion of ethanol and sugarcane production is expected to occur in Sao Paulo and its surrounding states because of its favorable growing environment and

---

[4] Value converted from original source assuming 80 metric tons of ethanol per ha and a ton = 32 gallons.

[5] Actual numbers reported are 376–396 CO_2-eq per cubic m with a range of 498 (worst case) to 282 (best case) CO_2-eq per cubic m of ethanol. Figure converted from original source assuming 1cubic m = 264.1 gallons.

[6] Actual number reported is 436 kg CO_2 eq per m3. Figure converted from original source assuming 1m3=264.

flat terrain, which makes mechanical harvesting possible. As of early 2008, there are about 350 mills and distilleries in Brazil producing both sugar and ethanol, with 20–30 more mills that are expected to start production in the next year (MAPA, 2008). Total area devoted to sugarcane plantations is expected to expand from the current 7.8 M ha to 14 M ha in 2020 while production is expected to reach 1 B tons, more than double the 2007/2008 level of 487 M tons (FNP, 2008). Brazil has vast amounts of land available to accommodate this expansion. Of Brazil's total land area of 850 M ha, 55% is under forests, 35% is pasturelands, and 7% is currently under agricultural production. Sugarcane plantations account for only 0.06% of the total land area or 2.3% of the total arable land (102 M ha) that is available for sugarcane expansion.

Cost of production, expansion capacity in Brazil, and share of US exports in the world market influence the position and slope of the excess supply curve of ethanol from Brazil; the latter in turn determines whether or not the United States is a price taker in the world market. The United States is the primary buyer of Brazilian ethanol exports; over 50% of exports went to the United States in 2005 while 30% of exports went to it in 2006 and 2007 (FNP, 2008). In the long run, Brazil's ethanol sector would probably be able to increase the number of ethanol plants and the area under sugarcane production, thus developing the capacity to accommodate a large quantity of import demand. In this case, Brazil's excess supply curve would be perfectly elastic and the United States would be a price taker. However, in the short run, there are likely to be capacity constraints so that the quantity of US import demand would be able to influence the world price.

15.2.1 Related Literature

Several studies have examined the welfare effects of biofuel policies, particularly a tax credit (Rajagopal et al., 2007; de Gorter and Just, 2008; Taheripour and Tyner, 2008; Gardner, 2007, Khanna et al., 2008) and a Renewable Fuels Standards (Gallagher et al., 2003; de Gorter and Just, 2007, 2008 and Tyner and Taheripour, 2008). The findings of these studies are being reviewed in Chapter 14 by Ando et al. of this book. Only a few studies have analyzed the market impacts of the ethanol tax credit and the import tariff. de Gorter and Just (2007) examined the impact of the tax credit and tariff in the presence of an ethanol use mandate. They show that the impact of removing the $0.51 per gallon ethanol tax credit and (or) the $0.54 per gallon import tariff is different when an ethanol mandate is binding and when it is not binding. Under a nonbinding mandate, a tax credit for ethanol is a subsidy to producers while under a binding mandate the tax credit is a subsidy to consumers. If the mandate in not binding, removing the tariff and the tax exemption increases imports by 7% relative to the baseline in which both policies are in place. Elobeid and Tokgoz (2008) also analyzed the effect of the subsidy and the tariff using a multimarket international ethanol model. They found that removing both the subsidy and the tariff decreases production and consumption by 10 and

2%, respectively. This leads to more than doubling of imports, all of which come from Brazil.

There are some key differences between this chapter and those described above. First, we define a miles production function where gasoline and ethanol are imperfect substitutes. This is unlike de Gorter and Just (2007) who define gasoline and ethanol as perfect substitutes and Elobeid and Tokgoz (2008) who define their relationship as dominant complements. By deriving demand for ethanol and gasoline from the demand for vehicle miles traveled, we can consider the effects of ethanol policy on the demand for miles and its feedback effects on the demand for gasoline and ethanol. Secondly, we consider the implications of the assumed trading relationship between the United States and Brazil. As discussed earlier, depending on the rate of expansion of Brazil's ethanol production relative to world import demand, the United States could be a price taker in the world ethanol market, or it can exert considerable influence on ethanol prices.

15.3 Welfare Effects of Biofuel Policies in the United States

15.3.1 Conceptual Framework

We now present a graphical analysis of the market impacts and welfare implications of trade restrictions on ethanol imports while recognizing the presence of an ethanol tax credit. The policies analyzed are a $0.51 per gallon subsidy for blending ethanol with gasoline and a $0.54 per gallon tariff on imported ethanol. We do not consider renewable fuel mandates in this analysis since the mandate is not binding in 2006.

In Fig. 15.2, the US market, with domestic supply (S_E^0) and demand (D_E^0) curves for ethanol, is on the left panel while the world market is on the right panel. We assume that the United States and Brazil are the only countries trading ethanol. The United States is a net importer while Brazil is a net exporter. The excess demand curve for the United States (ED^{US}) is the difference between the domestic demand and supply and is the total demand in the world market, while the excess supply curve for Brazil (ES^B) is the total supply of ethanol to the world market. Figure 15.2 shows the effect of a subsidy only. A subsidy (σ) shifts the domestic demand to the right by the size of the subsidy. This decreases the consumer price to (P_σ^D) and increases the producer price to (P_σ^S), with the difference in P_σ^D and P_σ^S being σ. The shift in the demand curve also shifts the excess demand curve to ED^{US}_σ although vertical rise of ED^{US} is less than σ because the increase in ethanol producer price also increases domestic production. The shift in excess demand increases the producer price of ethanol in the world market to P_σ^S, which increases imports to M_σ. This implies that importers also benefit from the domestic subsidy.

The subsidy induces a shift from gasoline toward both types of ethanol, and this substitution effect tends to lower GHG emissions. However, the subsidy also lowers the cost of the blended fuel for consumers below that of nonintervention and this

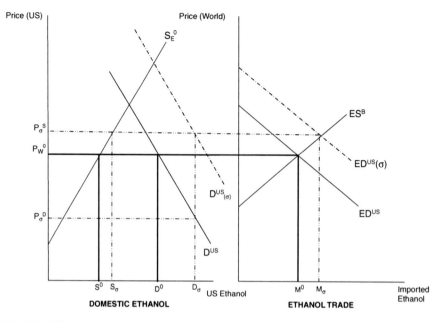

Fig. 15.2 Effect of a subsidy on domestic and world ethanol markets

increases demand for vehicle miles traveled (as in Khanna et al., 2008). The net impact of the subsidy on GHG emissions is, therefore, ambiguous.

If a tariff is imposed together with a subsidy, the benefit to importers is reduced or completely eliminated, depending on the size of the tariff. To keep the discussion simple, we consider only a specific tariff and not an ad-valorem tariff, but this presents no limitation to our analysis. In Fig. 15.3, with both a tariff and a subsidy in place, domestic demand ($D^{US}_{(\sigma)}$) and excess demand ($ED^{US}_{(\sigma)}$) both shift to the right, but there is a wedge the size of the tariff between $ED^{US}_{(\sigma)}$ and ES^B. Depending on the magnitude of the tariff and subsidy, the resulting consumer price in the United States could be higher or lower than the nonintervention price ($P_W{}^0$). As shown, $P_{\sigma\,t}{}^D$ the consumer price after a subsidy and tariff is lower than $P_W{}^0$. Domestic producers receive $P_{\sigma\,t}{}^S = P_{\sigma\,t}{}^D + \sigma$. This leads to an increase in domestic ethanol production from S^0 to $S_{\sigma\,t}$, and an increase in ethanol demand to $D_{\sigma\,t}$. Because of the tariff, ethanol exporters receive only $P_W{}^t$ which decreases imports to ($M_{\sigma\,t}$).

The subsidy and tariff policy induce substitution of corn ethanol for the less carbon-intensive sugarcane ethanol from Brazil. It also raises the price of ethanol above nonintervention levels, and thus induces substitution away from ethanol and toward gasoline. For both reasons, the substitution effect now works toward increasing GHG emissions. However, the increase in the price of domestic ethanol raises the price of the blended fuel to consumers and this will tend to reduce demand for

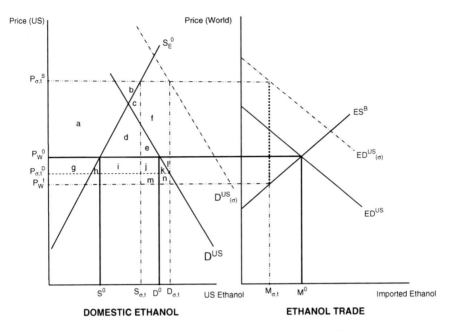

Fig. 15.3 Effect of a subsidy and tariff on domestic and world ethanol markets

miles and GHG emissions. Thus, the net impact of this policy on GHG emissions is ambiguous.

Figure 15.3 also shows the welfare effect for the United States of a subsidy and tariff policy as compared to a nonintervention scenario, where there is no subsidy or tariff. We find that the welfare effect of a tariff and a subsidy in the US market is ambiguous. Consumers gain $(g + h + i + j + k)$ and producers gain $(a + b)$, respectively, from the price change while the government spends $(a + b + c + d + e + f + g + h + i + j + k + 1)$ on subsidies and gets a tariff revenue of $(e + f + j + k + 1 + m + n)$. The net social surplus effect is positive if $(j + k + m + n) - (d + c) > 0$ and negative otherwise. For ethanol exporters, the subsidy received $(f + e)$ is less than tariff payments, which implies that the subsidy and tariff policy is welfare decreasing to ethanol exporters. In the case where the nonintervention price is lower than the domestic price with a subsidy and tariff, this ambiguity in welfare effect remains, although as a result of the subsidy and tariff, consumers will lose from the price increase and producers will have greater gains in producer surplus.

Since the welfare and GHG impacts of a subsidy and tariff are ambiguous, we develop a numerical model to quantify price, quantity, and GHG effects of the subsidy and tariff and nonintervention policy scenarios. We measure the change in welfare of a subsidy-only policy and a subsidy and tariff policy relative to nonintervention. This welfare measure does not include the external damages imposed by increased GHG emissions, as well as increased vehicle miles traveled in the form of congestion, traffic accidents, and air pollution. Inclusion of these external costs would further increase the deadweight loss associated with the subsidy and tariff biofuel policy. For more details of the model, as well as extensions to include valuations of GHG and miles externalities, see Lasco and Khanna (2008).

15.3.2 Empirical Model

In our analytical framework, we assume that consumers derive benefits from the consumption of miles and disutility from aggregate greenhouse gas emissions through its impact on air quality and global warming. The markets in our model are those for corn, domestic ethanol, imported ethanol, refined gasoline, and miles. We include the corn market to account for the effects of changing ethanol demand on the demand and price of corn, and thus on the ethanol supply curve; our welfare measurement is, however, limited to the miles and fuels markets. We use a constant elasticity of substitution (CES) production function for miles with gasoline and ethanol being imperfect substitutes but domestically produced ethanol and imported ethanol being perfect substitutes. The elasticity of substitution between gasoline and ethanol is set to 2. The scale and share parameters of the CES function are calibrated to 2006 market data. The level of GHG emissions is modeled as an additive function of carbon emissions per gallon of each fuel multiplied by the gallons consumed. The marginal cost of miles and the demand for ethanol and gasoline are derived within the model. Imported ethanol demand is defined as the excess demand in the domestic market. The rest of the supply and demand curves are assumed to have constant elasticity forms and are parameterized based on estimates available in the recent literature in this area and market data as explained below.

We assume a corn supply elasticity of 0.25 based on Lee and Helmberger (1985) and an ethanol supply elasticity of 1.5 based on Gallagher et al. (2003). Gasoline supply elasticity is assumed to be 0.25 based on the Department of Energy's estimates of gasoline demand and refining capacity (DOE, 2007). For the supply elasticity of imported ethanol, we use 2.70 as reported in Chapter 19 by Lee and Sumner, in this book. Corn demand elasticity is assumed to be −0.172 (USDA, 2008). The demand elasticity for miles is −0.40 (Parry and Small, 2005; Vedenov and Wetzstein, 2008).

We use 2006 market data to calibrate the model. The price of corn is $3 per bushel, which is the weighted average farm price reported by the USDA (2008). Ethanol and gasoline wholesale prices are $2.58 and $1.94 per gallon, respectively (Nebraska Ethanol Board, 2007). We add a markup of $0.30 per gallon and taxes of $0.38 per gallon to get the retail prices of ethanol and gasoline. In 2006, 12.5 B bushels of corn were produced, 17% (2.12 B bushels) of which went into the production of 4.86 B gallons of ethanol (RFA, 2007; USDA, 2008). RFA also reports that total ethanol imports for the same year were 0.65 B gallons, which brings total demand to 5.51 B gallons. According to the Department of Energy (2007), total gasoline input to motor fuels production was 112 B gallons. The US Federal Highway Authority (2007) also reported that miles driven in 2006 were 2,966 B miles. Emissions intensity of gasoline from "well to wheel" is 3.2 kg of C (kg C) per gallon, while for corn ethanol the value is 1.7 kg C per gallon (Farrell et al. (2006). Macedo et al. (2008) reported that "seed to factory gate" emissions of sugarcane ethanol are 0.44 kg C per gallon. Based on this, we assume that sugarcane ethanol has a "well to wheel" emission intensity of 0.60 kg C per gallon.

15.4 Numerical Simulation Results

15.4.1 Welfare Effects with Market Power in Ethanol Trade

We find that in the case where the United States has some market power, the combined effect of the subsidy and the tariff is to lower the price of ethanol by 3% and increase ethanol consumption by 6% (Table 15.1). This displaces gasoline consumption and its demand decreases by 0.09%, which given the fairly steep supply curve for gasoline lowers the gasoline price by 0.34%. Because of decreased fuel prices, the marginal cost of miles decreases, thus increasing miles consumption in the subsidy and tariff scenario by 0.19% from 2,960 to 2,966 B miles. Most of the increase in fuel use comes from ethanol rather than gasoline. With a subsidy and tariff most of the increased demand for ethanol is met by domestic production which increases by 9% while imports are reduced by 11%. GHG emissions are higher in the subsidy and tariff scenario by 0.08% than nonintervention. Even though ethanol has increased its fuel share (by 1%), the increase in miles traveled brought about by lower fuel prices increases overall fuel use by 0.17% or 207 M gallons. Thus, the benefits of reduced GHG emissions due to the substitution of ethanol for gasoline are more than offset by increased miles and fuels consumption due to the subsidization of fuel. The deadweight loss associated with the subsidy and tariff compared to nonintervention is $3.20 B. These results indicate that the tariff and subsidy are welfare decreasing relative to nonintervention and provide justification for removing the tariff.

We also examine the effects of removing the tariff without removing the $0.51 per gallon ethanol subsidy on welfare, prices, and quantities (see column 3, Table 15.1). We find that keeping the subsidy is even more welfare decreasing than keeping both the tariff and subsidy in place. Welfare would now decrease by $3.47 B compared to $3.20 B when the tariff is implemented together with the subsidy. This is because the removal of the tariff causes tariff revenues to be forfeited, while at the same time increasing the cost of subsidization through inducing a drop in ethanol prices (−6%), which in turn increases total demand to 5.79 B gallons (11% higher than nonintervention). The drop in ethanol prices leads to an increase in miles consumption of 0.36% from 2,960 B to 2,971 B miles, which in turn increases miles externalities. This policy does not change GHG emissions much (0.03%) because the decline in emissions due to substitution of ethanol (domestic and imported) for gasoline offsets the increase in emissions due to greater miles traveled.

Table 15.1 Effect of a subsidy and tariff policy on price, quantities, welfare, and GHG emissions

		Market power			Price taker		
	Unit	Nonintervention (1)	Subsidy and tariff (2)	Subsidy only (3)	Nonintervention (4)	Subsidy and tariff (5)	Subsidy only (6)
Quantity Ethanol							
Domestic supply	B gallons	4.48	4.86 / *8.58*	4.79 / *7.06*	4.36	4.86 / *11.39*	4.36 / *0.00*
Imports	B gallons	0.73	0.65 / *−11.04*	1.00 / *36.10*	1.26	0.65 / *−48.05*	3.70 / *193.95*
Total demand	B gallons	5.21	5.51 / *5.82*	5.79 / *11.15*	5.62	5.51 / *−1.90*	8.06 / *43.38*
Gasoline	B gallons	112.10	112.00 / *−0.09*	111.91 / *−0.16*	111.97	112.00 / *0.03*	111.29 / *−0.61*
Miles	B miles	2,960.29	2,966.00 / *0.19*	2,971.11 / *0.37*	2,967.98	2,966.00 / *−0.07*	3,009.04 / *1.38*
Consumer prices							
Ethanol	$/gallon	2.84	2.75 / *−3.17*	2.67 / *−5.85*	2.72	2.75 / *1.10*	2.21 / *−18.75*
Gasoline	$/gallon	2.63	2.62 / *−0.34*	2.61 / *−0.65*	2.62	2.62 / *0.11*	2.55 / *2.41*
Welfare change	B $		−3.20	−3.47		−3.51	−2.90
GHG emissions	M mt C	366.78	367.07 / *0.08*	366.90 / *0.03*	367.00	367.00 / *0.00*	366.00 / *−0.27*

Note: Numbers in italics are percentage changes.

15.4.2 Welfare Effects with United States as a Price Taker in Ethanol Trade

Unlike the case where the United States has market power, the subsidy and the tariff together slightly increase the price of ethanol by 1% when we assume that the United States is a price taker. This is because the tariff does not decrease the world price of ethanol, since the United States no longer has any influence on the world price (i.e., the slope of Brazil's excess demand curve is flat instead of being upward sloping). This leads to a 2% decrease in ethanol demand and an increase in gasoline demand of 0.03%. Because the prices of both fuels increase, there is a decrease in miles consumption of 0.07%. The result is that GHG emissions are unchanged from nonintervention to the subsidy and tariff case. As expected, imports also decline more significantly (48%) and domestic production increases by a higher percentage (11%) from nonintervention to the subsidy and tariff scenario than in the case with market power.

When the tariff is removed without removing the subsidy, welfare decreases by $3.51 B. This is a larger welfare decrease than in the previous case with market power due to the fact that there are no longer any terms of trade benefits from the tariff. Thus, instead of offsetting the deadweight loss from the subsidy, the tariff further adds deadweight losses to the economy. This is confirmed when we look at the effect of removing the tariff while leaving the subsidy in place. Column 6 in Table 15.1 shows that the welfare decrease from nonintervention to the subsidy case is $2.90 B, which is less than the welfare loss when both the tariff and subsidy were in place.

Our results[7] are fairly consistent with similar studies (de Gorter and Just, 2007; Elobeid and Tokgoz, 2008) in terms of the direction of the price and quantity responses although the magnitude of the impacts in the ethanol market differ considerably. This difference is primarily in the response of ethanol demand to the removal of the subsidy and the tariff. We find that this impact is modest (−6%) and relatively similar to the estimate of Elobeid and Tokgoz (−2%). In contrast, de Gorter and Just report a 90% decrease in ethanol demand with the removal of the subsidy and the tariff. The difference in these results can be explained by the difference in assumptions about the elasticity of substitution between ethanol and gasoline. If the elasticity of substitution is high, we expect a greater response from the ethanol market should ethanol price increase (for example, as we move from the subsidy and tariff to nonintervention). de Gorter and Just assume that gasoline and ethanol are perfect substitutes while Elobeid and Tokgoz assume a dominant complementary relationship. Since we assume that gasoline and ethanol are imperfect substitutes, the magnitude of our result falls between the two.

Our sensitivity analysis shows that the response in the ethanol market is dependent on the assumed relationship between gasoline and ethanol (as demonstrated

---

[7]We use the results where the US has market power since De Gorter and Just and Elobeid and Tokgoz both assume an upward sloping excess supply curve.

above). We find that our results are robust to varying assumptions on other parameter values. Detailed results of this analysis are available from the authors.

15.5 Conclusions and Policy Implications

Our findings show that a combined subsidy and tariff increases ethanol demand and domestic production in the United States, while restricting ethanol imports from Brazil. This creates deadweight losses in the economy and does not help mitigate climate change, since GHG emissions are increased, relative to nonintervention. Moreover, the increase in miles consumption increases miles externalities, which would further reduce welfare. These results suggest that it would be welfare increasing to remove the subsidy and the tariff and to import more ethanol from Brazil. In this analysis, we have considered the effects of these policies on social surplus and GHG emissions. One of the most important rationales for protecting the domestic ethanol industry may be to enhance energy security. In this case, gasoline (which is nonrenewable and largely imported) should be taxed and domestically produced ethanol should be subsidized. Parry et al. (2007) report that a tax of $0.12 would internalize the negative external effect of gasoline consumption related to energy security. In this case, a $0.12 subsidy given to ethanol for replacing gasoline is still much lower than the current subsidy of $0.51, so that our conclusion remains valid, although the welfare gains from removing the subsidy and tariff will be somewhat diminished.

Increased imports from the United States, as well as other countries like the European Union and Japan will necessitate rapid expansion of Brazil's ethanol sector. In terms of available land, Brazil has the capacity to increase sugarcane and ethanol production. However, massive investments are needed. McKinsey & Co. (2007) estimates that roughly $2.36 B in new investment is needed for every 1 B gallon of additional production in Brazil. Investment is needed to not only build mills, but also to improve transportation infrastructure. Brazil is also home to several biodiversity hotspots like the Amazon rainforest, the Pantanal, and the cerrado grasslands, and there is concern that expanding sugarcane production will lead to land use change that will result in increased GHG emissions and loss in biodiversity (Searchinger et al., 2008). Smeets et al. (2008) and Macedo (2005) published a comprehensive report assessing the sustainability of Brazilian ethanol production. Their studies show that there are no major adverse direct impacts of ethanol production. However, indirect impacts on land use change and other environmental indicators are harder to assess.

At the moment, the cost of production of ethanol in Brazil is lower than those in any other country and its environmental impacts are minimal. With decades of experience and vast natural resources, Brazil is poised to supply huge amounts of ethanol to the world market. The United States, as well as other countries could benefit from liberalizing ethanol markets and importing more ethanol from Brazil. However, expanding ethanol production to meet greater world demand presents a tremendous opportunity, as well as a challenge to the Brazilian ethanol and sugarcane industry. The sector must ensure that the growth of its ethanol and sugarcane industries will not be at the expense of worsening environmental quality and undesirable land use change. The long-term success and viability of this industry will depend on its ability to continue increasing efficiency and manage its growth in an environmentally sustainable manner. Future research is needed to quantify the magnitude of indirect impacts and ascertain whether the indirect cost of expanding sugarcane production in Brazil outweighs the benefits of using ethanol as fuel.

References

ANFAVEA (Associação Nacional dos Fabricantes de Veículos Automotores). 2008. Website accessed June 2008. http://www.anfavea.com.br/tabelas2007.html.

de Gorter H and Just DR (2007) "The Economics of US Ethanol Import Tariffs with a Consumption Mandate and Tax Credit." Working paper, Department of Applied Economics and Management, Cornell University.

de Gorter H and Just DR (2008) "The Welfare Economics of a Biofuel Tax Credit and the Interaction effects with Price Contingent Farm Subsidies." Am J Agr Econ 90:in press.

Department of Energy (2007) Website accessed June 2007. http://www.eia.doe.gov/emeu/mer/petro.html.

de Vera ER (2008) "The WTO and biofuels: the possibility of unilateral sustainability requirements." Chic J Int Law 8: 661–680.

Elobeid A and Tokgoz S (2008) "Removing distortions in the US ethanol market: What does it imply for the United States and Brazil?" Am J Agr Econ 90: 918–932.

Farrell A, Plevin R, Turner B, Jones A, O'Hare M, and Kammen D (2006) "Can Ethanol Contribute to Energy and Environmental Goals?" Science 311:506–508.

FAO (2008) "Bioenergy Policy, Markets and Trade and Food Security," Technical Background Document, Food and Agriculture Organization of the United Nations, Rome, HLC/08/BAK/7, June 2008.

Federal Highway Authority (2007) Website accessed June 2007. http://www.fhwa.dot.gov/ohpi.

FNP (FNP Consultoria & Comércio) (2008) Anuário da Pecuária Brasileira. São Paulo Brazil

Gallagher P, Shapouri H, Price J, Schamel G, and Brubaker H (2003) "Some longrun effects of growing markets and renewable fuel standards on additives markets and the US ethanol industry." J Policy Model 25 (6–7): 585–608.

Gardner B (2007) "Fuel Ethanol Subsidies and Farm Price Support" J Agric Food Industrial Org 5(2) : 1–20

IEA (Instituto de Economia Agricola) (2008) Website accessed July 2008. http://www.iea.sp.gov.br/out/index.php

Khanna M, Ando A, and Taheripour F (2008) "Welfare Effects and Unintended Consequences of Ethanol Subsidies." Rev Agr Econ 30 (3) : 411–421

Lasco C and Khanna M (2008) "Biofuels Trade Policy in the Presence of Environmental Externalities." Working paper, Department of Agricultural and Consumer Economics, University of Illinois Urbana Champaign.

Lee D and Helmberger P (1985) "Estimating Supply Response in the Presence of Farm Programs." Am J Agr Econ 67(2): 193–203.

Liska AJ, Yang HS, Bremer VR, Klopfenstein TJ, Walters DT, Erickson GE and Cassman KG (2009) "Improvements in Life Cycle Energy Efficiency and Greenhouse Gas Emissions of Corn-Ethanol." J Ind Ecol 00(0) :1–17.

Macedo IC, Leal MRLV and da Silva JEAR (2004) "Assessment of greenhouse gas emissions in the production and use of fuel ethanol in Brazil." Brazil: Secretariat of the Environment of the State of Sao Paulo, p. 32.

Macedo IC (Editor) (2005) Twelve studies on Brazilian sugarcane agribusiness and its sustainability. UNICA, São Paulo, Brazil.

Macedo IC, Seabra JEA, and Silva JEAR (2008) "Green house gases emissions in the production and use of ethanol from sugarcane in Brazil: The 2005/2006 averages and prediction for 2020." Biomass Bioenerg 32: 582–595.

MAPA (Ministério da Agricultura) (2008) Website accessed June 2008. http://www.mapa.gov.br

McKinsey & Company (2007) Positioning brazil for biofuels success. The McKinsey Quarterly special edition: Shaping a new agenda for Latin America.

NASS (National Agricultural Statistics Service) (2008) Land Values and Cash Rents 2008 Summary. United States Department of Agriculture Sp Sy 3 (08).

Nebraska Ethanol Board (2007) Ethanol and Unleaded Gasoline Average Rack Prices. Website accessed July 2006. http://www.neo.state.ne.us/statshtml/66.html.

Oliveria MED, Vaughan BE, and Rykiel EJ (2005) "Ethanol as fuel: Energy, carbon dioxide balances, and ecological footprint." Bioscience 55(7): 593–602.

Papageorgiou A (2005) Ethanol in Brazil. Working Paper, PRIMEA (European Commission).

Parry I and Small K (2005) "Does Britain or the United States Have the Right Gasoline Tax?" Am Econ Rev 95(4): 1276–1289.

Parry IWH, Walls M and Harrington W (2007) Automobile Externalities and Policies. Discussion Paper, Resources For the Future.

Rajagopal D, Sexton SE, Roland-Holst D, and Zilberman D (2007) "Challenge of Biofuel: Filling the Tank Without Emptying the Stomach?" Environ Res Lett 2: 1–9.

Rask K (1995) "The Social Costs of Ethanol Production in Brazil: 1978–1987." Econ Dev Cult Change 43(3):627–649.

RFA (2007) Renewable Fuels Association Ethanol Industry Overview. Website accessed June 2007. http://www.ethanolrfa.org

UNICA (2008) Website accessed June 2008. http://www.unica.com.br

USDA (2008) USDA Feed Grains Database. Website accessed July 2006. http://www.ers.usda.gov/Data/FeedGrains.

Vedenov D and Wetzstein M (2008) "Toward an Optimal US Ethanol Fuel Subsidy." Energ Econ 30:2073–2090.

Searchinger T, Heimlich R, Houghton RA, et al. (2008) "Use of US croplands for biofuels increases greenhouse gases through emissions from land-use change." Science 319(5867): 1238.

Smeets E, Junginger M, Faaij A, Walter A, Dolzan P, and Turkenburg W (2008) "The sustainability of Brazilian ethanol – An assessment of the possibilities of certified production." Biomass Bioenerg 32(8):781–813

Taheripour F and Tyner W (2008) "Ethanol Subsidies, Who Gets the Benefits?," in Joe Outlaw, James Duffield,, and Ernstes (eds), Biofuel, Food & Feed Tradeoffs, Proceeding of a conference held by the Farm Foundation/USDA, at St. Louis, Missouri, April 12–13 2007, Farm Foundation, Pak Brook, IL, 91–98.

Tyner W and Taheripour F (2008) "Policy Analysis for Integrated Energy and Agricultural Markets in a Partial Equilibrium Framework," Paper Presented at the Transition to a Bio-Economy: Integration of Agricultural and Energy Systems conference on February 12–13, 2008 at the Westin Atlanta Airport planned by the Farm Foundation.

Tokgoz S and Elobeid A (2006) "Policy and Competitiveness of US and Brazilian Ethanol." Iowa Ag Review 12(2) : 6–7,11.

von Lampe M (2006) "Agricultural Market Impacts of Future Growth in the Production of Biofuels." Working Party on Agricultural Policies and Markets, AGR/CA/APM (2005) 24.

Wu M, Wang M, and Huo H (2007) "Life-cycle energy and greenhouse gas emission impacts of different corn ethanol plant types." Environ Res Lett: 2.

Chapter 16
Food and Biofuel in a Global Environment

Gal Hochman, Steven Sexton, and David Zilberman

16.1 Introduction

As the human population and incomes continue to grow while the petroleum reserve is declining and the concerns about environmental damage are increasing, the world community is challenged to supply ever-greater quantities of energy without harming the environment.[1] The world primary energy demand is predicted to more than double between 2006 and 2030, and yet most oil-producing countries will reduce output over that period (IEA 2008). OPEC's share of total oil production will grow from 41 to 51%, and the oil production growth will need to come mostly from the small number of countries that own remaining oil resources. Whether they will make the necessary increases in capacity is uncertain, however, given a number of geopolitical, capital, and national policy constraints (IEA 2008). In response to the challenge posed by energy demand growth, energy supply constraints, and environmental risk, and driven by policies in the developed world, a new energy paradigm is emerging. It is characterized by the development of a relatively clean and renewable bioenergy alternative that not only affects energy prices, but also food prices, environmental preservation, and patterns of trade.

In this chapter, we use a general equilibrium trade model to analyze the economics of renewable energy, particularly bioenergy alternatives, while emphasizing the interactions between energy, food, and the environment. Using the proposed framework, it can be shown that growth, globalization, and capital flows increase demand for energy, leading to a decline in food production and loss of environmental land. The decline in food supply causes food prices to rise. Building on these results, policies to address the greenhouse gas (GHG) impacts of energy production

G. Hochman (✉)
Department of Agricultural and Resource Economics, University of California, Berkeley, CA, USA
e-mail: galh@berkeley.edu

[1]Needless to say, if the "consensus" on anthropogenic climate change is wrong, then oil is not very environmentally damaging; therefore, much of the energy-related environmental concerns are mitigated.

M. Khanna et al. (eds.), *Handbook of Bioenergy Economics and Policy*,
Natural Resource Management and Policy 33, DOI 10.1007/978-1-4419-0369-3_16,
© Springer Science+Business Media, LLC 2010

are discussed, as is the impact of biofuel production on the allocation of land among its three principal uses (food production, biofuel, and nature). Specifically, we derive four policy recommendations: The first two present roles for correctly pricing carbon and land, the third suggests that introduction of biofuels may result in increased food security problems, whereas the fourth emphasizes the importance of technology policies that will increase the productivity of both biofuel and food crops. We show that a tax on GHG emissions may be welfare decreasing if it is not accompanied by a policy to price the benefit of environmental land (i.e., the provision of environmental services). If land use is not properly valued in the marketplace, a carbon tax will induce too much (relative to social optimality) conversion of natural land to production. Given the costs of biodiversity loss associated with land conversion, it may be best to impose no tax if a tax on land conversion is not also introduced.

The remainder of this chapter is divided into three sections. Section 16.2 shows how energy should be incorporated into trade models and discusses its importance in the context of the new energy paradigm. Section 16.3 discusses policy implications of the modeling presented in Section 16.2 and Section 16.4 concludes.

16.2 Trade

Current analysis of biofuel policies in the context of trade (especially the implications of trade protection on domestic biofuel producers) builds on partial equilibrium modeling. Partial equilibrium models are used to explain the implications of import tariffs, given domestic policies that support biofuel production (e.g., de Gorter and Just 2008). Although the normative implications of biofuel policies can be elegantly illustrated using partial equilibrium models, important trade-related issues are overlooked. In particular, the study of international trade focuses on the gains from trade, the pattern of trade, protection, international policy coordination, and international capital markets. Most of these themes cannot be addressed using a partial equilibrium analysis. All are important in the context of bioenergy production. They can be studied using a general equilibrium framework, which is the focus of this chapter.

Countries engage in international trade because they are different from each other. There are several ways to model these differences (e.g., differences in technology or in resource endowments); we chose to model them as heterogeneity in resource endowments. The framework developed in this chapter is, however, robust to more complicated modeling environs.

Modeling differences in resources among countries suggest that a country's comparative advantage is influenced by the resource endowments of other countries, i.e., the relative abundance of production factors and the technology of production, i.e., the relative intensity with which various production factors are used in production of different goods. To illustrate this point, assume a two-country by two-goods by two-factors economy, and further assume country H has a higher ratio of input 1 to input 2 than country F. In other words, country H is abundant in input 1 relative to country F, and country F is abundant in input 2 relative to country H. Furthermore, assume good 1 requires a higher ratio of input 1 to input 2 in production compared to good 2; that is, good 1 is input 1 intensive and good 2 is input 2 intensive. Country H, the input 1-abundant country, therefore, has a comparative advantage in production of good 1. Country F, on the other hand, is input-2 abundant and has a comparative advantage in production of good 2. In this context, countries tend to export goods that are intensive in factors with which the countries are relatively abundantly endowed. For instance, Brazil, a country rich in fertile land for growing sugarcane, exports sugar ethanol to the world. In 2007, Brazil exported 3.5 billion gallons of ethanol, a quantity expected to grow to 4 billion in 2008 (Reuters 2008).

The gains from trade are not distributed evenly. Typically, there are winners and losers. In the factor-abundant model, for example, the owners of a country's abundant factor win from trade, whereas the owners of a country's scarce factor lose from trade. Continuing with our example of Brazil, the owners of the sugar plantations in Brazil win from trade. Typically, those who gain from trade are much less concentrated, informed, or organized than those who lose. This led the United States to levy tariffs on sugar-based ethanol from Brazil – even though consumers can benefit because sugar-based ethanol is not only more efficient than corn-based ethanol but also profitable at lower oil prices (holding biofuel crop prices constant).

This kind of traditional general equilibrium analysis, as we have seen, is capable of providing some insights into the consequences of a new energy paradigm characterized by bioenergy. Given the importance of energy to the world community – its importance is second only to food – it is necessary that these models be extended with a proper understanding of the role of energy in trade (Fig. 16.1).

To understand the implications of emerging energy trends, we need a general equilibrium framework that incorporates household production à la Becker (1965) and Lancaster (1966). After all, households are responsible for a significant share of energy demand (Vandenbergh and Steinemann 2007).[2] The framework should introduce energy as a ubiquitous factor in production and define utility over an untraded convenience characteristic and environmental commodity.

---

[2] A conservative estimate attributes 22% of all US energy consumption to households and 32% of all US (and 8% of all world) carbon emissions to individuals. The amount of energy consumed by autos, motorcycles, and total energy use for transportation data are taken from Table 2.7, "Highway Transportation Energy Consumption by Mode" (*Transportation Energy Data* Book Edition 26-2007, Oak Ridge National Laboratory). The numbers are used to compute the fraction of total transportation energy consumed by residential users (approximately 55%). Total residential energy consumption is determined by summing residential sector consumption and 55% of transportation sector consumption according to Table 2.1a, "Energy Consumption by Sector: 1949–2006" (*Annual Energy Review* 2006, Energy Information Administration). Data were taken for the year 2001. Calculation of the fraction of residential transportation energy consumption excluded light vehicles, which include light trucks, SUVs, and minivans. Therefore, this estimate is believed to provide a lower bound on the fraction of total energy consumed by residences.

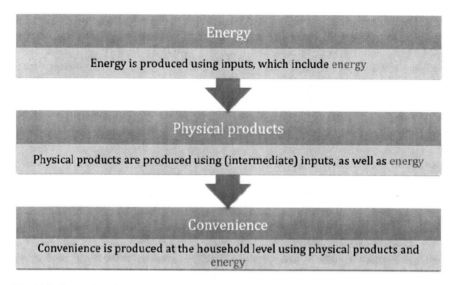

Fig. 16.1 Energy is a ubiquitous factor

To this end, and following Hochman et al. (2008a), we assume consumers derive utility from food, an environmental amenity (henceforth, the environment), and a characteristic called convenience, which is produced within the household in a production function that combines manufactured goods and energy. The household production function produces utility units from capital goods and energy (henceforth, referred to as convenience goods). This latter assumption introduces the consumption of energy at the household level and models energy as a ubiquitous good (e.g., petroleum is used to derive utility units from a car, while energy is also used to produce the car itself). The consumer utility function ranks a collection of characteristics and only indirectly ranks manufactured goods through the characteristics they possess. This structure also introduces a nontraded good, i.e., the convenience good. We assume the environment is increasing in land allocated to the environment (a local and a global public good) and decreasing in GHG emissions (a global public bad).[3]

We employ the Heckscher–Ohlin model as a clear and simple way to introduce household production in a general equilibrium trade framework. We assume a two-country model, with identical constant returns-to-scale production technologies and identical homothetic preferences. Both countries are endowed with land and capital. Land is used for environmental preservation and can be converted to production at a cost. Energy can be produced using either capital and labor or land and labor, using fossil fuel or biofuel technology, respectively. To produce agricultural and capital

---

[3]The environment offers local benefits in the form of open space and recreational opportunities. It also offers global benefits in the form of existence and option values associated with biodiversity.

products, land, capital, and energy are needed.[4] Both countries may use land taxes and emissions taxes to internalize externalities related to land conversion and energy consumption, respectively. Consumers also derive utility from food.

To simplify our exposition, and without loss of generality, we assume that H is capital abundant and F is land abundant, and country endowments are sufficiently similar, i.e., both countries produce both goods in equilibrium, and country H is rich in capital.

This framework can be used to generalize the Stolper–Samuelson theorem (Samuelson 1953) to show that government policies alter factor prices. Therefore, equality of the price of traded goods across countries is not sufficient for factor price equalization. Furthermore, countries that set lower taxes on land, de facto export the land services (Hochman et al. 2008a). In addition, if energy and capital are complements at both the household level and the industrial level, then both the residential demand and the industrial demand for energy increase with trade. To this end, industry demand is expected to surpass transportation demand for energy before 2010 and become the second largest end-use sector, after the combined residual services and agriculture sectors (IEA 2008).

In the analysis presented in Hochman et al. (2008a), the capital and the manufactured goods are assumed to be homogeneous. Switching to energy-efficient capital is therefore impossible. If we introduce heterogeneity, however, these assumptions may not be natural ones, and energy efficiency can be seen to offset the capital-induced increase in energy demand and weaken the complementarity of capital and energy. In periods of high-energy prices, therefore, there exists a strong rationale for investment in energy efficiency. In fact, if some manufactured goods are more energy efficient than others, capital and energy may become substitute inputs. Households, for example, may switch to hybrid cars or cars with smaller engines.[5] At the production level, firms might use more energy-efficient machinery. It is interesting to note that Pindyck and Rotemberg (1983) showed that capital and energy are complements. In Pindyck (1979), however, energy and capital are substitutes.

Growing demand for energy, together with the latest surge in oil prices, has spurred a search for new technologies and yielded a first generation of biofuel technologies. To this end, Brazil has managed to push these changes to extremes and

---

[4]The economic structure builds on work (originated by Samuelson 1953; and extended by Melvin 1968; Drabicki and Takayama 1979; Dixit and Norman 1980; Deardorff 1980; Dixit and Woodland 1982; and many others) that attempts to apply the law of comparative advantage to more than two goods and more than two factors. It generalizes the results of two-by-two theory to many-goods, many-factors models by making the same general assumptions as in the standard two-by-two theory and looking for weaker results (see, for instance, Jones and Scheinkman 1977; Dixit and Norman 1980; and Deardorff 1982). It investigates the accuracy of the classical trade theorems: the Rybczynski theorem, the Stolper–Samuelson theorem, the Heckscher–Ohlin theorem, and the factor price equalization theorem.

[5]Demand for Toyota Prius hybrids increased significantly during the latest surge in oil prices, partly at the expense of SUV and light trucks, which led Toyota to convert light truck production lines in the United States to produce the Toyota Prius hybrids (http://blog.wired.com/cars/2008/06/rising-gas-pric.html).

recently became an exporter of energy (in the 1970s, Brazil imported 75% of its oil). Moreover, in recent years, we also witnessed a surge in production of corn for ethanol, especially in the United States. In 2005, 97 ethanol plants in 21 states produced a record high of 3.904 billion gallons of ethanol – an increase of more than 14% over 2004 and up an incredible 139% from 2000 (Renewable Fuels Association 2006). As of February 2006, the annual capacity of the US ethanol sector stood at 4.3 billion gallons; plants under construction or expansion are likely to add another 2 billion gallons.

According to our model, growth in capital raises the aggregate supply of the manufactured good and energy and reduces aggregate supply of food (the price of food consequently rises). This conclusion is consistent with a number of empirical observations on global capital markets and the energy market. An important part of globalization in recent years has been the ongoing rise in foreign direct investment (FDI). UNCTAD (2000) reports that from 1979 to 1999, the ratio of world FDI inflows to global gross domestic capital rose from 2 to 14%. FDI to China increased from US \$11.7 billion annually during 1985–1995 to US \$46.8 billion in 2001. Similar, albeit smaller, changes in FDI flows to India are documented as well. FDI to India summed to US \$393 million in FY 1992–1993 and surged to US \$4,222 million in FY 2002–2003.

FDI, along with government investment, contributed to lift the overall investment in China, a major contributor to China's growth. China's growth led to a sharp increase in demand for energy and forced China to become a major importer of oil. Similar changes in demand for energy can also be documented in India, among other places. These global changes were followed by changes in oil prices, which reached record highs (more than US \$147) during summer 2008. Our theoretical prediction of rising food prices is also confirmed by leading newspapers around the world.

The linkage between energy, food, and the environment can be lessened by technical change. Technical changes in the production of biofuel, such as development of high-yield energy crops and economical conversion of cellulosic biomass to ethanol, would lessen the land constraint. The second generation of biofuel, using cellulosic feedstock, promises to reduce the competition for land between food and energy production and reduce overall demand for land through productivity gains. Agricultural biotechnology unambiguously reduces the land constraint and attenuates the impacts of biofuel adoption on food supply and land allocations. These results augur for a new commitment to agricultural biotechnology to address the new energy and environmental challenges. Policies to advance biotechnology would be consistent not only with heightened environmental concern in the developed world, but also address food security in the developing world (see Hochman et al. 2008a,b; Khanna et al. 2008).

When deriving policy, we assume (perhaps naively) that governments maximize social welfare and are able to cooperate and optimally choose policies to address environmental externalities, with global as well as local ramifications. These assumptions provide a normative benchmark for socially desirable policies.

16.3 Policy Considerations

Our model generates specific policy recommendations that maximize countries' social welfare subject to market-clearing conditions. These policy recommendations include a carbon tax, a land tax, a food security policy, and creating an environment that promotes technological innovations. Although this chapter offers clear and stark policy recommendations, implementation may be limited by equity and political considerations. Thus, in the following sections, we present the model's policy recommendations and then discuss some of the issues related to its implementation.

16.3.1 Climate Change Policy

The model suggests policies that protect the environment and improve food security. Before we introduce these polices, note that globalization that increases trade and capital flows, creates greater demand for energy. Because energy production and consumption emit GHG emissions, globalization contributes to global climate change. This may explain opposition to globalization among some environmental groups. We do not support this perspective, given the welfare gains from freer trade in goods and knowledge attained by billions of people in developing countries. The market failure associated with free trade can be corrected. The efficient response to the negative externalities associated with energy production is to impose a Pigouvian tax on polluters that equals the marginal cost of pollution, not to restrict free trade.

Economists nearly universally agree that the first-best response to anthropogenic climate change is the imposition of a global carbon tax on each unit of emissions equal to the marginal externality cost of carbon emissions (e.g., Mankiw 2007; Metcalf 2007; Glaeser 2008; Nordhaus 2008; Shapiro et al. 2008). Such a tax, known as a Pigouvian tax, would improve social welfare by reducing carbon emissions while creating no deadweight loss (Pigou 1938). It would also generate government revenue that could be used to reduce other taxes that do cause distortions, such as income taxes (e.g., Goulder 1994).

Furthermore, a carbon tax and a well-designed system of cap and trade are the only policies among those considered in the literature that can provide at once appropriate incentives throughout the energy market without choosing technology winners and losers (Fischer and Newell 2004). These policies encourage traditional energy producers to reduce the emissions intensity of fuels, renewable energy producers to increase output and invest in R&D to lower costs, and consumers to conserve. As Baumol and Oates (1971) note, the Pigouvian tax is the least-cost mechanism to correct an externality.

A system of tradable carbon-emission permits can be equivalent to a carbon tax in terms of efficiency and incidence if the permits are auctioned by the government and if the quota is set to equate demand and private marginal cost plus marginal externality cost. Europe instituted a system of tradable permits in 2005 with a quota of permits that gradually declines to impose greater and greater pressure for carbon-emissions mitigation. However, by 2007, the carbon quota was still higher than total emissions, and the price of a carbon-emission permit had plummeted to near zero. Emissions in Europe had also not fallen as firms faced insufficient incentives to invest in mitigation activities. Legislation to implement carbon trading in the United States has been repeatedly introduced, though by 2008 it had not garnered sufficient political support to near approval.

The Kyoto Protocol was intended to address the global nature of carbon emissions and climate change with a global consensus for emissions reductions. However, important countries like the United States did not sign on to the Kyoto Protocol, and important developing economies, like those of China and India, were exempted from the provisions of the agreement. Furthermore, many countries that were signatories to the Protocol have failed to meet their emission-reduction obligations. The international community is presently considering a successor to the Kyoto Protocol, which will expire in 2012. While it is unlikely that a truly global agreement for carbon-emissions reductions will be achieved (at least until the consequences of climate change become more apparent), a second-best policy would include the most industrialized economies to mitigate the potential for leakage.

Though, in theory, a carbon tax can efficiently accomplish goals to reduce carbon emissions and boost renewable energy production, the implementation of emission pricing is complicated by a number of issues. First, the true social cost of carbon emissions may be estimated with error. If the Pigouvian tax exceeds the true cost of emissions, emissions will be too low. If the tax is set too low, emissions will be too high. When the optimal level of emissions is uncertain, it may be better to use standards rather than financial incentives to minimize the deadweight loss associated with errors in determining the truly optimal level of emissions (Weitzman 1974).

Second, it is administratively difficult to regulate nonpoint sources of pollution. A first-best carbon tax would regulate all point and nonpoint sources of pollution, but this requires monitoring of emissions from every factory smokestack and every vehicle tailpipe. Although proxies can be used for true emissions (e.g., fuel consumption), they are also information intensive and can induce inefficiencies. Taxes that target the more easily monitored point sources of pollution, like factories, and ignore nonpoint sources of pollution induce considerable deadweight loss. Fowlie et al. (2008) show, for instance, that failure to regulate vehicle emissions of nitrogen oxide and sulfur dioxide raises the cost of meeting emission-reduction targets by as much as 10%; relatively cheap emission reductions from vehicles are lost and instead relatively costly reductions from industry are required because of asymmetric regulation. Industry is easier to regulate than agriculture, for instance, but Antle et al. (2001) have proposed methods for pricing carbon emissions and sequestration on the farm.

Some of the implementation problems can be reduced with improvements in monitoring technologies and through learning-by-doing in regulation, thus reducing the implementation cost of carbon tax (Millock et al. 2002). Moreover, Metcalf

(2009) shows that a carbon tax swap can be constructed, which will be distributional neutral, and thus politically viable (see also Metcalf and Weisbach 2009).

Third, carbon taxes have historically been politically unpopular (Fischer and Newell 2004; Owen 2004; Palmer and Burtaw 2005). Consumers and producers are generally opposed to paying for something that historically has been free. Industries in the United States, recognizing that climate policy is coming, have sought to influence the form it takes. Firms oppose carbon taxes because they take resources out of industry; they prefer carbon-trading regimes with permits allocated to industry for free (Buchanan and Tullock 1975). The reluctance to introduce "painful" polices to reduce GHGs stems partially from the uncertainty about the likelihood of climate change. The willingness to introduce stricter polices will increase in response to observed changes in climatic conditions in years to come.

While the desirable outcome is pricing carbon correctly, the reality is different. Currently, the most popular tools to address carbon emissions are mandates, setting a lower bound on the aggregate levels of biofuels used in transportation combined with the renewable fuel standards (RFS). The RFS sets an upper bound on the GHG emission per gallon of biofuel, based on life cycle analysis (LCA), which is used to measure the carbon footprints throughout the entire life cycle of the energy. Unlike carbon taxes that are typically based on emissions in consumption (i.e., combustion), LCA considers emissions over the entire life cycle of the energy, i.e., from "well to wheel" and "tractor to tailpipe" (e.g., Rajagopal et al. 2008a, b). The combination of mandates and RFS aims to increase fuel security to assure that biofuel reduces GHG emissions relative to fossil fuel that it replaces.

In 2005, Congress passed the Energy Policy Act that adopted gradually increasing mandates for biofuel production. By 2008, with the realization that all biofuels are not created equal, the renewable fuel standard (RFS) was amended to slow the mandated growth of conventional biofuels (i.e., corn ethanol) and adopted standards for advanced biofuels and cellulosic biofuels. The former offer limited GHG savings relative to fossil fuel and compete most intensely with food production for land. While the amended RFS appropriately acknowledges the heterogeneity in biofuels and offers the greatest incentive for expansion of the best technologies in 2008, it imposes no incentive for development of technologies beyond those needed to meet the threshold requirements for advanced and cellulosic biofuels. More generally, any policy based on upper-bound standards is inefficient because it offers no incentive for achieving better outcomes than required by the upper bound, even if they can be achieved at low cost (Hochman and Zilberman 1978). Admittedly, the current RFS does provide incentives for cost reductions, which are important to the viability of biofuels.

Related to fuel taxes are fuel standards that require energy supplies to meet carbon intensity standards. California is pursuing the Low Carbon Fuel Standards (LCFS) in its effort, codified by legislation, to reduce GHG emissions to 1990 levels by 2020. The California standard, modeled after a similar policy in the United Kingdom requires producers of transportation energy to track and report the "global warming intensity" (GWI) of their fuels (measured as the production of global warming materials per unit of energy) and reduce the GWI over time (Farrell and Sperling 2007). The determination of GWI is based on LCA. To achieve the GWI reductions with least cost, the LCFS permits producers to reduce the GWI by producing cleaner fuels (such as renewable fuels), by purchasing credits from traditional energy firms that exceed their required GWI reductions, or from renewable energy producers. Despite the introduction of a market for GWI reduction credits, the LCFS is inferior to a fuel tax in that it provides no incentive for emission reductions beyond those prescribed by policy. This can alter the R&D effort of firms. The LCFS also does not provide an incentive for conservation among consumers. In violation of the principle of targeting, the LCFS essentially taxes carbon intensity, when carbon emissions have been linked to climate change (Holland et al. 2007). The LCFS, therefore, is associated with greater excess burden than a carbon tax or even a fuel tax. Moreover, biofuels produced on land converted to production from natural habitat (which emits carbon to the air) should not be taxed the same as biofuels produced on existing farmland (e.g., Searchinger et al. 2008). Thus, payments for carbon sequestration do not generally result in optimal levels of biodiversity preservation.

A number of other policies have been employed by governments to reduce carbon emissions. A full discussion of all of them is beyond the scope of this chapter. We mention several in passing, however, to demonstrate the variety of policy responses and the efficiency implications. The United States has subsidized domestic ethanol production dating to the energy crisis of the 1970s, when domestic energy production was a foremost security concern. Given the consensus finding that conventional biofuels offer an 18% GHG emissions savings relative to fossil fuels (Farrell et al. 2006), subsidies represent an alternative to a carbon tax. However, such subsidies should account for heterogeneity in carbon savings across production methods and uses, much like a carbon tax should monitor actual carbon emissions.

Governments also regulate energy efficiency in an effort to reduce emissions (i.e., U.S. Corporate Average Fuel Economy (CAFE) standards for automobile manufacturers, energy-efficiency requirements for new buildings, and subsidies and tax credits for the purchase of energy-efficient technologies by firms and consumers). The inefficiency of such policies should be clear: They do not provide incentives for advancing fuel technology or for conservation. CAFE standards may, in fact, affect no reduction in carbon emissions as consumers respond to a reduction in the cost of travel (produced by greater fuel efficiency) by demanding more travel (Austin and Dinan 2005).

16.3.2 Land-Use Policy

The second major finding of the model presented above is that even an optimal tax on GHG emissions may reduce social welfare. A carbon tax, for instance, would make biofuels relatively more attractive because they are associated with lower carbon emissions than fossil fuels. Carbon policy would, therefore, increase demand for biofuels. However, biofuels, while relatively clean (in some cases only marginally) compared to fossil fuels, impose other costs on the environment. In particular, biofuel is a land-intensive technology, and production has led to the destruction of natural habitat, including tropical forest. Consequently, unless climate change policy is implemented along with a land policy that values environmental services and biodiversity, both will be made scarcer as biofuels expand. If biodiversity loss is a greater threat than climate change, as some suggest it is, then a carbon tax alone could reduce welfare by increasing demand for biofuels without appropriately taxing the negative externality associated with their production—the loss of natural lands. A land tax or, alternatively, payment for environmental services, would internalize the externality. This finding is consistent with the theory of the second best, which states that correcting one market distortion in the presence of multiple distortions can reduce welfare, especially if political and distributional, as well as economic considerations, are introduced.

If any carbon policy is implemented, it will put additional pressure on natural habitat, which is already threatened by the expansion of biofuel production. Unless nature is properly valued for the services it provides, including biodiversity preservation, carbon sequestration, water purification, waste assimilation, soil erosion prevention, etc., it will be converted to productive uses (e.g., food and fuel) at a rate that is too high relative to social optimality. There is currently no systematic and global mechanism for monetizing natural habitat; however, conservation organizations have succeeded in purchasing land or the development rights to land that is deemed ecologically significant. In addition, countries around the world do set aside lands for preservation and regulate land use through zoning ordinances. For instance, the United States pays farmers to retire land in sensitive ecosystems, and Costa Rica pays landowners to preserve or reforest sensitive watersheds.

However, biodiversity preservation is both a local and global public good, and therefore subject to free-rider problems and underprovision by landowners who do not appropriate the full benefits of their conservation effort (Perrings and Gadgil 2003). Without coordination, individuals and countries will not provide conservation beyond the level that is in their own self-interest, a level that is below the socially optimal level for an unregulated public good. At the local level, biodiversity sustains the productivity of ecosystems, enabling them to provide economically important local services such as the hydrological cycle – including flood control, water supply, waste assimilation, nutrient recycling, soil conservation and regeneration, and crop pollination (Daily 1997). At the global level, public benefits include carbon sequestration and preservation of the global gene pool. The extinction of any species within a region involves a largely irreversible loss of genetic information that is an intergenerational global public good (Sandler 1999). Therefore, just as a global policy to address climate change is needed, so too is a global policy for environmental preservation required to ensure optimal land allocations. To derive the global, as well as the local social cost from allocating land to production, we note that consumers benefit from natural land, a local public good, whereas reallocating natural land to production releases carbon to the air, a global public bad.

The conversion of natural habitat to agricultural use could rival global warming in its environmental and social impact. Tilman et al. (2001) forecasted expansion of agricultural land use by hundreds of millions of hectares by 2050. It would constitute a loss of natural land equal in size to the United States and would be concentrated in Latin America and sub-Saharan Africa. It could lead to a loss of a third of remaining tropical and temperate forests, savannas, and grasslands. Their projection did not assume significant biofuel production and did assume a continuation to 2050 of yield gains similar to those of the Green Revolution. In other words, high demand for biofuel and diminishing productivity growth could mean even more natural land is at risk.

Optimal land conversion can be achieved with a Pigouvian tax on the conversion of natural land to other uses. In general, and abstracting from the framework presented in the paper, a first-best land tax should recognize heterogeneity in land values, based on the biodiversity at stake and the potential to provide ecosystem services, including carbon sequestration. It should also account for heterogeneity in environmental impacts based on land use. Even within agricultural uses, for instance, there is significant variation in environmental impacts depending on which crops are grown and which farm practices are adopted. Low-tillage methods would continue to provide soil carbon sequestration, though less than natural habitat. Pastureland provides environmental benefits relative to land planted to grow crops. In addition, the taxes should recognize the irreversibility of ecosystem damage (such as species extinction) and the fact that productive land can only be restored to natural habitat at great cost. Policy must, therefore, account for the appropriately discounted value of future benefits from natural land.

An analogous financial instrument is the subsidization of environmental services. Such subsidies would have the same efficiency implications as a land tax, but would require government outlays to landowners who preserve their lands. The tax, in contrast, generates government revenues without deadweight loss. The revenues could be used to reduce distortionary taxes. Biodiversity preservation could be achieved most effectively through payment for (subsidization of) conserving natural habitat. However, biodiversity can also be protected indirectly through payments for other environmental services, such as carbon sequestration (Kiss 2004).

As with most environmental public goods, the value of biodiversity preservation is unknown. Though attempts have been made to empirically determine values for biodiversity and ecosystem services, such estimates are insufficient for policymaking (Perrings and Gadgil 2003). Given this uncertainty, imposition of a land tax will necessarily be associated with error. In such circumstances, and as in the case of uncertainty about carbon-emission costs, it may be more efficient to adopt quantity-based regulations rather than price-based mechanisms (Weitzman 1974). Significant heterogeneity in the value of natural lands across regions further complicates the implementation of efficient policy. Nevertheless, as tropical forest is destroyed to make room for energy crops, it is clear additional policy action is needed. The highest value natural habitat is generally found in developing countries, and yet it is most valued by the relatively affluent in industrialized countries. As Perrings and Gadgil (2003) note, poverty can undermine biodiversity. Concern for the environment and intergenerational bequests of the genetic commons is generally subordinate to more immediate livelihood concerns such as access to food, shelter, and water. Given the greater value placed on environmental preservation by wealthy countries, an international consensus on biodiversity may demand transfers from developed countries to developing countries that house the most valuable environmental lands (Lipper et al. 2005).

In sum, pricing both carbon and land correctly, while internalizing environmental externalities associated with biofuel production, makes the whole debate of indirect land use irrelevant. In particular, and given the complexity and difficulties in accurately measuring the indirect land-use effect, creating a pricing mechanism that correctly reflects the cost from carbon and land use is a more promising path to take by practitioners.

16.4 Food Policy

The third result of the Hochman et al. (2008a) model is that rising demand for energy (from increasing trade and capital flows, income growth, or otherwise), amid fossil fuel supply constraints, raises food prices and reduces food availability. This is because increased demand for energy raises demand for biofuels, which in turn raises demand for land to produce energy crops. Energy crops compete for land with food production (at least for the time being), raising the rental rate of land and increasing the cost of food production. This can be considered a shift back of the food-supply curve, which increases price and reduces output. As a result of this linkage between food and energy markets, policies may be needed to protect food supplies in the event of food emergencies, such as that witnessed in 2008.

Perhaps the most dire consequence of biofuel production is the pressure it imposes on the food system. Whereas elevated carbon emissions and biodiversity loss have negative effects that will play out over decades and centuries, rising food prices and reduced food production means people go hungry now. In 2008, biofuel production contributed to record-high food prices that induced the first food crisis in more than three decades and marked a dramatic end to more than a generation worth of declining food prices. Deadly riots ensued as a result of that crisis. To some extent, biofuel policies trade food in the stomach for fuel in the tank. They benefit energy consumers and hurt consumers of food, particularly the poor who devote a large share of their incomes to food purchases (Sexton et al. 2009). The ability to mitigate food impacts will be crucial for the success of biofuels. Fortunately, policies can help ease the food-vs.-fuel trade-off. We discuss some of these policies now, reserving a discussion of R&D for the subsequent section.

Biofuel policies should be flexible and adjust to food market conditions. Biofuel mandates and subsidies could be tied to food inventories in order to avert food crises. Wright (2008) showed that low inventories lead to price-increasing behaviors such as export restrictions in producing countries, increased storage, and speculative activity. Inventories can be used to reduce variations in prices and prevent unanticipated shortages that result from weak harvests (due to weather, for instance).

Recent price increases in food and fuel reflect record-low inventories and the expectation of greater demand for food crops due to biofuel mandates. Such mandates are expected to reduce future supply, which makes storage cheap and further reduces food availability in the present.

Tyner and Taheripour (2008) proposed that biofuel subsidies be tied to the price of oil in recognition of the increasing competitiveness of biofuels as oil prices rise. Subsidies would fall as oil prices rise. Likewise, biofuel subsidies and mandates could be reduced as food inventories decline. While providing increased protection against food crisis, this would create a less certain market for biofuels and could slow capital investment and innovation. The cost imposed on the biofuel industry increases further if the price of staple crops is correlated with the price of energy.

16.4.1 Policy for Biofuel and Agriculture R&D

In Section 16.2, we showed that improvements in biofuel technology could reduce demand for land, reducing pressure on food production and environmental preservation – the two predominant uses of land today. Together, these findings suggest investment in plant biosciences can have beneficial impacts for food security and the environment. Improvements in biofuel technology (improving the intensive margins in agricultural production) can improve both climate change outcomes and biodiversity outcomes by reducing the carbon and land intensity of biofuel production. Similarly, agricultural biotechnology reduces the land intensity of traditional farming and can free land in food production for energy crop production, reducing the demand for land and thereby mitigating the loss of natural lands.[6]

Until recently, and for more than 30 years, relative food prices have been declining and food supplies sufficient to produce incremental reductions in the number of starving people throughout the world. In fact, food prices fell 75% in real terms from 1974 to 2005 (*The Economist* 2007), allowing an entire generation to grow up in an era of cheap food and yielding a sense of complacency about agricultural production and plant science. Human hunger, it seemed, had been defeated by the Green Revolution.

In 2008, however, the world awoke to a pressing challenge for agriculture to feed a growing population that is also growing wealthier, and to provide an energy alternative to oil. Some suggest agricultural production will need to double in the first half of the twenty-first century strictly to feed the world (Johnson 2000; United Nations 2001). Genetic engineering of traditional agricultural and new energy crops offers opportunities to increase agricultural production without land expansion and without more intensive use of inputs.

---

[6]Other policies, not addressed in this chapter, include policy-inducing adoptions of energy-saving technologies that may affect the degree of complementarity between capital and energy. We elected to abstract from these policies since in this chapter goods are homogeneous, not heterogeneous.

FAO and IFPRI rely on yield gains to provide 60% of production growth over the next few decades in order for supply of staple crops to meet demand. Yet, yield gains have slowed down, not just in developed countries that saw large yield gains from the Green Revolution, but also in developing countries. Yield growth in cereals, for instance, averaged 3% per year during the 1960s, but fell to only 1.3% in the 1990s. It is expected to fall further still over the next decade. Even sustaining current yields will be difficult as yield increases from incremental fertilizer applications are falling and pesticide resistance is growing (FAO 2008; von Braun 2007; Ruttan 2002).

The tools of biotechnology – molecular- and marker-assisted selection – have produced a first generation of genetically modified (GM) staple crops. Since transgenic crops were commercialized in the mid-1990s, yield gains have been observed from the United States to South Africa, and from Argentina to China (Qaim and Zilberman 2003; Thirtle et al. 2003; Ismael et al. 2001; Traxler et al. 2001; Marra et al. 2002; Qaim et al. 2006). Increased productivity in food and feed is important in order to avert future food crises, but so too is improved productivity of biofuels. The first-generation biofuels have made clear the obstacles that must be overcome in order for biofuels to provide a real solution to global food, energy, and environmental challenges. First, they will need to compete less intensely for staple food crops to avert future food crises. Second, biofuels will need to offer greater GHG emissions reductions. And third, biofuels will need to be less land intensive to avoid negative impacts on food and global warming and to prevent the loss of biodiversity, which some have estimated to be more costly today than global warming (Mooney and Hobbs 2000).

Emerging plant science can convert cellulosic plants like switchgrass and Miscanthus into liquid fuel, make use of agricultural residues, and turn marginal land into productive land. It cannot yet scale up these processes in a commercially viable way. Cellulosic ethanol promises to resolve the most significant problems associated with existing biofuels. Whereas only 500 ethanol gallons of harvestable corn grains are extracted from each acre of corn grain, 1,700 ethanol gallons of harvestable biomass is produced on each acre of Miscanthus. Thirteen hundred gallons of cellulosic ethanol can be produced from each acre of Miscanthus. Only 450 gallons of corn ethanol are yielded per acre of corn. Under a hypothetical scenario of 35 billion gallons of ethanol production, corn ethanol would demand one-quarter of all harvested cropland in the United States. Miscanthus would need less than one-tenth (Heaton et al. 2008). With dedicated energy crops that can be grown on marginal lands and that are, in fact, more productive on lands where traditional crops are less productive, the next generation of biofuels would permit the entire harvest of staple crops to be used for food and feed (Khanna 2008).

Cellulosic ethanol can also overcome the challenge of biofuels to offer significant GHG emission reductions relative to fossil fuels. By utilizing a greater share of harvested plants, by utilizing crops that produce more biomass per acre, by reducing input intensity of feedstock production, and by increasing efficiency of depolymerization and fermentation, the next generation of biofuels can greatly reduce the carbon intensity of biofuel production. Corn ethanol emissions are at best 25% below gasoline emissions, but Miscanthus emits 89% less emissions. Corn stover, which uses the residue of corn food and feed production, causes only 18% as much pollution as gasoline (Khanna 2008).

Despite the need for improved plant science, federal spending for R&D has declined as a percent of GDP since 1975. It fell from 1.27% of GDP in 1975 to a low of 0.88% in 2001, before recovering marginally through 2005. It has since fallen again to 1% of GDP in the 2009 federal budget (American Association for the Advancement of Science, 2008). Public funding for agricultural research has fallen in recent years – in some cases in nominal as well as real terms (Alston et al. 1998). Given the modern tools of plant science, the returns to agricultural research can be as large as they were during the Green Revolution.

The capacity to introduce new varieties also depends on regulation. Regulation not only slows the adoption of technology ex post, but also reduces innovation ex ante. A precautionary approach that places uncertain impacts in the future ahead of certain hunger in the present should be reconsidered. To this end, Europe's de facto ban on GMO from 1998 to 2004 severely hampered innovation and introduction of new traits (Graff et al. 2009; Paarlberg 2001).

16.5 Conclusion

Extending general equilibrium trade models to incorporate the household model can capture many of the features of the changing energy industry and help to better understand policies for ensuring sufficient energy supplies and preservation of the world environmental commons. The framework introduced in this chapter also permits modeling of interactions between different sources of energy, e.g., biofuels and fossil fuels. For instance, the use of general equilibrium models allows us to better understand the impact of land-use change, and therefore the merits of biofuels. It extends traditional trade models to incorporate energy, a ubiquitous factor of production that is going to have significant impact on economies around the world, especially given predicted future supply constraints and demand growth.

This work illustrates the importance of identifying all externalities associated with energy production and of developing a pricing mechanism that will internalize them. It also highlights the importance of supporting R&D in second-generation biofuel (what is also called advanced biofuels) and biotech to mitigate both the surge in energy and food prices. We have identified a number of policies that can respond to challenges of the new energy paradigm – the threat of climate change, the loss of natural habitat, and the end of an era of secure food supplies. The breadth of policy responses can be explained by the range of market failures associated with these challenges and by the relative political palatability of various interventions. The first-best responses to carbon emission and agricultural expansion are particularly untenable from a political economy perspective.

As other mechanisms are pursued, a few principles should be followed. First, following the Principle of Targeting, policies should target as explicitly and narrowly as possible to the sources of market failures. If emissions of automobiles cause external social costs, emissions should be taxed, not production of energy or a particular production or end-use technology. Second, policies should account for heterogeneity in costs and benefits of various energy technologies and natural habitats. Third, "market opening" policies, supporting development of new promising technologies and motivated by dynamic efficiency considerations, should be temporary and seek to support technologies that will eventually compete on their own in the market. Fourth, governments should support R&D activities to provide basic knowledge to develop new agricultural and biofuel technologies. The regulations of these technologies should be based on their social opportunity costs, namely, performance relative to existing technologies. Thus, at times, we need to take some small risks to avoid bigger environmental and economic risks.

References

Alston JM, Pardey PG and Roseboom J (1998) Financing agricultural research: International investment patterns and policy perspectives. World Dev 26:1057–1071.

American Association for the Advancement of Science (2008) Historical trends in R&D. Accessed at: http://www.aaas.org/spp/rd/guihist.htm.

Antle JM, Capalbo SM, Mooney S, Elliott ET, Paustian, K (2001) Economic analysis of agricultural soil carbon sequestration: An integrated assessment approach. J Agric Resour Econ 26:344–367.

Austin D, Dinan T (2005) Clearing the air: The costs and consequences of higher CAFE standards and increased gasoline taxes. J Environ Econ Manag 50(3):562–582.

Baumol WJ, Oates WE (1971) The use of standards and prices for protection of the environment. J Econ 73: 42–54.

Becker GS (1965) A theory of allocation of time. The Econ J 75(299):493–517.

Buchanan J, Tullock G (1975) Polluters profits and political response: Direct control versus taxes. Amer Econ Rev 65:139–147.

Daily G, ed. (1997) Nature's services: Societal dependence on natural systems. Island Press, Washington DC.

Deardorff AV (1980) The general validity of the law of comparative advantage. J Politi Econ 88(5):941–957.

Deardorff AV (1982) General validity of the Heckscher-Ohlin Theorem. Amer Econ Rev 72(4):683–694.

de Gorter H, Just DR (2008) 'Water' in the US ethanol tax credit and mandate: Implications for rectangular deadweight costs and the corn-oil price relationship, NBER. Available at SSRN: http://ssrn.com/abstract=1071067.

Dixit A, Norman V (1980) Theory of international trade: A dual, general equilibrium approach. Cambridge University Press, Cambridge.

Dixit A, Woodland A (1982) The relationship between factor endowments, and commodity trade. J Inter Econ 13:201–214.

Drabicki JZ, Takayama A (1979) An antinomy in the theory of comparative advantage. J Inter Econ 9:211–223.

Farrell A, Sperling D (2007) A low-carbon fuel standard for California. Institute for Transportation Studies, University of California, Davis.

Farrell AE, Plevin RJ, Turner BT, Jones AD, O'Hare M, Kammen DM (2006) Ethanol can contribute to energy and environmental goals, Science 311(5760):506–508.

Fischer C, and Newell R (2004) Environmental and technology policies for climate change, RFF Discussion Paper 04–05. Resources for the Future, Washington, DC.

Food and Agriculture Organization of the United Nations (FAO) (2008) Food price index. Food and Agriculture Organization of the United Nations, Washington, DC.

Fowlie M, Knittel CR, Wolfram CD (2008) Sacred cars? Optimal regulation of stationary and non-stationary pollution sources. NBER Working Paper Series #14504.

Glaeser EL (2008) The folly of 'fixing' energy price hikes. The Boston Globe (August 1).

Goulder L (1994) Environmental taxation and the double dividend: A reader's guide. NBER Working Paper No. 4896. National Bureau of Economic Research, Cambridge.

Graff GD, Hochman G, Zilberman D (2009) The political economy of agricultural biotechnology policies, AgBioForum.

Heaton EA, Dohleman FG, Long SP (2008) Meeting US biofuel goals with less land: The potential of Miscanthus. Global Change Biology 14(9):2000–2014.

Hochman G, Sexton S, Zilberman D (2008a) The economics of biofuel, trade, and the environment, The Canadian Economic Meetings, Vancouver, BC.

Hochman G, Sexton SE, Zilberman D (2008b) The economics of biofuel policy and biotechnology. J Agric Food Ind Org 6(2):Art. 8.

Hochman E, Zilberman D (1978) Examination of environmental policies using production and pollution microparameter distributions. Econometrica 46(4):739–760.

Holland SP, Knittel CR, Hughes JE (2007) Greenhouse gas reductions under low carbon fuel standards? Center for the Study of Energy Markets. Paper CSEMWP-167. http://repositories.cdlib.org/ucei/csem/CSEMWP-167.

International Energy Agency (2008) World Energy Outlook 2008.

Ismael Y, Beyers L, Lin L, Thirtle C (2001) Smallholder adoption and economic impacts of Bt cotton in the Makhathini Flats, South Africa. Paper presented at the 5th ICABR International Conference on Biotechnology, Science and Modern Agriculture: A New Industry at the Dawn of the Century, Ravello, Italy.

Johnson DG (2000) Population food and knowledge. Amer Econ Rev 90:1–14.

Jones RW, Scheinkman J (1977) The relevance of the two-sector production model in trade theory. J Politi Econ 85:909–935.

Khanna M (2008) Economics of biofuel production: Implications for land use and greenhouse gas emissions. Presented at Sustainable Biofuels and Human Security Conference, University of Illinois, Urbana-Champaign.

Khanna M, Hochman G, Rajagopal D, Sexton S, Zilberman D (2009) Sustainability of food, energy, and environment with biofuels. CAB Reviews: Perspectives in Agriculture, Veterinary Science, Nutrition and Natural Resources, 4 (028): 10 pp.

Kiss A (2004) Making biodiversity conservation land use priority. In: McShane T, Wells M (eds.) Getting biodiversity projects to work: Towards more effective conservation and development. Columbia University Press, New York, NY.

Lancaster KJ (1966) A new approach to consumer theory. J Politi Econ 74(2):132–157.

Lipper L, Cooper J, Zilberman D (2005) Synthesis chapter: Managing plant genetic diversity and agricultural biotechnology for development. In: Cooper J, Lipper LM, Zilberman D (eds.) Agricultural biodiversity and biotechnology in economic development, Springer, New York.

Mankiw NG (2007) One solution to global warming: A new tax. The New York Times (September 16).

Marra M, Pardey P, Alston J (2002) The payoffs to transgenic field crops: An assessment of the evidence. AgBioForum 5(2):43–50. Available on the World Wide Web: http://www.agbioforum.org.

Melvin JR (1968) Production and trade with two factors and three goods. Amer Econ Rev 58:1248–1268.

Metcalf G (2007) A green employment tax swap: Using a carbon tax to finance payroll tax relief. Policy Brief, The Brookings Institution, Washington, DC.

Metcalf GE (2009) Designing a carbon tax to reduce U.S. greenhouse gas emissions. Rev Environ Econ and Policy Advance Access. Published online on January 7, 2009.

Metcalf GE, Weisbach DA (2009) Design of a carbon tax, U of Chicago, Public Law Working Paper No. 254. Available at SSRN: http://ssrn.com/abstract=1324854.

Millock K, Sunding D, Zilberman D (2002) Regulating pollution with endogenous monitoring. J Environ Econ Manag 44:221–241.

Mooney HA, Hobbs RJ, eds. (2000) Invasive species in a changing world. Island Press, Washington DC.

Nordhaus W (2008) A question of balance: Weighing the options on global warming policies. Yale University Press, New Haven, CT.

Owen AD (2004) Environmental externalities, market distortions and the economic of renewable energy technologies. The Energy J 25:127–156.

Palmer K, Burtaw D (2005) Cost-effectiveness of renewable electricity policies. Energy Econ 27:873–894.

Paarlberg RL (2001) The politics of precaution: Genetically modified crops in developing countries, Johns Hopkins University Press, Baltimore.

Perrings C, Gadgil M (2003) Conserving biodiversity: Reconciling local and global public benefits: In Kaul I, Conceicao P, le Goulven K, Mendoza RL (eds.) Providing global public goods: Managing globalization. Oxford University Press, Oxford.

Pigou AC (1938) The economics of welfare. 4th ed. Weidenfeld and Nicholson, London.

Pindyck RS (1979) The structure of world energy demand. MIT Press, Cambridge, MA.

Pindyck RS, Rotemberg JJ (1983) Dynamic factor demands and the effects of energy price shocks. Amer Econ Rev 73(5):1066–1079.

Qaim M, Subramanian A, Naik G, Zilberman D (2006) Adoption of Bt cotton and impact variability: Insights from India. Rev Agric Econ 28:48–58.

Qaim M, Zilberman D (2003) Yield effects of genetically modified crops in developing countries. Science 299:900–902.

Rajagopal D, Hochman G, Zilberman D (2008a) Regulation of GHG emissions from biofuel blended energy, Farm Foundation Conference, St. Louis, MO.

Rajagopal D, Hochman G, Zilberman D (2008b) A simple framework for regulation of biofuels, 28th USAEE/IAEE North America Conference, New Orleans, LA.

Reuters UK (2008) http://uk.reuters.com/article/oilRpt/idUKB53020120080305.

Renewable Fuels Association (2006) http://www.ethanolrfa.org/.

Ruttan V (2002) Productivity growth in world agriculture: Sources and constraints. J Econ Perspective 16:161–184.

Samuelson PA (1953) Prices of factors and goods in general equilibrium. Rev Econ Studies 21: 1–20.

Sandler T (1999) Intergenerational public goods: Strategies, efficiency, and institutions. In: Kaul I, Grunberg I, Stern MA (eds.) Global public goods: International cooperation in the 21st Century. Oxford University Press, New York.

Searchinger T, Heimlich R, Houghton RA, Dong F, Elobeid A, Fabiosa J, Tokgov S, Hayes D, Yu TH (2008) Use of U.S. croplands for biofuels increases greenhouse gases through emissions from land use change. Science, 319(5867):1238–1240.

Sexton S, Rajagopal D, Hochman G, Roland-Host D, Zilberman D (2009) Welfare analysis, CA Agric.

Shapiro R, Pham N, Malik A (2008) Addressing climate change without impairing the U.S. economy. U.S. Climate Task Force, Washington, DC.

Thirtle C, Beyers L, Ismael Y, Piesse J (2003) Can GM technologies help the poor? The impact of Bt cotton in Makhathini Flats, KwaZulu-Natal. World Devel 31: 717–734.

Tilman D, Fargione J, Wolff B, D'Antonio C, Dobson A, Howarth R, Schindler D, Schlesinger WH, Simberloff D, Swackhamer D (2001) Forecasting agriculturally driven global environmental change. Science 292:281–284.

The Economist (2007) The end of cheap food. London (December 6).

Traxler G, Godoy-Avila S, Falck-Zepeda J, de J. Espinoza-Arellano J (2001) Transgenic cotton in Mexico: Economic and environmental impacts. Paper presented at the 5th ICABR International Conference on Biotechnology, Ravello, Italy.

Tyner W, Taheripour F (2008) Biofuels, policy options, and their implications: Analyses using partial and general equilibrium approaches, J Agric Food Ind Organ 6(2), Art. 9.

UNCTAD (2000) http://www.unctad.org/Templates/StartPage.asp?intItemID=2068.

United Nations (2001) World population prospects: The 2000 revision. United Nations Department of Economic and Social Affairs, ESA/P/WP 165. Available at http://www.un.org/esa/population/wpp2000.htm_.

Vandenbergh MP, Steinemann AC (2007) The carbon neutral individual, New York Univ Law Rev 82:10–12.

Von Braun J (2007) World food situation: New driving forces and required actions. Food Policy Report. International Food Policy Research Institute, Washington, DC.

Weitzman M (1974) Prices and quantities. Rev Econ Stud 44:511–518.

Wright B (2008) Speculators, storage, and the price of rice. ARE Update 12:7–10.

Chapter 17
Meeting Biofuels Targets: Implications for Land Use, Greenhouse Gas Emissions, and Nitrogen Use in Illinois

Madhu Khanna, Hayri Önal, Xiaoguang Chen, and Haixiao Huang

Abstract This article develops a dynamic micro-economic land use model to identify the cost-effective allocation of cropland for traditional row crops and perennial
grasses and the mix of cellulosic feedstocks needed to meet predetermined biofuel targets over the 2007–2022 period. Yields of perennial grasses obtained from
a biophysical model and together with county level data on costs of production for
Illinois are used to examine the implications of these targets for crop and biofuel
costs, greenhouse gas emissions, and nitrogen use. The economic viability of cellulosic feedstocks is found to depend on their yields per acre and the opportunity cost
of land. The mix of viable cellulosic feedstocks varies spatially and temporally with
corn stover and miscanthus coexisting in the state; corn stover is viable mainly in
central and northern Illinois while miscanthus acres are primarily located in southern Illinois. Biofuel targets lead to a significant shift in acreage from soybeans and
pasture to corn and a change in crop rotation and tillage practices. The biofuel targets assumed here lead to a reduction in greenhouse gas emissions, but an increase
in nitrogen use.

Biofuels are increasingly being viewed as the center piece in any strategy for energy
independence, stable energy prices, and greenhouse gas (GHG) mitigation in the
United States. A key challenge to the expansion of biofuel production is the allocation of limited agricultural land between crops and biomass to meet the needs for
food, feed, and fuel and its potential to raise the prices of food/feed crops. The share
of corn being used for ethanol production has increased from 10 to 28% between
2004/2005 and 2007/2008, and despite an unprecedented increase by 15% in the
acreage under corn in 2007 relative to 2005, corn prices reached record high levels
in 2007 that were twice as high as those in 2005 (Bange, 2007).

M. Khanna (✉)
Department of Agricultural and Consumer Economics, Energy Biosciences Institute, University of
Illinois, Urbana-Champaign, IL, USA
e-mail: khanna1@illinois.edu

M. Khanna et al. (eds.), *Handbook of Bioenergy Economics and Policy*,
Natural Resource Management and Policy 33, DOI 10.1007/978-1-4419-0369-3_17,
© Springer Science+Business Media, LLC 2010

Energy policy in the United States initially sought to promote production and use of first-generation biofuels, corn ethanol, through mandates and tax credits; this has changed due to concerns about the implications of expanding demands for corn ethanol for food prices, as well as the greater potential of cellulosic biofuels to mitigate climate change. The recently enacted Energy Independence and Security Act of 2007 places greater emphasis on the next generation of biofuels and mandates that 21 of the 36 billion gallons of ethanol be advanced biofuels that reduce GHG emissions by at least 50% relative to baseline levels.

Unlike the current generation of biofuels based on a single feedstock, that is corn, cellulosic biofuels can be produced from several different feedstocks including crop residues, woody biomass, and perennial grasses. The use of crop residues, being by-products of crop production, does not create a food–fuel competition for land but can impact soil productivity unless they are harvested sustainably with a substantial portion of the residue left on the field. Currently available corn stover (with 40% residue collection) could meet about a third of the advanced biofuels mandate for 2022 in the United States (Perlack et al., 2005) necessitating reliance on other sources, such as perennial grasses to meet the rest of the mandate.

Two perennial grasses, switchgrass (*Panicum virgatum*) and miscanthus (*Miscanthus x giganteus*), have been identified in particular as among the best choices as dedicated energy crops in the United States (Heaton et al., 2004; Lewandowski et al., 2003). These grasses have higher yields than others, provide high nutrient use efficiency, and require growing conditions and equipment similar to those for corn, making them compatible with conventional crop cultivation. They can provide a larger volume of biofuels per acre and lower life cycle GHG emissions per gallon of fuel than corn ethanol, and thus alleviate the competition for land. Moreover, switchgrass and miscanthus can be grown on marginal lands with no adverse impacts on yields (unlike in the case of corn). These grasses also have the potential to reduce soil erosion and chemical run-off compared to the row crops they may displace due to their low chemical input needs and root structure.[1]

This article develops a dynamic micro-economic land use allocation model that determines the profit-maximizing land use choices to meet a targeted level of corn ethanol and cellulosic ethanol (from corn stover, miscanthus, and switchgrass) over the 2007–2022 horizon, while taking into account the spatial heterogeneity in yields, costs of production, and land availability within a region. Spatially heterogeneous yields of switchgrass and miscanthus are obtained from a biophysical crop growth model and used to examine the heterogeneity in the viability of biofuels from alternative feedstocks across geographical locations and the mix of feedstocks that is likely to be economically viable. A second purpose of this article is to examine the

---

[1] There have been some concerns that miscanthus, as an introduced species, might be an invasive plant. However, most varieties used for biofuel production (like *Miscanthus x Giganteus*) are sterile hybrids and do not produce seed.

impact of these biofuel targets for the price of food crops that will be displaced from cropland and for the cost of producing biofuels to meet given mandates. The diversion of corn needed to meet the target for corn ethanol is expected to raise the prices of both corn and other competing commodities, and thus the cost of production of corn ethanol. Rising corn prices would also raise the opportunity costs of land to be converted to energy crops, and thus the costs of producing cellulosic biofuels.

This article also investigates the effects of biofuel targets on nitrogen use and life-cycle GHG emissions. Biofuels from different feedstocks differ in their nitrogen requirements, energy-balance, and life-cycle emissions. While corn ethanol reduces GHG emissions relative to gasoline, the production of corn is nitrogen and carbon intensive compared to perennial grasses. Reliance on current-generation biofuels, therefore, poses a trade-off between reducing GHG emissions and potentially increasing nitrate run-off and causing water quality problems.

The model is used to examine the economic and environmental implications of biofuels targets over the period 2007–2022 employing county-level data for Illinois. Restricting the regional coverage to Illinois is mainly because of data availability, but despite its limitations this approach is justifiable given that the State currently produces 17% of corn and 19% of the ethanol in the United States and has the climatic and soil conditions conducive to the production of herbaceous perennials that can be used as feedstocks for cellulosic biofuels. We also estimate the nitrogen use and life-cycle GHG emissions associated with biofuels from different feedstocks based on the crop patterns generated by the model and the county-specific production practices in Illinois. The next section describes the related literature. The economic model is described in Section 3 followed by a description of the dedicated energy crops being considered here and the data and assumptions underlying the numerical simulation. Results of the numerical simulation are presented in Section 5 followed by conclusions in Section 6.

17.1 Related Literature

The dynamics of agricultural land use changes have been examined by several studies. Foremost among these are the studies based on the Forest and Agricultural Sector Optimization Model (FASOM) which is a multiperiod, price-endogenous, spatial market-equilibrium model of land allocation between agricultural crops and forests. The model is run on a decadal time step. Biophysical relationships that quantify the growth of timber and the sequestration of carbon in forests and land are included. Alig et al. (1997) apply this model to investigate the allocation of land among 39 crop and livestock activities and forests across five regions in the United States to achieve given carbon sequestration targets, while Alig et al. (2000) examine the land use implications of producing hardwood short rotation woody crops on cropland for the US pulp and paper sector and its impact on the agricultural and forest sectors in the United States.

McCarl et al. (2000) apply FASOM to examine the competitiveness of electric power generation using bioenergy from milling residues, whole trees, logging residues, switchgrass, and short-rotation woody crops instead of coal while disaggregating the United States into 11 homogenous regions. McCarl and Schneider (2001) expand this model into the ASMGHG model to investigate competitiveness of various carbon-mitigation strategies that include soil sequestration, biofuel crops, and afforestation at alternative carbon prices across 63 regions in the United States. They find that at low carbon prices, soil carbon sequestration through a change in cropping practices is competitive, while at high carbon prices, abatements are achieved mainly through use of biomass for power generation and conversion of land to forests.

Another dynamic agricultural sector model used to analyze allocation of cropland in the United States is POLYSYS (Ugarte et al., 2003). The model includes various traditional and energy crops and investigates land use impacts of exogenously set bioenergy prices. It has a finer regional disaggregation than FASOM, with 305 agricultural statistical districts as defined by the USDA, and provides annual estimates of changes in economic outcomes. Walsh et al. (2003) apply POLYSYS to examine the potential for using CRP land to produce bioenergy crops at various bioenergy prices and find that switchgrass is more competitive than woody bioenergy crops and annual farm income and crop prices would increase due to bioenergy crop production.

A few studies examine the environmental effects of the ethanol mandate. Using POLYSYS, English et al. (2008) show that the corn-ethanol mandate will lead to major increases in corn production in the Corn Belt, shifting soybeans and wheat production to the southeast and shifting cotton westward over the period 2007–2016 (assuming that cellulosic biofuels are not feasible over this period). Fertilizer use and soil erosion will increase significantly while soil carbon sequestration will decline. Malcolm (2008) uses Regional Environment and Agriculture Programming Model (REAP), a partial-equilibrium model of the US agricultural sector consisting of 50 regions, to quantify the extent to which substitution of crop-residue-based cellulosic ethanol for corn ethanol reduces soil erosion and nutrient deposition.

The dynamic land use allocation model developed here differs from the models used in the studies mentioned above. We incorporate spatial and temporal heterogeneity in returns to land at county level rather than much broader regions considered in those studies, and the optimal mix of competing cellulosic feedstocks – corn stover, miscanthus, and switchgrass – is examined to meet the ethanol mandates. Due to the perennial nature of miscanthus and switchgrass, we use a multiperiod dynamic rolling horizon model. The model generates a time path of the costs of meeting the biofuel mandate and examines its sensitivity to assumptions about the costs of producing cellulosic feedstocks. A biophysical model of energy crop yields and life-cycle analysis of carbon emissions is integrated with the land use model to examine the environmental implications of land use changes to meet the specified targets.

17.2 The Model

A dynamic spatial-optimization model is developed to analyze market prices, socially optimal land use strategies, and production and consumption of various row crops and perennial crops while meeting specific targets for ethanol production in Illinois over the 16-year planning horizon of 2007–2022. The annual crops considered here are corn, soybean, wheat, and sorghum, while the perennial crops considered are alfalfa, switchgrass, and miscanthus. Since Illinois is a major producer of corn and soybeans, a significant change in the crop pattern in this region is likely to alter the market prices of these two commodities. Therefore, when determining the optimum resource allocation, the model incorporates market-equilibrium prices for corn and soybeans as endogenous variables. This is done by using a conventional approach where the sum of consumers' and producers' surplus is maximized subject to demand–supply balances, resource-availability constraints, and technical constraints underlying production possibilities in Illinois (see, Takayama and Judge, 1971, and McCarl and Spreen, 1980, for a rigorous presentation of this methodology and a review of studies that used this approach). Consumers' behavior is represented by constant elasticity demand curves for soybeans and for traditional (nonethanol) uses of corn, both specified regionally, while the prices of wheat, sorghum, and alfalfa are fixed at their base year levels. The parameters of the regional constant elasticity demand curves for corn and soybeans are computed based on national demand elasticity estimates of these commodities and their base year consumption levels (quantities sold at the farmgate) using the method in Kutcher (1972). When computing the producers' surplus, returns from commodity sales and the costs associated with production of row crops and perennial crops, costs of land conversion between perennial and row crops, and the processing costs of both corn ethanol and cellulosic ethanol are incorporated in the objective function of the model. In addition, returns from the sales of coproducts of biofuel production (such as distiller's dried grains with solubles (DDGS) and electricity, a by-product of cellulosic biofuel production) are included in the producers' surplus, with the price of DDGS linked to the price of corn. The production costs of row crops vary with alternative management practices (rotations and tillage choices) while the costs and the yields of perennials vary with the age of the perennials. Yields of all crops are assumed constant over time here.

The model determines optimal allocation of agricultural land simultaneously in all of the 102 counties in Illinois that are heterogeneous in their crop productivity and related costs, including the costs of producing biofuels (due to differences in feedstock costs), across various crops, rotations, and management practices while satisfying the county-level land-availability constraints, policy constraints (ethanol targets), and various technical constraints underlying the row crop rotation choices and dynamics of perennial crop production. The annual targets for corn ethanol and cellulosic ethanol for Illinois are assumed to be proportional to those set by the renewable fuels mandate, based on the current share of Illinois in national ethanol production.

Our modeling approach has three salient features. First, the perennial nature of switchgrass and miscanthus requires consideration of year-to-year changes in crop yields and costs, thus multiyear production plans. Switchgrass has a lifetime of 10 years while miscanthus has been observed to have a lifetime of 20 years. The production of both crops involves high establishment costs and low yields in the initial years and requires farmers to make long-term planting decisions to recover the initial sunk costs. We assume that farmers have a 10-year planning horizon and make decisions based on expected prices (assumed to be the same as the previous year's prices) in each year, the dynamics of crop yields and costs, and the demands for corn and biomass that are consistent with the ethanol targets. The aggregate of individual crop-production decisions together with the crop-demand functions and ethanol-production targets determine the equilibrium prices realized in the market. We assume that farmers observe these realized prices at the end of the first year of the 10-year planning horizon and use these to reoptimize their decisions for the subsequent 10 years. This "rolling horizon" approach incorporates considerations of the initial fixed costs of producing perennial grasses, their annual costs in subsequent years, and their lifetimes. A longer time horizon is unlikely to influence the model results significantly since the present (discounted) value of net benefits received beyond 10 years is likely to be small.

Second, our model allows cropland availability to respond endogenously to the increase in demand due to the steady increase in the ethanol production targets. Marginal lands not being utilized currently may be converted to crop land, with the extent of conversion depending on the variations in crop prices over time. Using the rolling-horizon approach we treat agricultural land supply as "semi-endogenous." Specifically, we solve a 10-year market-equilibrium model for each year of the 2007–2022 period assuming that land availability in each county is fixed during each run of the model. However, the county-level land availability is varied between successive runs based on an estimated land supply elasticity that reflects the responsiveness of cropland expansion to changes in an expected crop price index. The endogenously determined first-year prices for corn and soybeans determine the overall crop price index and the cropland available for subsequent runs of the model. In this iterative procedure, we first solve the model using the base-year (2007) land availability and ethanol targets for 2007–2022. Then, using the 2007 prices determined endogenously and observed prices prior to 2007, we compute the expected crop price index for 2008, update the land availability accordingly, and solve the model again considering the ethanol targets for the next 16 years (i.e., 2008–2023). This is repeated for each year of the planning horizon with the ethanol targets beyond 2022 being set at their levels in 2022.

The third feature of the model is the limited flexibility for changes in optimal crop patterns. To prevent unrealistic changes in land use, we incorporate a combination of historical and "synthetic" acreage patterns into the land allocation for each row crop. Observed historical acreages can be used under "normal" conditions to guide the potential planting behavior for row crops as in McCarl (1982) and Önal and McCarl (1991). Since we are considering further increases in the production of corn and planting new bioenergy crops in order to meet mandatory cellulosic ethanol targets, unprecedented land use patterns are likely to occur in the near future. To ensure that the model can generate results which are consistent with farmers' planting history and potential future trends, we incorporate both historical and synthetic' acreage patterns (crop mixes, each mix being a vector of crop acreages in a given year) in the model. The synthetic' crop mixes included in the model are generated a priori based on estimated acreage supply elasticities (both own price and cross price elasticities) and considering a set of price vectors in which crop prices (for corn, soybeans, and wheat only) are varied systematically. In addition, we impose a constraint that governs the dynamics of land conversion between perennials and row crops. These constraints are partly imposed by the allowable crop rotation possibilities and partly by limits imposed on the extent to which land can be converted from conventional to conservation tillage and from row crops to perennial grasses.

17.3 Data

We estimate rotation and tillage specific costs of production in 2007 prices for four row crops – corn, soybeans, wheat, and sorghum – and three perennial grasses – alfalfa, switchgrass, and miscanthus. The three perennial grasses have lifetimes of 5, 10, and 20 years, respectively. Application rates for nitrogen, potassium, phosphorus, and seed for the four row crops and for alfalfa vary with yields per acre (University of Illinois Extension, 2002), as do the costs of drying and storage of crops (FBFM, 2003). Costs of producing row crops and alfalfa are obtained from the Farm Business and Farm Management data (FBFM, 2007). County-specific, 5-year (2002–2006) historical average yield per acre for each row crop is obtained from National Agricultural Statistics Service (USDA/NASS, 2008a) and used to construct these costs for each of the 102 counties in Illinois. Observed yields per acre are assumed to be those under a corn–soybean rotation, which is the dominant rotation practiced in Illinois. Corn yield per acre under a continuous corn practice is assumed to be 12% lower than under a corn–soybean rotation. Costs of machinery operation, depreciation, and interest vary across the northern, central, and southern regions of Illinois and are obtained from the FBFM data for various years (FBFM, 2003; FBFM, 2007, 2008). The per acre costs of labor, building repair and depreciation, and overhead (such as farm insurance and utilities) are excluded from these costs of production since they are likely to be the same for all crops and would not affect the relative profitability of crops. These are, therefore, part of the opportunity costs of using existing farm land, labor, and capital to produce bioenergy crops.

Corn stover yield for each county and each rotation is obtained from corn yields assuming a 1:1 ratio of dry matter of corn grain to dry matter of corn stover and 15% moisture content in the grain (Sheehan et al., 2003). Corn stover yields range from a low of 2.25 t dm per acre (metric tons of dry matter per acre) in southern Illinois to a high of 4 t dm per acre in northern and central Illinois. In the absence of long-term observed yields for switchgrass and miscanthus, a crop productivity model MISCANMOD is used to simulate these yields in Illinois using GIS data on climate, soil moisture, solar radiation, and growing degree days, as described in Khanna et al. (2008). Harvestable yields of miscanthus and switchgrass are estimated to be lower in northern Illinois (9.8 t dm per acre and 4.4 t dm per acre, respectively) than in southern Illinois (12.1 t dm per acre and 5.8 t dm per acre, respectively). This pattern of yield is in contrast to that observed for corn and corn stover. This is because solar radiation and growing degree days which are more abundant in southern Illinois are critical determinants of biomass yield while soil quality is more important for corn yields.

Agronomic data indicate that miscanthus does not yield harvestable biomass in the first year; it provides 50% of its maximum yield in the second year, and 100% of yields from the third year onwards for its remaining life. For switchgrass, we assume that 50% of the maximum yield can be harvested in the first year and full yield can be obtained in the second year and onwards. We also assume that 33% of the peak yield is lost during harvest of miscanthus, but there are no harvest losses for switchgrass (unlike Khanna, 2008; Khanna et al., 2008). Harvested switchgrass and miscanthus have moisture contents of 15 and 20%, respectively.

In estimating the costs of producing miscanthus and switchgrass, we rely on agronomic assumptions about fertilizer, seed, and pesticide application rates for switchgrass and miscanthus described in Khanna et al. (2008), while updating the costs of inputs using 2007 prices. Miscanthus is planted using rhizomes and planting costs are estimated as $1,000 per acre. Costs of harvesting switchgrass and miscanthus (i.e., mowing, raking, baling, and staging) are obtained from the FBFM data (FBFM, 2007, 2008) and from Duffy (2007). Costs of mowing/conditioning and raking in Illinois are $14.2 and $4.5 per acre, respectively, while the cost of staging is $2.75 per bale (with a weight of 950 lbs). Baling costs for switchgrass and miscanthus are based on current estimates of the cost of baling hay. The cost of baling hay with a yield of 1.18 metric tons per acre is estimated to be $20.5 per acre in Illinois. We consider a high-cost scenario in which baling costs of switchgrass and miscanthus increase proportionately with yield. In the low-cost scenario the fixed costs of baling (tractor and implement overhead) are estimated to be $14.3 per acre and invariant with yield. The variable costs of baling include costs of fuel, lube, and labor which depend on the biomass yield to be baled. These are estimated to be $5.25 per metric ton (FBFM, 2008). We also consider a high- and a low-cost scenario for storage of biomass; the former with storage in an enclosed building and the latter with storage in the open field on crushed rock covered by tarp. Storage costs are estimated to be $18.37 per metric ton in the former case (Duffy, 2007) and $3.22 per metric ton in the latter case (Brummer et al., 2000). Loss of biomass during storage is assumed to be 2% and 7% in the high- and low-cost scenarios, respectively.

The costs of producing corn stover include the cost of fertilizer that needs to be applied to replace the loss of nutrients and soil organic matter due to removal of residue from the soil. The costs of replacement fertilizer are obtained by assuming that removal rates of N, P, and K are 7.72, 1.76, and 16.76 pounds, respectively, per dry metric ton of stover removed as estimated by Sheehan et al. (2003). In addition, corn stover collection will involve a second pass through the field using commercial equipment after harvesting the corn grain. The costs of mowing, raking, baling, and staging are determined for a high-cost and low-cost case using similar assumptions as described above. Similar to Malcolm (2008), we assume that 50% of the residue can be removed from fields if corn is produced using no-till continuous corn rotation and 30% can be removed if conventional till is practiced. These estimates are more conservative than those in Khanna (2008). In addition, we consider a scenario of high stover yield, in which 70% of residue can be removed from fields if corn is produced using no-till continuous corn rotation and 50% can be removed using conventional till while other cost items remain the same as in the low-cost scenario.

The estimates of breakeven cost of production of cellulosic feedstocks under average yield conditions in Illinois are shown in Table 17.1. The opportunity costs of land are the foregone profits from a corn–soybean rotation on that land. In the case of corn stover, the opportunity cost of land is estimated under the assumption that demand for corn stover leads to a switch from a corn–soybean rotation to continuous corn with 12% lower corn yields and 40 lbs per acre greater fertilizer applications in the absence of nitrogen fixation by soybeans (University of Illinois Extension, 2002). The costs of producing these feedstocks vary considerably due to spatial

Table 17.1 Farmgate costs of production of cellulosic feedstocks in Illinois

Cost items ($/Acre)	Switchgrass		Miscanthus		Corn stover		
Scenario	High cost	Low cost	High cost	Low cost	High cost	Low cost	High yield
Fertilizer	66.7	66.7	29.8	29.8	11.85	11.85	16.59
Chemicals	7.7	7.7	0.5	0.5	–	–	–
Seed	7.0	7.0	70.8	70.8	–	–	–
Interest on operating inputs	5.7	5.7	7.1	7.1	0.83	0.83	1.16
Preharvest machinery	14.1	14.1	11.0	11.0	–	–	–
Harvesting	117.4	82.5	277.5	151.6	55.0	52.3	60.15
Storage	77.3	14.6	199.3	37.6	29.8	5.6	7.86
Annualized total operating cost ($/acre)	296.3	198.3	595.9	308.4	97.5	70.6	85.8
Annualized deliverable yield (t dm/acre)[a]	3.5	3.3	8.5	8.1	1.4	1.3	1.79
Breakeven cost ($/t dm)	84.5	59.6	70.1	38.2	70.2	55.1	47.91
Opportunity cost of land ($/t dm)[b]	125.8	132.6	51.9	54.7	61.5	64.8	46.26
Breakeven cost inc. land ($/t dm)	210.3	192.2	122.0	92.9	133.6	119.9	94.03

[a]Deliverable yield at the farmgate estimated after including losses during harvest and storage. Yield losses during storage are assumed to be 7% in the low-cost scenario and 2% in the high-cost scenario.
[b]Opportunity cost of land is estimated assuming a price of $5 per bushel for corn and $12 per bushel for soybeans and a yield of 145 bushels/acre for corn and 50 bushels/acre for soybeans with a corn–soybean rotation.

differences in their yields, as well as differences in the costs of land. The costs of corn and corn stover are lower in the northern and central regions of Illinois, while the lowest costs for miscanthus prevail in the southwestern and southern regions of Illinois. The per unit cost of producing switchgrass in Illinois is extremely high compared to miscanthus.

Ethanol yield from corn grain is 2.8 gallons of denatured ethanol per bushel of corn. Based on pilot demonstrations, cellulosic biofuel yield from an n th-generation stand alone plant is estimated as 87.3 gallons per metric t dm of biomass (Wallace et al., 2005). Because of its high deliverable yield (average annualized value of 8.5 t dm per acre), miscanthus produces 86% more ethanol than corn per unit of land (with a yield of 145 bu/acre under a corn–soybean rotation), more than twice as much as switchgrass, and five times as much as corn stover.

The cost of conversion of corn grain to ethanol is obtained from a dry mill ethanol plant simulator developed by Ellinger (2008), which simulates the performance of a 100-million gallon capacity plant over a 7-year period. The cost is estimated to be $0.69/gallon in 2007 prices with adjustments based on Wu (2008). A coproduct credit for DDGS is included assuming that 17.75 lbs of DDGS is produced per bushel of corn used for ethanol. The nonfeedstock costs of producing cellulosic ethanol are estimated to be $1.46 per gallon for a 25-million gallon capacity plant operating 330 days a year in 2007 prices (Wallace et al., 2005).

The costs of biofuel production from alternative feedstocks are reported in Table 17.2 . Ignoring the opportunity cost of land, corn ethanol has the highest feedstock cost while ethanol from miscanthus has the lowest. When the opportunity cost of land is included, miscanthus is still the cheapest feedstock at farmgate, but switchgrass becomes the most expensive. Under average conditions, the cost of ethanol production from corn is estimated as $1.99/gal while the cost of cellulosic ethanol varies between $2.61/gal and $3.96/gal, with miscanthus ethanol being the cheapest and switchgrass ethanol the most expensive.

Table 17.2 Cost of production of biofuels from alternative feedstocks (in $/gallon)[a]

Feedstock	Feedstock cost[b]		Opportunity cost of land				Total cost		Feedstock cost at farmgate[c]	
	High cost	Low cost	High cost	Low cost	Non feed-stock cost	Copro duct credit	High cost	Low cost	High cost	Low cost
Corn	1.78	1.78	–	–	0.69	0.48	1.99	1.99	1.78	1.78
Corn stover	1.03	0.84	0.70	0.74	1.46	0.12	3.08	2.92	1.53	1.37
Switchgrass	1.18	0.89	1.44	1.52	1.46	0.12	3.96	3.71	2.41	2.20
Miscanthus	1.01	0.65	0.59	0.63	1.46	0.12	2.95	2.61	1.40	1.06

[a] Due to space limitations, costs of biofuel from corn stover in the high-yield scenario are not reported in this table. These costs are feedstock cost, $0.55/gal; opportunity cost of land, $0.53/gal; total cost, $2.42/gal; and feedstock cost at farmgate, $0.87/gal.

[b] This cost includes transportation cost but excludes opportunity cost of land.

[c] This cost excludes transportation cost but includes opportunity cost of land.

To obtain the demand functions for corn and soybeans faced by Illinois producers, we use short-run national demand and supply price elasticities estimated by various sources. For corn, we use the demand elasticity of −0.16 (OECD, 2001) and supply elasticity of 0.2 (Gardner, 1976). The corresponding estimates for soybeans are −0.594 (USDA/ERS, 2007) and 0.45 (Gardner, 1988), respectively. The share of Illinois in the US corn and soybean production in 2007 is 17.1 and 14.9%, respectively (USDA/NASS, 2008a). The commodity prices and production quantity in 2007 (excluding the amount of corn used for ethanol) are used to estimate the parameters of the demand functions for corn and soybeans. For wheat, sorghum, and alfalfa, the farmgate prices are assumed to be exogeneous and remain constant throughout the planning horizon at their 2007 values observed in Illinois (USDA/NASS, 2008b).

We use the data on total planted acres by county- and state-level prices for corn, soybeans, sorghum, wheat, and alfalfa for 1995–2007 to estimate the relationship between cropland acreage and the lagged value of the Laspeyres crop price index for each county (with 1995 as the base year). We determine the elasticity of crop-specific acreage responses with respect to own and cross prices for corn and soybeans for each of the nine crop-reporting districts (CRD) in Illinois. In the estimation procedure, we incorporate the current and lagged regional acreages, the lagged state-level crop prices, a time trend, and the national commodity stock levels in December of the previous year. The crop acreage response elasticities estimated thereby for each CRD are then used for determining the land supply in the counties belonging to that CRD.

We consider six most commonly practiced rotation choices in Illinois and two tillage choices for the row crops. Methods used to determine the costs of production of each crop under conservation tillage are described in Dhungana (2007). County-specific historical acres under each crop (crop mixes) for the period 1995–2007 are obtained from USDA/NASS (2008b) and used to set bounds for the allocation of land among crops in each county. We also use simulated (synthetic') crop mixes for each county, which are generated by assuming different combinations of crop prices increased by 50 and 100% over their 2007 levels and using the estimated crop specific elasticities mentioned above.

We conduct a life-cycle analysis of the above-ground CO_2 equivalent emissions (CO_2e) generated from biofuels production using different feedstocks; emissions of the major GHGs are converted to equivalent levels based on their 100-year global-warming potential (IPCC, 2001). We include the CO_2e generated not only from various inputs and machinery used on the farm in the production of each feedstock and the energy used to produce and transport those inputs to the farm, but also from the energy used to transport the feedstock to a biorefinery and the energy used to convert the feedstock to biofuel. Specifically, inputs for feedstock production include fertilizers (e.g., nitrogen, phosphorous, and potassium), herbicides, and insecticides. Energy used in the production of biofuel feedstock includes the direct consumption of gasoline, diesel, liquefied petroleum gas, and electricity, and the indirect consumption of energy embodied in farm equipment such as tractors and plows. Similarly, CO_2e generated during the biorefinery phase accounts for the energy used to convert the feedstock to fuel and the energy embodied in build-ings and equipment in the biorefinery. CO_2e is obtained by aggregating the CO_2 emissions from the energy used and the GHG emissions induced from the use of the inputs such as nitrogen and lime. For more details regarding the assumptions and parameters used in our life-cycle analysis for biofuel feedstock production, see Dhungana (2007); for biofuel conversion, see Farrell et al. (2006). CO_2e from corn stover ethanol is estimated using an incremental emissions approach as in Wu et al. (2006). Specifically, life-cycle emissions arising from stover harvesting and addi-tional chemical application as a result of stover removal are evaluated. If demand for corn stover, miscanthus, or switchgrass leads to a switch away from the baseline corn–soybean rotation to alternative land uses, the change in emissions due to this is also incorporated.

Finally, the annual corn and cellulosic biofuel production targets for Illinois are assumed to be 20% of their respective annual national ethanol mandates. These tar-gets are specified for each year of the planning horizon (e.g., for 2022 the respective mandates are 3 billion and 4.2 billion gallons).

17.4 Results

We simulate land use decisions in Illinois between 2007 and 2022 under four sce-narios: no biofuel targets (baseline), biofuel targets with high costs of feedstock production (high cost), biofuel targets with low costs of feedstock production (low cost), and biofuel mandates with high corn stover removal rates and low costs of feedstock production (high stover yield) (see Table 17.3). Imposing biofuel targets has three types of effects on land use. First and foremost, it increases the demand for land, which in turn increases the cropland brought into production relative to the baseline. Second, the mandate leads to a conversion of land from food crops to bio-fuel crops. Third, the biofuel targets and the resulting demand for corn stover lead to a significant change in the tillage and rotation choices for crop production. More specific results are given below.

Under all three scenarios with biofuel targets, we find that the total land use increases by about 5% by 2022. The results also show an increase in the percentage of land under corn (from 47 to 53–55%), a decrease in the percentage of land under soybeans (from 45 to 29%), wheat (15% reduction), and pasture (44% reduction). Of the total corn produced, 56% would be used to produce ethanol and 14% of the cropland would be diverted to produce miscanthus by 2022 under the high- and low-cost scenarios. The land under miscanthus would be lower, but still significant (10% of the total cropland) under the high stover yield scenario. Switchgrass would not be produced under any of the scenarios we analyzed because of its yield and cost disadvantage compared with miscanthus. The biofuels target results in the use of 100% of the available corn stover for cellulosic biofuel production in 2022 under all scenarios.

Table 17.3 Effect of biofuel targets on land use, crop production, and the environment

	Variables (values calculated for 2022)	Nonethanol baseline	High cost	Low cost	High stover yield
Land use	Total land (M acres)	22.04	23.10	23.09	23.13
	Land under corn (%)	47.83	53.46	53.32	55.27
	Land under soybeans (%)	45.42	28.32	28.50	29.37
	Land under wheat (%)	3.34	2.89	2.83	3.24
	Land under pasture (%)	3.09	1.74	1.74	1.84
	Land under stover (%)		53.46	53.32	55.27
	Land under miscanthus (%)		13.50	13.53	10.18
	Land under conservation tillage (%)	27.55	58.96	58.61	61.49
	Land under corn–soybean rotation (%)	80.05	27.02	29.48	30.58
	Land under corn–corn rotation (%)	7.21	37.25	35.95	36.96
Crop production, consumption (M bushels)	Corn production	1,709.24	1,927.1	1,923.61	1,983.11
	Corn consumption (non ethanol use)	1,709.24	855.67	852.18	911.68
	Soybeans	449.83	283.71	285.69	292.44
Prices in 2022 (in 2007 dollars)	Corn ($/Bu)	4.22	6.04	6.09	5.85
	Soybean ($/Bu)	10.59	11.35	11.35	11.31
	Corn ethanol ($/gallon)		2.16	2.17	2.09
	Cellulosic ethanol ($/gallon)		4.08	2.99	2.54
Volume of ethanol	Corn (B gallons)		3.00	3.00	3.00
	Stover (B gallons)		1.53	1.53	2.21
	Miscanthus (B gallons)		2.67	2.67	1.99
Environmental effects	**Greenhouse gas emissions (M tons)**	0.84	0.39	0.39	0.38
	Energy equivalent fuel emissions	0.76	0.26	0.26	0.26
	Corn production	0.08	0.10	0.11	0.11
	Stover production		0.007	0.007	0.015
	Miscanthus production		0.008	0.008	0.005
	Nitrogen use (1,000 tons)	13.39	16.76	16.75	16.91
	Corn production	12.99	15.79	15.78	15.88
	Stover production		0.39	0.39	0.49
	Miscanthus production		0.20	0.20	0.15

The trends in acreage under corn, stover, and miscanthus under the low-cost scenario are shown in Fig. 17.1. We find that miscanthus and corn stover would be used conjunctively to produce biofuels. Specifically, 36% of the cellulosic target in 2022 would be produced from corn stover. Assumptions about corn stover removal rates have a significant impact on the trends in allocation of acreage among cellulosic feedstocks. In this case, stover production begins in 2010 and is used to meet 83%

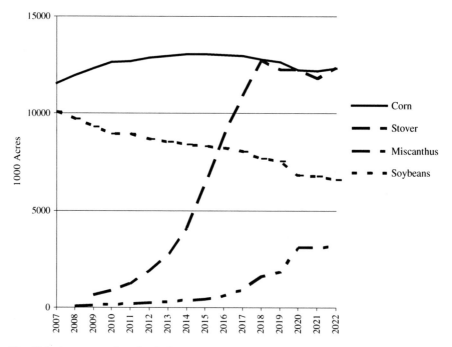

Fig. 17.1 Acreage trends under the low-cost scenario

the cellulosic target in 2015 and 53% of the cellulosic ethanol in 2022. Miscanthus production starts in 2010 and is used to meet 17% of the cellulosic target in 2015 and 47% in 2022.

The diversion of land to biofuel production affects the prices of both corn and soybeans because of the reduced acreage and production of these commodities for food and feed uses, as shown in Fig. 17.2. As compared to the baseline, biofuel targets lead to an increase in total production of corn from 1.7 billion bushels in the baseline to about 1.9 billion bushels, a decrease in food and feed uses by about 50% for corn and 36% for soybeans, an increase in corn prices from $4.22 to $6.09 per bushel (by 44%), and an increase in the price of soybeans by 7% from $10.60 to $11.40 per bushel. Corn and soybean prices in the high stover yield scenario are very similar to those in the other scenarios. The cost of producing cellulosic biofuels differs in the three scenarios due to differences in the share of biofuels from corn stover vs. miscanthus. For instance, the cost of corn ethanol in 2022 is $2.17 per gallon and that of cellulosic ethanol is $2.99 per gallon in the low-cost scenario and $4.08 per gallon in the high-cost scenario. With the high stover yield, the costs of producing corn ethanol and cellulosic ethanol are lower, $2.09 and $2.54 per gallon, respectively.

The cropland under corn–soybean rotation decreases from 80 to 29% in the low-cost scenario, and 27% in the high-cost or the high stover-yield scenarios, while the

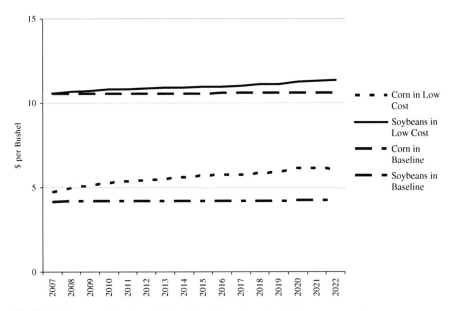

Fig. 17.2 Price trends in the no biofuels baseline and under the low-cost scenario

land under continuous corn increases from 7 to 36, 37, and 37% in the low-cost, high-cost and the high stover-yield scenarios, respectively. We also see an increase in the land under conservation tillage, which allows a larger percentage of corn stover to be collected, from 28% in the baseline to about 59% in both the low- and high-cost scenarios with the mandate in 2022. The land under conservation tillage increases from 27 to 61% in the high stover-yield scenario, leading to a reduction in the land allocated to miscanthus compared to the other two scenarios.

We find considerable spatial variability in the acres devoted to cellulosic feed-stocks across counties and over time. Under the low-cost scenario, in 2015, 90% of the corn acreage would be in the central and northern counties while corn stover would be collected in 73 of 102 counties (6.3 million acres). In contrast, by 2022, the land under corn is reduced in 33 southern counties (by 884 thousand acres) due to the increase in the cellulosic biofuel target which in turn increases the acreage of miscanthus in that region, as shown in Fig. 17.3. Under the high stover-yield sce-nario, in 2015, corn stover would be collected from 4.6 million corn acres in 52 central and northern Illinois counties. Under all three scenarios, corn stover is col-lected from the entire corn acreage in 2022. Under the low-cost scenario, toward the end of the planning horizon, 67 of the 102 Illinois counties would allocate about 14% of their total cropland to miscanthus production, which expands pri-marily in the southern counties, from 1.3 million acres in 2015 to 2.5 million acres in 2022.

Finally, we estimate that the cumulative GHG emissions (2007–2022) from pro-duction of corn and soybeans and the use of energy equivalent gasoline in the

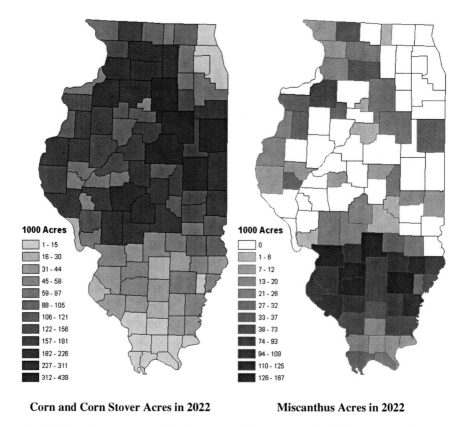

1000 Acres

	1 - 15
	16 - 30
	31 - 44
	45 - 58
	59 - 87
	88 - 105
	106 - 121
	122 - 156
	157 - 181
	182 - 226
	227 - 311
	312 - 439

1000 Acres

	0
	1 - 6
	7 - 12
	13 - 20
	21 - 26
	27 - 32
	33 - 37
	38 - 73
	74 - 93
	94 - 109
	110 - 125
	126 - 167

Corn and Corn Stover Acres in 2022   **Miscanthus Acres in 2022**

Fig. 17.3 Spatial heterogeneity in land use with biofuel targets under the low-cost scenario

absence of biofuel targets is about 0.84 million metric tons. The total emissions over the same period with biofuel targets are estimated as 0.39 million metric tons, 54% lower than the baseline. This reduction is generated primarily by the displacement of gasoline by ethanol which more than offsets the increase in emissions due to greater corn production. The flip side of this environmental benefit is the increased use of nitrogen in agricultural production, which may have adverse implications for water quality. While the total GHG emissions are halved, nitrogen use would be increased by 25% relative to the baseline level because of several reasons, including (i) the expansion of corn acres, (ii) conversion of land from corn–soybean rotation with conventional tillage to the more carbon-intensive continuous corn, and iii) removal of corn stover that has to be compensated by increased use of nitrogen fertilizers. Expansion of perennial crop acreage to meet the biomass demand of the cellulosic ethanol industry adds very little to nitrogen use due to its low requirements for nitrogen.

17.5 Conclusions

This chapter examines the implications of biofuel production targets up to 2022, mandated by the Energy Independence and Security Act, for the allocation of land among food and fuel crops and the resulting impacts on crop prices. Although the study has a somewhat narrow regional focus, the main conclusions are likely to be valid for US agriculture as a whole, commodity markets, and environmental costs/benefits. We find that biofuel targets lead to a significant shift in the acreage from soybean, wheat, and pasture to corn, and a change in crop rotation and tillage practices. Despite an increase in corn production by 12%, the biofuel targets considered here result in significantly higher corn and soybean prices due to the diversion of about 56% of the corn produced to ethanol production. Among cellulosic feedstocks, we find that corn stover is likely to play an important role in meeting the cellulosic biofuel targets in Illinois mainly due to its relatively low cost of production and high yields in this region. All of the available corn stover that can be sustainably harvested is, however, insufficient to meet the biofuel target; this creates demand for miscanthus as an inevitable alternative source of bioenergy due to comparative advantage of this biomass crop vis-a-vis switchgrass in the case of Illinois. There is considerable spatial variability in the allocation of land to food and fuel crops across Illinois, with much of the corn stover production occurring in central and northern Illinois while miscanthus production occurs mostly in southwestern Illinois. Finally, our analysis highlights the trade-offs involved in relying on biofuels, particularly the current generation of biofuels, in terms of climate change mitigation and water quality improvements. Increased biofuel production reduces GHGs by 54%, but it increases nitrogen use by 25% relative to the baseline. In contrast, cellulosic biofuels from grasses, such as miscanthus, offer the potential for carbon emissions reduction with minimal increases in nitrogen applications.

Acknowledgments The authors gratefully acknowledge funding from CSREES\US Department of Agriculture, US Department of Energy and the Energy Biosciences Institute, University of California, Berkeley.

References

Alig R, Adams D, McCarl B, Callaway JM, and Winnett S (1997) "Assessing Effects of Mitigation Strategies for Global Climate Change with an Intertemporal Model of the U.S. Forest and Agriculture Sectors." Environ Res Econ 9: 259–274.

Alig R, Adams M, McCarl B, and Ince PJ (2000) "Economics Potential of Short-Rotation Woody Crops on Agricultural Land for Pulp Fiber Production in the United States." Forest Prod J 50 (5): 67–74.

Bange GA (2007) "The Situation and Outlook for World Corn, Soybean, and Cotton Markets." Presentation to National Grain and Oils Information Center, Beijing, China.

Brummer E, Burras C, Duffy M, and Moore K (2000) "Switchgrass Production in Iowa: Economic Analysis, Soil Suitability, and Varietal Performance." Ames, Iowa: Iowa State University,

Prepared for Bioenergy Feedstock Development Program, Oak Ridge National Laboratory, Oak ridge, TN

Dhungana BR (2007) "Economic Modeling of Bioenergy Crop Production and Carbon Emission Reduction in Illinois." PhD Dissertation, University of Illinois.

Duffy M (2007) "Estimated Costs for Production, Storage, and Transportation of Switchgrass." Dept. Agr Econ, Iowa State University, Ames, Iowa.

Ellinger P (2008) Ethanol Plant Simulator. Dept. Agr Econ, University of Illinois at Urbana-Champaign, Urbana, IL.

English B, Ugarte DG, Menard R, and West T (2008) "Economic and Environmental Impacts of Biofuel Expansion: The Role of Cellulosic Ethanol." Paper presented at the Integration of Agricultural and Energy Systems Conference, Atlanta, GA.

Farrell AE, Plevin RJ, Turner BT, Jones AD, O' Hare M, and Kammen DM (2006) "Ethanol Can Contribute to Energy and Environmental Goals." Science 311 (27): 506–509.

FBFM (2003) "Farm Economics Facts and Opinions." Farm Business and Farm Management Newsletters. Department of Agricultural Economics, University of Illinois at Urbana-Champaign.

FBFM (2007) "Farm Economics Facts and Opinions." Farm Business and Farm Management Newsletters. Department of Agricultural Economics, University of Illinois at Urbana-Champaign.

FBFM (2008) "Farm Economics Facts and Opinions." Farm Business and Farm Management Newsletters. Department of Agricultural Economics, University of Illinois at Urbana-Champaign.

Gardner B (1988) *Economics of Agricultural Policies*. New York: Macmillan Publishing company.

Gardner BL (1976) "Futures Prices in Supply Analysis." Am J Agr Econ 58 (1): 81–84.

Heaton EA, Clifton-Brown J, Voigt T, Jones MB, and Long SP (2004) "Miscanthus for Renewable Energy Generation: European Union Experience and Projections for Illinois." Mitig Adapt Strat Glob Change 9: 433–451.

IPCC (2001) "Climate Change 2001: The Scientific Basis." Third Assessment report of the Intergovernmental Panel on Climate Change. New York: Cambridge University Press.

Khanna M (2008) "Cellulosic Biofuels: Are They Economically Viable and Environmentally Sustainable?" Choices, 3rd Quarter 23, (3): 16–21.

Khanna M, Dhungana B, and Clifton-Brown J (2008) "Costs of Producing Miscanthus and Switchgrass for Bioenergy in Illinois." Biomass Bioenerg 32 (6): 482–493.

Kutcher GP (1972) "Agricultural Planning at the Regional Level: A Programming Model of Mexico's Pacific Northwest." Ph. D dissertation, University of Maryland.

Lewandowski I, Scurlock JMO, Lindvall E, and Christou M (2003) "The Development and Current Status of Perennial Rhizomatous Grasses as Energy Crops in the U.S. and Europe." Biomass Bioenerg 25 (4): 335–361.

Malcolm S (2008) "Weaning Off Corn: Crop Residues and the Transition to Cellulosic Ethanol." Paper Presented at the Transition to A BioEconomy: Environmental and Rural Development Impact, Farm Foundation, St. Louis, MO.

McCarl B and Schneider U (2001) "The Cost of Greenhouse Gas Mitigation in U.S. Agriculture and Forestry." Science 294: 2481–2482.

McCarl BA (1982) "Cropping Activities in Agricultural Sector Models: A Methodological Approach." Am J Agr Econ 64 (4): 768–772.

McCarl BA, Adams DM, Alig RJ, and John TC (2000) "Analysis of Biomass Fueled Electrical Power Plants: Implications in the Agricultural and Forestry Sectors." Ann Oper Res 94: 37–55.

McCarl BA and Spreen TH (1980) "Price Endogenous Mathematical Programming as a Tool for Policy Analysis." Am J Agr Econ (62): 87–102.

OECD (2001) "Market Effectsof Crop Support Measures."

Önal H and McCarl BA (1991) "Exact Aggregation in Mathematical Programming Sector Models." Can J Agr Econ 39: 319–334.

Perlack RD, Wright LL, Graham RL, Stokes BJ, and Erbach DC (2005) "Biomass as Feedstock for a Bioenergy and Bioproducts Industry: The Technical Feasibility of a Billion-Ton Annual Supply." DOE/GO-102005-2135, ORNL/TM-2005/66, Oak Ridge National Laboratory, Oak Ridge, Tennessee.

Sheehan J, Aden A, Paustian K, Killian K, Brenner J, Walsh M, and Nelson R (2003) "Energy and Environmental Aspects of Using Corn Stover for Fuel Ethanol." J Ind Ecol 7: 117–146.

Takayama T and Judge GG (1971) *Spatially and TemporalPrice and Allocation Models.* Amsterdam: North Holland Publishing Co.

Ugarte DG, Walsh ME, Shapouri H, and Slinsky SP (2003) "The Economic Impacts of Bioenergy Crop Production on U.S. Agriculture." U.S. Department of Agriculture.

University of Illinois Extension (2002) "Illinois Agronomy Handbook, 23rd Edition." University of Illinois, College of Agricultural, Consumer and Environmental Sciences, Cooperative Extension Service.

USDA/ERS (2007) Commodity and Food Elasticities: Demand Elasticities from Literature, http://www.ers.usda.gov/Data/Elasticities/query.aspx.

USDA/NASS (2008a) U.S. & All States County Data – Crops, http://www.nass.usda.gov/QuickStats/Create_County_All.jsp.

USDA/NASS (2008b) U.S. & All States Data – Crops, http://www.nass.usda.gov/QuickStats/Create_Federal_All.jsp.

Wallace R, Ibsen K, McAloon A, and Yee W (2005) "Feasibility Study for Co-Locating and Integrating Ethanol Production Plants from Corn Starch and Lignocellulosic Feedstocks." NREL/TP-510-37092 Revised January Edition: USDA/USDOE/NREL.

Walsh ME, Ugarte DGdlt, Shapouri H, and Slinsky SP (2003) "Bioenergy Crop Production in the United States: Potential Quantities, Land Use Changes, and Economic Impacts on the Agricultural Sector." Environ Res Econ 24 (4): 313–333.

Wu M (2008) "Analysis of the Efficiency of the U.S. Ethanol Industry 2007." Argonne National Laboratory, Report Delivered to Renewable fuels Association.

Wu M, Wang M, and Huo H (2006) "Fuel-Cycle Assessment of Selected Bioethanol Production Pathways in the United States." Argonne National Laboratory Report, ANL/ESD/06-7.

Chapter 18
Corn Stover Harvesting: Potential Supply and Water Quality Implications

L.A. Kurkalova, S. Secchi, and P.W. Gassman

Abstract Corn stover is a likely bioenergy feedstock, but whether farmers would find harvesting stover profitable depends on its price as well as on the prices of corn and other agricultural commodities. We analyze the potential supply of corn stover and the associated changes in crop rotations and tillage practices for a major US corn production region, the state of Iowa. Using remote-sensing crop-cover maps and digitized soils data as inputs to integrated economic, geographical, and environmental models, we simulate the land use changes in response to alternative crop and corn stover prices. The ensuing changes in soil erosion, nitrogen runoff, and phosphorus runoff are predicted with the Environmental Policy Integrated Climate model. We find that the amount and the location of corn stover available for sale at a viable stover market are greatly influenced by the primary product (crop) markets. High corn stover prices can significantly affect land use, which in turn may incur significant soil erosion as well as nitrogen and phosphorus runoff in the region.

Keywords Corn stover · Land use change · Economic analysis · Market prices · Soil erosion · Nitrogen runoff · Phosphorus runoff

18.1 Introduction

With its vast, productive agricultural land, Iowa is posed to play a major role in the bioeconomy. While the long-term bio-based production of energy largely depends on technologies still under development, a potential source of biomass in the state that could be used with such technologies is corn stover, comprised of corn stalks, cobs, and leaves left in the field after grain harvest. To date, only large-scale, rough estimates are available on the amount of corn stover that could be brought to market (Gallagher et al., 2003; Perlack et al., 2005; English et al., 2005, 2006a, b; Nelson

L.A. Kurkalova (✉)
North Carolina A & T State University, Greenboro, NC, USA
e-mail: lakurkal@ncat.edu

M. Khanna et al. (eds.), *Handbook of Bioenergy Economics and Policy,*
Natural Resource Management and Policy 33, DOI 10.1007/978-1-4419-0369-3_18,
© Springer Science+Business Media, LLC 2010

et al., 2005; Graham et al., 2007). Moreover, many of these studies merely calculate the total amount of corn residue that could be produced in a geographic area, without regard for how changes in the prices of key agricultural commodities such as corn and soybeans would change the composition of crops, or how corn stover prices would affect the profitability of alternative cropland uses. Such issues would affect both the ability of farmers to participate in the corn stover market and have substantial implication on environmental indicators such as water quality.

If corn stover becomes a widely tradable commodity, water quality may suffer for several reasons. The collection of corn stover precludes crop residue recycling (Smil, 1999). Agronomists and soil scientists have long argued that the recycling of crop residues by leaving them to decay on field surfaces after the harvest or by incorporating them into soil by tillage operations provide valuable environmental benefits including nutrient recycling, maintaining soil carbon levels, control of nutrient runoff, and prevention from water and wind erosion (Johnson et al., 2006). Nitrogen and phosphorus moving from fields to surface water when sediment is transported through runoff and soil erosion are the leading contributors to reduced water quality. The ability of the residues to control erosion by intercepting raindrops, thereby reducing surface runoff and preventing soil particles detachment is well recognized (Smil, 1999; Johnson et al., 2006). While the existing assessments of the potential of harvesting of crop residue for bioenergy production note the associated potential reduction in soil productivity and environmental risks, they focus on containing soil erosion within tolerable levels (Gallagher et al., 2003; English et al., 2005; Nelson et al., 2005; Graham et al., 2007) and/or on maintaining the soil organic carbon levels (Wilhelm et al., 2004; McCarl, 2008), and do not assess the potential water quality impacts.

There are additional reasons to be concerned about water quality, which are associated with the potential expansion of corn acreage. A viable corn stover market would increase the profitability of corn relative to other crops, thus increasing the acreage under corn. Since corn is typically associated with the high nutrient use and loss (e.g., Balkcom et al., 2003; Randall et al., 2003), the expansion of corn acres might lead to worsening nutrient runoff. Also, if the expansion of corn acreage comes through a higher incidence of corn in the rotations on the land that is already in row crop production, than the commonly higher nitrogen fertilizer application rates (Hennessy, 2006) and tillage intensity (Werblow, 2007) under corn monoculture would lead to even more soil erosion and more pressure on water quality. Most early assessments of corn stover supply ignored this concern by treating stover as an existing resource and assuming that stover prices do not alter farmers' decisions on which crops to grow or what tillage practices to use (Gallagher et al., 2003; Nelson et al., 2005; Graham et al., 2007.

In this paper, we link economic, geographical, and environmental models to assess the spatial distribution of corn production, corn residue availability, and associated water quality indicators under alternative corn stover, corn, and soybean prices. Importantly, rather than relying on a representative farm approach, we use the micro (field) level data to capture the spatial heterogeneity of soils, and hence the growing conditions and the potential nutrient runoff and soil erosion impacts in a major US crop production region, the state of Iowa. Like the work of English et al. (2005, 2006a, b), we investigate the effect of alternative corn stover prices on cropland uses, but we build our estimates by using a GIS-based modeling system with much finer spatial resolution. In contrast with previous work that assumed homogeneous units of analysis of the size of a state, crop reporting district, or county, the grid size of our analysis is of a field-scale, with a 30 by 30 m resolution.

We complement previous assessments of corn stover potential for biomass production that focused on the stover supply, overall impacts on the rural economy, and soil erosion by specifically investigating nutrient (nitrogen and phosphorus) runoff. This is an important addition to the understanding of the overall environmental impact of bioenergy production as the pollution from agricultural production in Iowa has been identified as one of the factors contributing to hypoxic conditions in the Gulf of Mexico (EPA Science Advisory Board, 2008).

In the following section we present our data and methodology. After describing the economic and environmental quality models, we present the results on several crop and stover price scenarios. We summarize the results and policy implications in the last section.

18.2 Data and Methods

18.2.1 Data

This work builds on previous efforts to use GIS-based data to assess the environmental consequences of expanding biofuels production and changing agricultural land use in Iowa (Secchi et al., 2008; Kurkalova et al., 2009). We construct the state cropland use baseline and simulate changes in it under several scenarios using the US Department of Agriculture (USDA) National Agricultural Statistics Service (NASS) remote-sensing crop-cover maps, which have been published yearly since 2002 (U.S. Department of Agriculture, National Agricultural Statistics Service, 2002–2006). We combine 5 years of data, 2002–2006, the last year available, to construct historical rotations using a 30-meter square grid as the basis of the analysis. The cropland data are supplemented by the GIS-based soil layer Soil Survey Geographic Database (SSURGO) (U.S. Department of Agriculture, Natural Resources Conservation Service, 2006) and the Corn Suitability Rating (CSR). The CSR is an index from 0 to 100 that measures land's productivity in crop production. CSR data were obtained from the Iowa Cooperative Soil Survey (2003). Secchi et al. (2008) provide further details on the data construction.

Figure 18.1 shows the current use of agricultural land in Iowa. Given the historical farming practices in the area, our analysis considers three mutually exclusive crop rotations: continuous corn (CC), corn–corn–soybeans (CCS), or corn–soybeans (CS). Continuous soybean is usually not practiced in our study area due to the significantly increased likelihood of crop failure because of serious problems with soybean nematodes – *Heterodera glycines* (see, e.g., Koenning et al., 1995), and

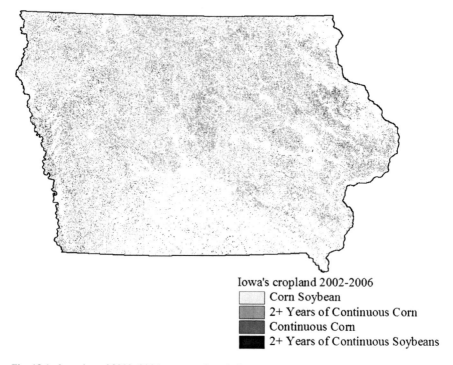

Iowa's cropland 2002-2006
☐ Corn Soybean
▨ 2+ Years of Continuous Corn
▩ Continuous Corn
■ 2+ Years of Continuous Soybeans

Fig. 18.1 Location of 2002–2006 crop rotations in Iowa

thus is not considered in this analysis. While continuous corn is also likely to contribute to a higher probability of pest infestation (e.g., by corn rootworm), the resulting yield reduction is not as severe as for continuous soybeans (see, e.g., Pikul et al., 2005). Reflecting these tendencies, the state cropland baseline based on the 2002–2006 GIS data shows 69.4% of cropped land in CS and 2.1% of cropland in continuous soybean, with the rest of the area in the rotations with two or more years of continuous corn (Fig. 18.1).

Three mutually exclusive tillage systems: conventional, mulch, or no-till are modeled in the study. Conventional tillage usually involves the use of plowing and is defined as any tillage practice that leaves less than 30% residue cover after planting. Mulch tillage involves the use of tillage tools such as chisels, field cultivators, disks, sweeps, or blades. No-till leaves the soil undisturbed from harvest to planting except for nutrient injection and planter passes. As the field-level data on historical tillage use are not available, we make assumptions as follows. Due to the conservation compliance provisions mandated by the Federal government, all highly erodible land (HEL) is assumed to be in no-till. Because reduced tillage is known to have significant impacts on yields in continuous corn (see the discussion below), we assume conventional tillage for all non-HEL land in CC rotation. Finally, given the historical wide-spread use of mulch tillage in many portions of the state (see, e.g., Kurkalova et al., 2009), we assume mulch till on the rest of the land in the baseline.

Corn stover can be used as livestock bedding and winter feed for beef cows. As no historical data are available on the amount of corn stover collection for these purposes in Iowa, we assume no stover collection in the baseline.

18.2.2 Economic Modeling

We treat the years 2002–2006 as the baseline and compare farmers' rotation and tillage choices under alternative corn, soybean, and corn stover prices. In the economic modeling of farmers' choices, we assume that profit maximization is the driving force in farmers' behavior and for simplicity abstract from the impact that the choices have on variance of production.

The field-by-field farmers' choices are determined by comparing multiple-year cumulative expected net returns. The expected net returns are the difference between expected revenue and the expected costs of production. The expected revenue is the product of crop price and expected yield, plus the revenues from selling corn stover if a famer decides to participate in the corn stover market. Since the crop prices modeled in the study are on the higher side of the range observed historically (see the discussion below), we do not consider any additions to the expected revenues from government commodity payments, which have been a part of farmers' income in the times of depressed commodity prices.[1]

Following previous research, maximum possible stover production is estimated to be equal to corn grain mass produced (Gallagher et al., 2003; Johnson et al., 2006; Graham et al., 2007). There is considerable discussion in the agronomic literature about the levels of residue removal that are sustainable from the soil organic matter and soil productivity perspective (e.g., see the discussion in Johnson et al. 2006 and Wilhelm et al., 2004). Ongoing research indicates the acceptable, sustainable level of residue removal depends on numerous factors such as minimum residue requirements, yield, tillage management, and cropping system. Accordingly, previous large-scale assessments assumed that only a proportion of all available stover is available for collection. Gallagher et al. (2003) estimate that 50% corn stover harvest would marginally result in tolerable (5 ton/acre or less) soil erosion on the Class IIIe soil, and subsequently exclude the soils of this class from the assessment. A 100% stover collection is assumed on the land with no soil erosion concerns (Class I) and the land with minor soil erosion concerns (Class II). Graham et al. (2007) note that the equipment presently available for stover collecting would not allow for more than 75% of the stover to be collected. After imposing soil-specific, erosion-consideration constraints on stover collection, Graham et al. (2007) estimate that some 68% of stover in Iowa and Minnesota can be collected without causing

---

[1] The assumption of irrelevance of government commodity payments is consistent with previous economic assessments of the impact of biofuels on agriculture which either did not account for the payments (Gallagher et al., 2003; English et al., 2005), or found that government payments diminish rapidly as commodity market prices increase in response to the increased demand for biomass feedstocks (De La Torre Ugarte et al., 2007).

intolerable soil erosion under 1995–2000 corn management practices. English et al. (2005) assume that only 45% of stover could be collected on the non-HEL land. Given the variety of the estimates of sustainable stover-collection percentages in the literature, we assume a flat, 50% residue removal rate in this study and leave a refinement of this assumption to future studies.

It must be noted that several previous studies, including English et al. (2005), assume no stover collection on HEL because of conservation compliance, which requires farmers to implement basic soil conservation practices to receive government subsidies. However, several recent publications questioned the effectiveness of conservation-compliance implementation, as well as the ability of the government to enforce the policy (Giannakas and Kaplan, 2005; Perez, 2007). Provided the uncertainty about the magnitudes of the government subsidies that the farmers found in noncompliance would lose, as well as the future of the conservation-compliance policy, we chose not to treat HEL differently from other soils.

Given the current instability in the commodity markets, we use two sets of prices in the analysis: Food and Agricultural Policy Research Institute (FAPRI) long-term projections (FAPRI Staff, 2007), with corn prices of $153.54/MT ($3.9/bushel) and soybean prices of $360.09/MT ($9.8/bushel), and the Chicago Board of Trade (CBOT) futures contract prices for the Fall of 2010 from the early summer of 2008 (http://www.cbot.com/), with corn at $259.04/MT for corn ($6.6/bushel) and $540.13/MT for soybean ($14.7/bushel).

Unlike traditional crops, corn stover does not currently have a national or even a regional market. The prices at which corn stover bales have been occasionally sold in the state at hay auctions in 2007 (Edwards, 2007) imply the corn stover price of approximately $50/MT. In the scenarios below detailed, we consider the prices of stover ranging from $5/MT to $100/MT.

The expected crop yield is modeled as the maximum potential yield adjusted for tillage and rotation effect. Following Secchi et al. (2009), the maximum potential yields are approximated as $2.25*CSR$ bu/ac for corn and $0.67*CSR$ bu/ac for soybeans. Both the crop previously grown on the field and the tillage affect yields, but the exact magnitude of the effect depends on a multitude of factors, including geographic location, soils, and climatic factors. While it is well known that corn residue from reduced tillage practices can create problems for germinating and emerging plants, the literature is not uniform on the magnitude of the tillage effect on yield. For example, Kapusta et al. (1996) found no long-term impact of tillage on continuous corn yields. On the other hand, Katsvairo and Cox (2000) found consistently higher yields with a moldboard plow than with a chisel or ridge till regime. After analyzing long-term trends for plow, chisel, ridge, and no-till systems and continuous corn and corn–soybean rotations, Vyn et al. (2000) report that corn yields were 10 to 11% higher on average in the rotations across all tillage regimes, but the corn yield reductions due to continuous corn plantings were much higher in no-till corn (19%) than in plowed corn (5%). Similarly, Wilhelm and Wortmann (2004) report a 12% and a 5% decrease in corn yields due to a change from a plow to a no-till system in continuous corn and in corn–soybean rotation, respectively. Previous corn stover potential assessments assumed a reduced tillage yield reduction independent of rotation: for example, Gallagher et al. (2003) assumed a 9.5% reduction in corn yields due to a switch from conventional to mulch till.

In accordance with the results of agronomic studies conducted under conditions similar to our study area (e.g., Katsvairo and Cox, 2000), we assume the maximum potential yields for corn after soybeans and soybeans after corn, under both conventional and mulch till. No-till corn after soybeans and no-till soybeans after corn reduce the expected yield by 5 and 20% from the maximum potential yield, respectively. Soybeans provide a proven soil nitrogen benefit to corn grown in the following year. Therefore, growing corn after corn leads to a yield reduction across all tillage categories: the expected yield is assumed to be 95, 90, and 80% of the maximum potential yield for conventional, mulch, and no-till, respectively.

Assumptions about representative machinery operations, fertilizer applications, and costs of production are based on Duffy and Smith (2008). Levels of fertilization are assumed independent of tillage, but are affected by previous crop to partially counter the corn-after-corn yield reduction effect. We follow Duffy and Smith (2008) who report fertilizer rates for average Iowa farms, and assume that nitrogen (N), phosphate (P), and potash (K) are applied at the rates of 213, 62, and 50 kg/ha (190, 55, and 45 lb per acre respectively) for corn after corn, and at the rates of 151, 67, and 56 kg/ha (135, 60, and 50 lb/ac) for corn after soybeans, respectively. For soybeans after corn, P and K are applied at the rates of 45 and 84 kg/ha (40 and 75 lb/ac), respectively. We are well aware that fertilizer applications significantly affect yields, but an explicit modeling of varying fertilizer levels is beyond the scope of the study and is left for future analyses.

Following Edwards (2007), we account for the following components of the on-farm of corn stover removal: chopping and raking corn stalks, baling the stover, and replacing lost crop nutrients. Edwards (2008) provides the custom rates for chopping, raking, and baling under the assumption of diesel fuel price of ¢0.726 per liter ($2.75 per gal). To adjust for the current fuel price, we follow his advice and multiply the fuel use per acre by the change in the price of fuel since the survey was conducted. The estimated fuel consumption values per acre for the involved farm operations come from Hanna (2005).

Baling costs are customarily reported on per bale basis. Edwards (2007) reports that large round bales, typically 65 inch in diameter and 60 inch wide, contain only about 8 to 9 lbs of dry stover per cubic foot. Using the mid-range estimate of 8.5 lbs of dry stover per cubic foot, a round bale is estimated to contain about 979 lbs or 444 kg of dry stover per bale.

Extra nutrients removed by harvesting stover must be replaced for future crops. The rates of removal vary depending on the hybrid planted, yields obtained, and how the stover is harvested. Following Edwards (2007), we assume the average rates of 9.1 kgs of nitrogen, 2.7 kgs of phosphate, and 11.3 kgs of potash per ton of dry matter harvested (20, 5.9, and 25 lbs, respectively), and cost the nutrients removed via appropriate fertilizer prices. Kurkalova, 2009, "Modeling costs of crop production and stover collection in Iowa", unpublished, provides further details on the calculation of the on-farm cost of corn stover harvesting used in the analysis.

For each corn, soybean, and corn stover price level under consideration, the outcome of the economic model is the prediction of the rotation and tillage choices associated with the each year in rotation, as well as whether stover is harvested, for each spatial grid unit of the study area. The economic model predictions are subsequently used to assess the changes in soil erosion, nitrogen runoff, and phosphorus runoff, all at the edge of the field.

18.2.3 Modeling Environmental Impacts

We use the Environmental Policy Impact Climate (EPIC[2]) model (Izaurralde et al., 2006) to simulate the impact of corn stover removal at each spatial grid unit in the study. EPIC is a field-scale model that is designed to simulate drainage areas of up to 100 ha characterized by homogeneous weather, soil, landscape, crop rotation, and management system parameters. It operates on a continuous basis using a daily time step and can perform long-term simulations for hundreds and even thousands of years. We used historical monthly weather parameters from 26 weather stations that are available in the EPIC weather generator database for the state of Iowa. The weather parameters generated from these stations were allocated to each SSURGO soil on the basis of proximity. A wide range of crop rotations and other vegetative systems can be simulated with the generic crop growth routine used in EPIC. An extensive array of tillage systems and other management practices can also be simulated with the model.

The model consists of 10 major subcomponents: weather, hydrology, erosion, nutrients, soil carbon, soil temperature, plant growth, plant environment control, tillage, and economic budgets. EPIC has been applied for a wide array of crop production and environmental problems such as irrigation and climate change impacts on crop yields, bioenergy crop production, water and wind erosion, nitrogen, phosphorus, and/or pesticide losses in surface runoff and sediment, nitrate–nitrogen (NO_3–N) losses via subsurface tile drainage, and soil carbon sequestration (Gassman et al., 2005). While the focus of our analysis is on sediment and nutrient losses, which are the main causes of water quality impairment in Iowa (Secchi et al., 2007), we also report carbon results due to their importance in the overall discussion of the pros and cons of biofuel production.

EPIC was applied in this study to simulate 50% removal of corn biomass for cropping systems containing corn in Iowa. The biomass removal was performed in EPIC after corn harvest in October based on guidance provided by J.R. Williams (Personal communication, Blacklands Research and Extension Center, Texas AgriLIFE, Texas A&M University, Temple, TX). The impacts of the biomass removal on soil erosion and other environmental indicators were simulated over a 30-year time period using weather records generated internally in the model.

---

[2]Originally known as the Erosion Productivity Impact Calculator (e.g., see Williams et al., 1984; Williams, 1990).

18.3 Results and Discussion

Overall, we find that the primary product (corn and soybean) markets dictate to a large extent the consequences of the introduction of a corn stover market. CBOT prices entail much higher corn acreage than the FAPRI prices, even in the absence of corn stover markets: some 4.6 million acres are estimated to be in CC under the CBOT prices vs. no acres in CC under the FAPRI prices (Figs. 18.2 and 18.3). If the crop prices remain similar to those witnessed in the last couple of years as in the FAPRI forecast, than a corn stover market would result in no additional corn acreage, at least at the prices below $90/MT, because the stover prices would have to be quite high to force a shift from corn–soybean rotations to continuous corn, due to the differential returns of the two rotations. However, if the crop markets clear at the much higher prices as in the futures contracts traded at CBOT in the summer of

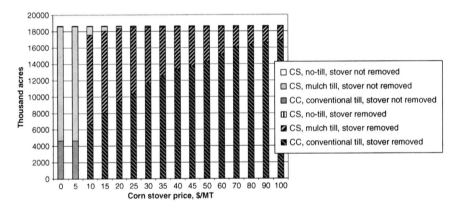

Fig. 18.2 Cropland in alternative farming systems under CBOT prices

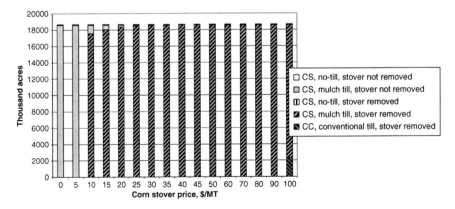

Fig. 18.3 Cropland in alternative farming systems under FAPRI prices

2008, then the crop acreage expansion brought about by the stover market is significant: some 5.8 million acres would convert from CS to CC rotation under the CBOT prices even at the stover price as low as $25/MT, because even low stover prices would be enough to move relatively less productive land into continuous corn, given the high relative corn prices. The corresponding corn and stover production patterns are also very different under the two alternative primary product prices (Fig. 18.4). The more corn is already produced (as under the CBOT prices), the lower is the marginal cost of supplying stover, i.e., the more stover would be supplied at a given price. Overall, however, the farm gate biomass prices that result in substantial production levels are at the lower end of those reported in previous studies, even recent ones (Toman et al., 2008; English et al., 2006a). This suggests that corn stover may be a lower cost feedstock than dedicated crops such as miscanthus or switchgrass – at least if the externalities we examine here are not taken into account.

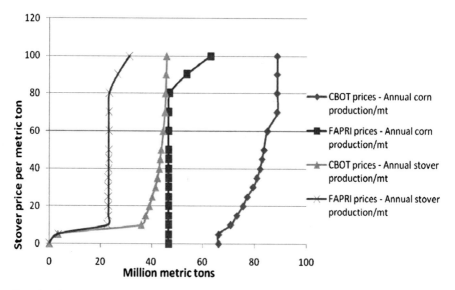

Fig. 18.4 Corn and corn stover supply

The novel, field-scale level of spatial detail used in the study permitted the comparison of the spatial distribution of residue availability under alternative crop market scenarios. Under the FAPRI forecast, the spatial distribution of saleable stover is relatively uniform across the state (Fig. 18.5). However, under the CBOT prices, not only more stover is estimated to be available in the state at any considered stover price, but also the spatial distribution of saleable stover is no longer uniform. Relatively more stover would be available in North-central Iowa (Des Moines lobe) as opposed to the relatively less productive lands along the Mississippi and Missouri rivers. These insights have big implications for the development of the ethanol production industry, especially for the efficient location of stover-processing facilities, as well as biomass transportation and storage costs.

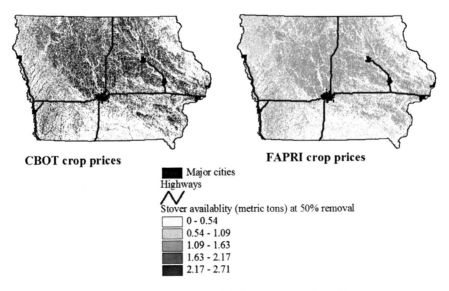

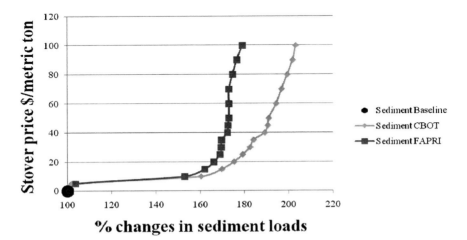

Fig. 18.5 Spatial distribution of corn stover availability at a price of $25/MT

Figures 18.6, 18.7, 18.8, and 18.9 illustrate supply curves corresponding to stover prices $0 to $100/MT for four important environmental indicators using the two sets of crop prices. Overall, they unequivocally show that corn stover removal has negative edge-of-field environmental effects, but they also provide several important, more specific insights on the possible environmental impacts of corn stover removal.

Fig. 18.6 Edge-of-field sediment loss supply curves corresponding to stover prices $0–$100

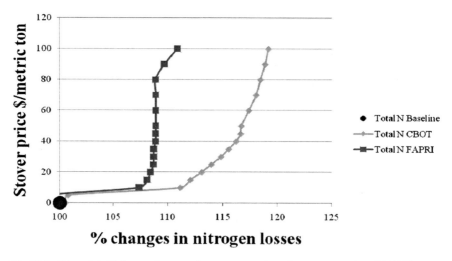

Fig. 18.7 Edge-of-field nitrogen loss supply curves corresponding to stover prices $0–$100

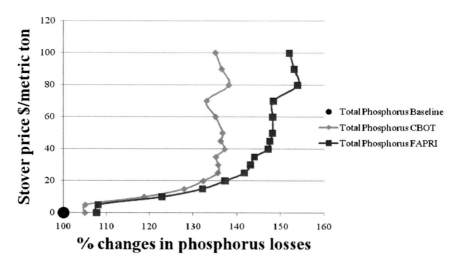

Fig. 18.8 Edge-of-field phosphorus loss supply curves corresponding to stover prices $0–$100

For farm gate corn stover prices above $20/MT, which would be close to the minimum level necessary to bring about substantial levels of corn stover production, all environmental indicators are substantially worse than in the baseline. However, the environmental indicators do not all follow the same trends. In particular, sediment and nitrogen losses (Figs. 18.6 and 18.7) are bigger for all stover price ranges under the CBOT crop price scenario, which is associated with much larger continuous corn acreage. On the other hand, phosphorus losses (Fig. 18.8) are bigger under the FAPRI price scenario. This is likely due to the assumptions in our modeling, which includes a relatively high level of phosphorus application in the soybean years

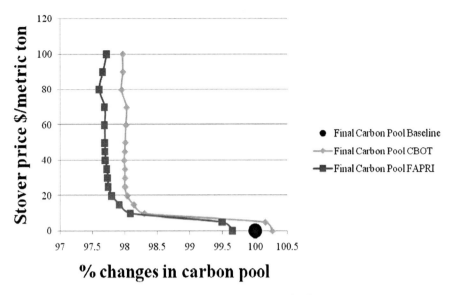

Fig. 18.9 Carbon supply curves corresponding to stover prices $0–$100

(45 kg/ha) and the lack of residue in the soybean years. This is important because phosphorus runoff is essentially controlled by precipitation, which makes available energy for transport of the nutrients, and the impact of precipitation is mediated by its timing and the amount of vegetative cover (Udawatta et al., 2004). The backward bend in the phosphorus losses at high levels of stover prices is related to this issue as well. Those high stover prices are associated with increases in corn production (see Fig. 18.4). Under CBOT prices, there is a sizable increase in corn production at $70/MT for stover, and the corn crop makes the use of the phosphorus fertilizer better than soybeans. Similarly, the slighter bent over $80/MT for stover is due to the need to replace nutrients for the increased stover harvest. With FAPRI prices, for the same reason, the backward bend in phosphorus losses corresponds to the large increase in corn production that occurs with stover prices over $80/MT.

Both phosphorus and nitrogen pollution can have significant impacts on water quality. Excess nitrogen in the Mississippi River system is argued to be a major cause of the oxygen-deficient "dead zone" in the Gulf of Mexico, while excess phosphorus is linked more often to more local, inland water bodies' oxygen deficiency (see e.g., EPA Science Advisory Board, 2008). Recent studies estimate that 43% of the nitrogen and 27% of the phosphorus flux to the Gulf of Mexico originate in the Upper Mississippi River Basin (Aulenbach et al., 2007), to which a large portion of Iowa rivers flows. Our results imply that the potential water quality impacts (and consequently, the potential remedies to alleviate the impacts) of the expansion of biofuel production would depend not only on which metric of water quality is considered (nitrogen or phosphorus content) but also on what crop prices are prevailing.

Final carbon stocks are higher under the CBOT scenario (Fig. 18.9). This is likely due to the fact that corn biomass – both above and below ground – is substantially higher than soybeans (Tufekcioglu et al., 2003). Therefore, even if half of the above ground corn biomass is removed, larger continuous corn acreages correspond to a bigger level of carbon sequestration than those achieved with a corn–soybean rotation associated with a partial removal of the corn biomass.

Comparing the supply curve for stover under CBOT prices and the corresponding sediment and nitrogen losses shows that these environmental attributes have a higher elasticity than the crops themselves. Alternatively, as one would expect, these results indicate a positive correlation between land productivity and environmental fragility. As higher corn stover prices induce more and more marginal land to be harvested for stover, the marginal environmental impact of that land increases.

On the other hand, the supply curve for carbon is very steep, following more closely the crop supply curves. This suggests that the levels of carbon sequestration are directly associated to the crop grown and the amount of biomass harvested rather than the soils and general physical characteristics of the land that the crops are grown on, though, of course, those do indirectly affect productivity.

18.4 Concluding Comments

The uncertainty in predicting the primary commodity prices translates into uncertainty in predicting farmers' choices of crops and rotations, and their water quality impacts. In future research, additional sources of stover supply uncertainty could be investigated. For example, one of the factors that might potentially impact the amount of harvested stover is associated with a short, approximately 3-week-long window for harvesting stover and the corresponding needs for storage (see, e.g., discussion in Petrolia (2008)) and for labor and machinery resources employed over a short time period only. The ambiguity in weather conditions during the stover harvest time would translate into uncertainty about the amount of stover that could be realistically collected. If the probability of weather conditions that preclude complete harvesting of the chosen percentage of stover is positive, then the expected value of the harvested stover (computed over the distribution of all possible weather conditions) is less than the amount we estimate in this study. Moreover, if the harvest time weather uncertainty varies spatially, as represented, e.g., by a spatially varying probability of severely adverse weather conditions during the stover harvest window, then the spatial distribution of the expected collectable stover may be different from the spatial distribution identified in our study.

Admittedly, the treatment of the corn price as exogenous in this analysis would not be justified if larger production regions or higher corn stover prices are considered. Future extensions of this work would need to take into account the positive impact of stover market on corn production and supply which in turn may affect corn prices.

Our analysis is also limited to edge-of-field impacts. While the correlation between these and in-stream water quality indicators is likely high, there is a need to use watershed-based approaches to better assess the regional and downstream effects of these landscape changes.

Finally, while most of the previous literature on use of biomass for energy production used aggregated data, our analysis shows that spatially explicit, fine scale analysis can directly link on-farm management choices to large-scale environmental outcomes. Our results also suggest that thorough assessments using microeconomics-based life-cycle analysis will be necessary to determine the environmental impacts of corn stover harvesting, and other biofuels in general. Economic drivers are important determinants of management choices that affect environmental indicators. Basing assessments of biofuels on past average impacts rather than marginal, economic-driven estimates is likely to provide inaccurate forecasts.

Acknowledgments We thank Todd Campbell for excellent computational assistance. This research was made possible in part by USDA-CSREES grant 2005-51130-02366, EPA collaborative agreement CR83371701-1, NSF CDI CBET grant 0835607, and USDA-CSREES grant 2009-10002-05149.

References

Aulenbach BT, Buxton HT, Battaglin WA and Coupe RH (2007) "Streamflow and nutrient fluxes of the Mississippi-Atchafalaya River Basin and subbasins for the period of record through 2005: U.S. Geological Survey Open-File Report 2007-1080", http://toxics.usgs.gov/pubs/of-2007-1080/index.html

Balkcom KS, Blackmer AM, Hansen DJ, Morris TF, and Mallarino AP (2003) "Testing soils and cornstalks to evaluate Nitrogen management on the watershed scale." *J Enviro Qual* 32: 1015–1024.

De La Torre Ugarte DG, English BC, Hellwinckel CM, Menard RJ, and Walsh ME (2007) "Economic implications to the agricultural sector of increasing the production of biomass feedstocks to meet biopower, biofuels, and bioproduct demands." Working paper, Research Series 08-01. Department fo Agricultural economics, University of Tennessee, Knoxville, TN.

Duffy M and Smith D (2008) "Estimated costs of crop production in Iowa – 2008." Iowa State University Extension, Ag Decision Maker Document FM-1712, revised January 2008. Available at http://www.extension.iastate.edu/agdm/crops/pdf/a1-20.pdf, accessed July 2008.

Edwards W (2007) "Estimating a value for corn stover." Iowa State University Extension, Ag Decision Maker Document FM-1698, revised December 2007.

Edwards W (2008) "2008 Iowa farm custom rate survey." Iowa State University Extension, Ag Decision Maker Document FM-1698, revised March 2008.

English BC, Menard RJ, De La Torre Ugarte DG and Walsh M (2005) "Economic impacts of ethanol production from maize stover in selected Midwestern States." in *Agriculture as a Producer and Consumer of Energy*. Edited by J. Outlaw, K.J. Collins, and J.A. Duffield, CABI Publishing, MA, pp 218–231.

English BC, De La Torre Ugarte D, JensenK, Hellwinckel C, Menard J, Wilson B, Roberts R and Walsh M (2006a) "25% Renewable Energy for the United States by 2025: Agricultural and Economic Impacts". Department of Agricultural Economics, University of Tennessee. Available at: http://www.agpolicy.org/ppap/REPORT%2025x25.pdf.

English BC, De La Torre Ugarte D, Walsh ME, Hellwinckel C, and Menard J (2006b) "Economic competitiveness of bioenergy production and effects on agriculture of the Southern region." *J Agri Applied Econ*, 38(2): 389–402.

EPA Science Advisory Board, Hypoxia Advisory Panel, (2008) "Hypoxia in the Northern Gulf of Mexico: An Update by the EPA Science Advisory Board."

FAPRI Staff (2007) FAPRI 2007 U.S. and World Agricultural Outlook. FAPRI Publications 07-FSR1, Food and Agricultural Policy Research Institute (FAPRI) at Iowa State University and University of Missouri-Columbia.

Gallagher P, Dikeman M, Fritz J, Wailes E, Gauther W, and Shapouri H (2003) "Supply and social cost estimates for biomass from crop residues in the United States." *Enviro Resource Econ* 24: 335–358.

Gassman PW, Williams JR, Benson VW, et al. (2005) Historical development and applications of the EPIC and APEX models. Working Paper 05-WP 397. CARD, Iowa State Univ., Ames, IA. Available at: http://www.card.iastate.edu/publications/synopsis.aspx?id=763.

Giannakas K and Kaplan J (2005) "Policy design and conservation compliance on Highly Erodible Lands." *Land Econ* 81(1): 20–33.

Graham RL, Nelson R, Sheehan J, Perlack RD, and Wright LL (2007) "Current and potential U.S. corn stover supplies." *Agron J* 99: 1–11.

Hanna M (2005) "Fuel required for field operations." Iowa State University Extension, Ag Decision Maker Document A3-37, revised October 2005. http://www.extension.iastate.edu/agdm/crops/pdf/a3-27.pdf, accessed July 2008.

Hennessy DA (2006) "On monoculture and the structure of crop rotations." *Am J Agricult Econ* 88(4): 900–914.

Iowa Cooperative Soil Survey (2003) Iowa Soil Properties and Interpretation Database. URL: http://icss.agron.iastate.edu/.

Izaurralde RC, Williams JR, McGill WB, Rosenberg NJ, and Quiroga Jakas MC (2006) "Simulating Soil C Dynamics with EPIC: Model Description and Testing Against Long-Term Data." *Ecolo Model* 192: 362–384.

Johnson JM-F, Reicosky D, Almaras R, Archer D, and Wilhelm W (2006) "A matter of balance: conservation and renewable energy." *J Soil Water Conserv* 61(4): 121A–125A.

Kapusta G, Krausz RF, and Mathews JL (1996) "Corn yield is equal in conventional, reduced and no-till after 20 years." *Agron J* 88 (5) 812–817.

Katsvairo and Cox (2000) "Economics of Cropping Systems Featuring Different Rotations, Tillage, and Management." *Agron J* 92 (3):485–493.

Kurkalova LA, Secchi S and Gassman P (2009) "Greenhouse gas mitigation potential of corn ethanol: accounting for corn acreage expansion." In *Proceedings of the 3rd National Conference on Environmental Science and Technology, Greensboro, NC, 2007*, Springer, 251–257.

Koenning SR, Schmitt DP, Barker KR, and Gumpertz ML (1995) "Impact pf crop rotation and tillage system on Heterodera glycines population density and soybean yield." *Plant Disease* 79, 282–286.

McCarl BA (2008) "Bioenergy in a Greenhouse Mitigating World." *Choices* 23(1): 31–33.

Nelson GR, Walsh ME, and Sheehan JJ (2005) "The supply of maize stover in the Midwestern United States." in *Agriculture as a Producer and Consumer of Energy*. Edited by J. Outlaw, K.J. Collins, and J.A. Duffield, CABI Publishing, MA, 2005, pp 195–204.

Perez M (2007) "Trouble downstream: upgrading Conservation Compliance." Report prepared by Environmental Working Group for the Mississippi River Water Quality Collaborative. Washington, DC, and Oakland, CA.

Perlack RD, Wright LW, Turhollow A, Stokes B, Erbach DC (2005) "Biomass as feedstock for a bioenergy and bioproducts industry: the technical feasibility of a billion-ton annual supply." Report prepared for the U.S. Department of Energy and the U.S. Department of agriculture, ORNL/TM-2005/66, Oak Ridge National Laboratory, Oak Ridge, TN, April.

Petrolia DR (2008) "An analysis of the relationship between demand for corn stover as an ethanol feedstock and soil erosion." *Rev Agricult Econ* 30: 677–691.

Pikul Jr. JL, Hammack L and Riedell WE (2005) "Corn yield, Nitrogen use, and Corn Rootworm infestation of rotations in the Northern Corn Belt." *Agron J* 97:854–863.

Randall GW, Vetsch JA and Huffman JR (2003) "Nitrate losses in subsurface drainage from a corn-soybean rotation as affected by time of nitrogen application and use of Nitrapyrin." *J Environ Qual* 32: 1764–1772.

Secchi S, Gassman PW, Williams JR and Babcock BA (2009) Corn-based ethanol production and environmental quality: A case of Iowa and the Conservation Reserve Program, *Environmental Management* 44:732–744 .

Secchi S, Kurkalova LA, Gassman PW and Hart C (2008) Land Use Change on the Intensive and Extensive Margin in a Biofuels Hotspot: The Case of Iowa, USA. – under review.

Secchi S, Gassman PW, Jha M, Kurkalova L, Feng HH, Campbell T and Kling CL (2007) The Cost of Cleaner Water: Assessing Agricultural Pollution Reduction at the Watershed Scale. *J Soil Water Conserv* 62(1).

Smil V (1999) "Crop residues: agriculture's largest harvest." *BioScience* 49(4): 299–308.

Toman M, Griffin J and Lempert RJ (2008) Impacts on U.S. energy expenditures and greenhouse-gas emissions of increasing renewable-energy use. RAND Technical report 384-1. URL: http://www.rand.org/pubs/technical_reports/2008/RAND_TR384-1.pdf.

Tufekcioglu A, Raich JW, Isenhart TM and Schultz RC (2003) Biomass, carbon and nitrogen dynamics of multi-species riparian buffers within an agricultural watershed in Iowa, USA. Agroforestry Syst 57(3): 187–198

Udawatta RP, Motavalli PP, and Garrett HE (2004) Phosphorus loss and runoff characteristics in three adjacent agricultural watersheds with claypan soils. J Environ Qual 33:1709–1719.

U.S. Department of Agriculture, Natural Resources Conservation Service (2006) Soil Survey Geographic (SSURGO) Database for Iowa. Available at: http://soildatamart.nrcs.usda.gov.

U.S. Department of Agriculture, National Agricultural Statistics Service, Research and Development Division (2002–2006) Cropland Data Layer (http://www.nass.usda.gov/research/Cropland/SARS1a.htm).

Vyn TJ, West TD, and Steinhardt GC (2000) "Corn and soybean response to tillage and rotation systems on a dark prairie soil: 25 year review." No. 196 (p. 1–10) in Proceedings, 15th Conf. of Int'l. Soil Tillage Research Org., July 2–6, 2000. Fort Worth, Texas. Available at: http://www.agry.purdue.edu/staffbio/ISTROConf.2000Manuscript.pdf.

Werblow S (2007) "More Corn: Is Conservation Tillage at Risk?" *Partners: Quarterly Publication of the Conservation Technology Information Center*. April, 25(2).

Wilhelm WW, Johnson JMF, Hatfield JL, Voorhees WB, and Linden DR (2004) "Crop and soil productivity response to corn residue removal: a literature review." Agron J 96(1): 1–17.

Wilhelm WW and Wortmann CS (2004) "Tillage and Rotation Interactions for Corn and Soybean Grain Yield as Affected by Precipitation and Air Temperature." Agron J 96: 425–432.

Williams JR (1990) The erosion productivity impact calculator (EPIC) model: a case history. *Phil Trans R Soc Lond* 329: 421–428.

Williams JR, Jones CA, and Dyke PT (1984)A modeling approach to determining the relationship between erosion and soil productivity. *Trans ASAE* 27(1): 129–144.

Part V
Economic Effects of Bioenergy Policies

Chapter 19
International Trade Patterns and Policy for Ethanol in the United States

Hyunok Lee and Daniel A. Sumner

Abstract The US trade policy for ethanol affects imports and all aspects of ethanol production and use. The paper reviews U.S. trade policy for ethanol and then examines the pattern of imports of ethanol. Despite high tariff barriers the U.S. is a major ethanol importer and we document the pattern of ethanol imports over the past decades. We then show how ethanol imports have responded to market conditions. We find that the demand for imports is likely to have been very elastic in recent years. Our econometric estimates show how ethanol imports have responded to market conditions. We find a significant supply elasticity for imports into the U.S. of about 3.0. Finally we use the forgoing analysis to discuss potential impacts of trade policy changes under alternative market conditions that depend crucially on domestic biofuel policies.

19.1 Introduction

Every facet of ethanol economics in the United States is dominated by government policy, and international trade is no exception. With some exceptions, imported ethanol has been subject to a 2.5% ad valorem tariff plus a specific duty of $0.54 per gallon. In the early 1980s, Congress imposed the specific duty to offset the benefit

H. Lee (✉)
Department of Agricultural and Resource Economics, University of California, Berkeley, CA, USA
e-mail: hyunok@primal.ucdavis.edu

Lee is a Research Economist in the Department of Agricultural and Resource Economics, University of California, Davis. Sumner is the director of the University of California Agricultural Issues Center and the Frank H. Buck, Jr. is Professor in the Department of Agricultural and Resource Economics, UC Davis. Both are members of the Giannini Foundation. We thank Dr. Sébastien Pouliot and Dr. William Matthews for contributions to some of the empirical analysis.

M. Khanna et al. (eds.), *Handbook of Bioenergy Economics and Policy*,
Natural Resource Management and Policy 33, DOI 10.1007/978-1-4419-0369-3_19,
© Springer Science+Business Media, LLC 2010

afforded by the tax credit to imports. The duty was set to expire at the end of 2008 before being extended to 2010 when it must be reconsidered.

As ethanol takes a more prominent role as a transportation fuel in the United States, the economic implications of the ethanol tariff become more significant and the debate surrounding the tariff is expected to intensify. The current political environment for ethanol is volatile in the United States and the tariff renewal is required by the end of 2010. Trade policy surrounding ethanol demands further analysis as the issue becomes more controversial. In this study, we provide some insight on the important trade policies, the recent patterns of imports, and the response of imports to price. We provide the first econometric evidence on the price elasticity of ethanol import supply to the United States and draw implications of that evidence for implications of policy change.

Recently, there has been a surge of studies surrounding ethanol issues, and the literature in this area has broadened and deepened. Four recent papers analyzing the impact of trade liberalization (and the elimination of the US tax credit) may be most relevant (Elobeid and Tokgoz 2008; Slaski 2008; de Gorter and Just 2008; Martinez-Gonzalez et al. 2007).

Elobeid and Tokgoz (2008) analyze a multimarket international model that includes refiners and ethanol producers and corn, as well as sugarcane input markets. Their results indicate that with the elimination of the US tariff alone, the US price decreases moderately with a large increase in US imports of ethanol. Slaski (2008) examines the impact of eliminating the tariff in the context of multimarkets which include oil refiners and ethanol and ethanol input markets (his simulation uses elasticities calculated by Elobeid and Tokgoz). His study concludes that with the elimination of the tariff little change is expected in the US ethanol price, but there would be a large change in US imports. de Gorter and Just (2008) obtained results similar to Slaski in their US–Brazil multimarket simulation.

Martinez-Gonzalez et al. estimate a system of Brazil export supply and US import demand equations. Their estimation indicates that the price elasticity of export supply from Brazil is very inelastic (0.024), which contrasts with other studies (more discussion is deferred in the later section). None of the previous studies have estimated an export supply to the US market that is relevant to understanding the impact of US price changes that affect exporters as would be caused by a change in the US tariff.

19.2 Summary of US Ethanol Policy

We will not review in detail the domestic policies surrounding ethanol. These have been summarized elsewhere including by the Congressional Research Service (Yacobucci 2007 and 2008a,b). Here, we will focus more on the recent events on domestic ethanol policy, which will then provide the context to the following discussion on international policy.

19.2.1 Domestic Policy

On the supply side of the ethanol market, the main policy is an array of federal, state, and even local subsidies for construction and operation of ethanol plants. Farm subsidy programs that apply to corn have also played a role by increasing the quantity of corn supplied and reducing the price of corn. The corn subsidy is less important when the price of corn is expected to be relatively high as has been true since 2007.

On the demand side, the federal excise tax credit paid to refineries for use of ethanol in their fuel blends was the most important policy for many years. Also important have been environmental rules that required the use of oxygenate additives for gasoline in some regions, along with other environmental policies that precluded the major alternatives to ethanol. Regulations that define "renewable" fuels to meet national environmental standards and California regulations that define "low-carbon" fuels to meet greenhouse gas reduction commitments are emerging as important domestic policies that may have important trade implications. Into the near future, however, the most important domestic policy is likely to be the federal renewable fuels mandates that require a specific amount of ethanol to be blended into gasoline in the United States.

The renewable fuels mandates were expanded and extended in the Energy Independence and Security Act of 2007 (EISA) (P.L. 110–140) that was signed into law in December. The "renewable" fuels mandate was raised to 36 billion gallons to be fully phased in by 2024. Particular fuels made from specified feedstocks have separate mandates. While most of the EISA provisions were scheduled to take effect in 2009, the mandates began in 2008 with a renewable biofuels mandate of 9 billion gallons. The minimum requirement for use of renewable biofuels grows to 10.5 billion gallons for 2009 and then jumps to 12 billion gallons for 2010. For the next 5 years, the mandate increases by 0.6 billion gallons per year to reach the plateau of 15 billion gallons in 2015. In addition to these mandates that apply to conventional biofuels, additional quantities apply to advanced biofuels (defined as other than corn-starch based ethanol) a growing proportion of which must be cellulosic biofuels.

In defining fuels to meet these mandates, the US Environmental Protection Agency (EPA) must determine that the fuels reduce total or "lifecycle" greenhouse gas (GHG) emissions. Higher reduction requirements are set for advanced biofuels and cellulosic biofuels. These rules, which are still in development, may become important for international trade, particularly if it is determined that imported fuels are at an advantage or disadvantage relative to domestically produced fuels. The law allows the Administrator of the EPA to waive a portion of the mandates if they are determined to cause significant economic or environmental harm or if domestically produced biofuels are inadequate to meet the mandate. This later provision related to production of biofuels could raise potential issues for compliance with international agreements.

The Food, Conservation, and Energy Act of 2008 (P.L. 110–246), also known as the 2008 Farm Bill, included some biofuels program modifications, but left in place or extended the basic policies. The energy title of the Farm Bill mainly includes a series of subsidies on the supply side including new funding and loan guarantees for biofuels processing. The most relevant trade provision requires the USDA to purchase sugar that would have been forfeited to the USDA under the loan rate provisions of the sugar program. The sugar is to be sold to eligible bioenergy producers or disposed. The United States typically has sufficient import limits to assure that the price of domestic sugar remains above the loan rate specified under the sugar program. However, enough sugar may enter under NAFTA to undermine the price support program and imply that the cost of sugar purchased by the USDA would be charged against the sugar program. Under the 2008 Farm Bill provision in the energy title, the purchase of sugar would still be costly, but the costs would be attributed to energy policy, not to the sugar program.

Important policies, including the excise tax credit modifications and the renewal of the ethanol import duties, were adjusted in the Farm Bill, but not in the Farm Bill "Energy" title. The excise tax credit began when the Energy Tax Act of 1978 created a partial excise tax exemption for motor fuel that included ethanol in its blend (P.L. 95–618). Since 1978, the tax exemption has ranged between 40 and 60 cents per gallon. The tax credit, known as the Volumetric Ethanol Excise Tax Credit (VEETC), was updated in the American Jobs Creation Act of 2004 (P.L. 108–357). The VEETC provided blenders with a federal tax refund of 51 cents per gallon of ethanol blended with gasoline, which was down from 54 cents per gallon previously. The 2008 Farm Bill reduced the tax credit from 51 cents per gallon to 45 cents per gallon starting in calendar year after annual production and import of ethanol reaches 7.5 billion gallons. The reduction takes effect in 2009 because ethanol production reached the 7.5 billion gallon trigger in 2008.

19.2.2 Trade Policy

As noted above, the United States levies a 2.5% ad valorem tariff and a specific duty of 54 cents per gallon of imported ethanol. The item subject to the duty is listed in the tariff schedules as "ethyl alcohol not for beverage purposes." The specific duty was established in 1980 but is not part of the regular tariff schedule in the Harmonized Tariff Schedule (HTS) classifications of the United States. This duty is found in Section 99 titled "Temporary Legislation Providing for Additional Duties" (http://www.usitc.gov/tata/hts/). The specific duty was not reduced on multilateral basis in 1995 as a part of the Uruguay round World Trade Organization (WTO) agreement. The classification of this duty is controversial internationally in part because other tariffs on agricultural products were reduced by an average of 36% starting in 1995 and the ethanol special duty was not adjusted. However, leaving the ethanol duty out of the tariff reductions for agricultural imports, even if accepted initially, is not transparently appropriate under the rules of the WTO.

The origin of the import duty on ethanol is tied directly to the tax exemption or credit for users of ethanol. Shortly after the tax exemption or credit for ethanol was enacted in 1978, the Internal Revenue Service ruled that the tax credit applied to ethanol from any national origin, not just domestically produced ethanol. That meant that gasoline blenders using imported ethanol would also be eligible for the tax credit. Because one of the initial motivations of the excise tax benefit for ethanol was to stimulate a domestic ethanol industry, Congress responded by amending the US tariff schedule in 1980 to include an additional duty on ethanol in the Omnibus Reconciliation Act of 1980 (P.L. 96–499). Congress explicitly imposed the added duty to offset any benefit that the tax credit would provide to imports and effectively singled out the domestic ethanol industry for subsidy. South Carolina Senator Hollings stated, "This duty increase (for ethanol) would offset what amounts to a 40 cents per gallon subsidy now available to imported alcohol as a result of the excise tax exemption and the 40-cent tax credit," (Congressional Record S19181, July 23, 1980). This clear history may be the basis for legal concerns about the special duty in a WTO context. The duty was set to expire at the end of 2008 before being extended in the Farm Bill.

The United States has created exemptions to the special import duty for ethanol for selected trade partners. For example, the US–Israel Free Trade (FTA) Agreement and the North American Free Trade Agreement allow ethanol to enter duty free so long as it has been produced solely with feedstock sourced within the FTA countries. However, these two exemptions have not been important to trade. More important, ethanol can enter duty free under specific provisions of certain preference programs, such as the Caribbean Basin Initiative (CBI) and the Andean Trade Preference Act. Under the CBI, ethanol from Caribbean countries avoid the secondary duty as long as the ethanol is produced using at least 50% of the feedstock (sugarcane) grown in the CBI countries. Ethanol shipped from the CBI countries may still be eligible for exemption from the special duty when made from less than 50% local feedstock if such imports comprise less than 7% of the US domestic ethanol market.

The typical case of CBI ethanol imports has been hydrous ("wet") ethanol produced in places such as Brazil or European countries that is reprocessed in a dehydration plant in a CBI country and then reshipped in dehydrated form to the United States. The US International Trade Commission is charged with determining the volume of duty-free ethanol imports allowed. The volume of CBI imports has typically been less than half the allowed duty-free limit, so this restriction has not been binding. For the past several years until late 2008, more ethanol has often arrived directly from Brazil than has arrived through the duty-free access from the CBI countries.

A second way that ethanol exports have avoided the high import duty is by the use of the duty drawback provisions that are available to US manufacturers who import materials used in production. In this case, some ethanol is used in blended fuels and processors qualify for the drawback if they export a "like-product." Jet fuel, whether it contained ethanol or not, was used to offset duty paid on ethanol because it was considered a like-product with ethanol used to blend into gasoline. (The US exports jet fuel in the jets as they leave US airports with full tanks.) Hard data on how much ethanol duty was refunded under these provisions have not been available. Nonetheless, the 2008 Farm Bill ended the provision that allowed fuel that does not contain ethanol to qualify for the duty drawback on October 1, 2008. Thus, starting in that month some ethanol imports that had previously been exempt from the $0.54 per gallon duty were now required to pay the duty. We discuss the import pattern surrounding the end of the duty drawback and the increase in the effective tariff after September 2008.

19.3 Expected Effects of the Tariff on the US Market Under Recent Market Conditions

The main focus of this section is to examine how the US ethanol imports may adjust in response to a change in US border policy. In a simple framework, a reduction in the tariff shifts down the supply curve for imports by reducing the wedge between the price paid by importer buyers and the incentive price received by import sellers. Thus, the resulting change in the market quantity mainly depends on the supply and demand elasticities in the import market. In this section, we first discuss how the market demand for ethanol imports may be related to other market variables, and then proceed to discuss on the supply curve. In the analysis of the latter, we will investigate empirically by actually estimating the price elasticity of supply curve for imports into the United States.

19.3.1 Price Elasticity of Import Demand: Relationship with Other Market Parameters

Effects of tariff policy depend on the reach of the tariff, given that there are policies that allow tariff avoidance. Effects also depend on the elasticity of demand for imported ethanol and elasticity of supply of ethanol from export sources. The nature of the demand function also determines the empirical strategy for estimating the elasticity of the supply of imports.

We expect the US demand for ethanol from import sources to be very elastic for several reasons. The intuition about the magnitude of the elasticity of the demand for imports can be gained from considering the import demand as an excess demand function derived from the total domestic demand for ethanol and the supply of domestically produced ethanol.

In most cases, imported ethanol and domestic ethanol are identical products that can be used in identical ways by gasoline blenders. Thus, we can treat domestic and imported products as perfect substitutes in a single market. Import demand quantity then is the total quantity of ethanol demanded minus the quantity supplied from domestic sources and $dQ_i = dQ_{tu} - dQ_{uu}$, which says that a change in the quantity imported, dQ_i, is the change in the total quantity demanded in the US market, dQ_{tu}, minus the change in the quantity supplied to the US market from US suppliers, dQ_{uu}. In elasticity terms, we may write the residual import demand elasticity facing import supplies, η_i, as

$$\eta_i = (Q_{tu}/Q_i)\eta_t - [(Q_{tu}/Q_i) - 1]\varepsilon_u,$$

where η_t is the total demand elasticity for ethanol and ε_u is the US domestic supply elasticity (see Perloff Chapter 8 among many other texts for a simple derivation). Note that Q_{tu}/Q_i is the inverse of the import share in the domestic ethanol market. The larger is the share of the market accounted for by imported ethanol the more imports face downward sloping demands and, therefore, the more price responds to a given percentage increase in imports. Also, the less elastic is either final demand or the supply response from competitors, the more price falls when imports expand.

Imports have increased in recent years, yet share of imports in US ethanol has remained quite small. The share of imports in the US ethanol market has varied between close to 0 and about 10% of the market. In 2006, imports were about 12% of sales of ethanol in the United States. That share dropped to about 6.5% in 2007 and about 8% in 2008. During the first half of 2008, ethanol imports were less than 5% of the US market, the share was more than 5% in August and September only to drop precipitously for the period October to December 2008. Based on these data, the inverse of the market share, $(1/S_i)$, is likely to be at least 10 and recent in the range of 20 or more. (As we note below, for purposes of considering a tariff policy change the market share of imports that actually pay the import duty is relevant.)

The domestic demand for ethanol, η_t, must be very elastic over the range in which ethanol substitutes for gasoline. That is, over the range of data or policy experiment when no strict ethanol mandate is binding, the demand for ethanol is determined by its residual role in the blended gasoline market. Since the share of ethanol in the total US fuel use is very small (about 3%), over this range the overall demand for ethanol will be almost perfectly elastic. When ethanol is mainly used as an oxygenate to meet environmental standards or if the federal mandate for a specific quantity of ethanol is binding, the demand function is less elastic. In the extreme case of a strictly binding mandate in all domestic geographic markets, the overall demand elasticity, η_t, is zero. For the sake of calculating the demand facing imported ethanol supplies, we can use two values of −5.0 and −0.1, to represent elastic and inelastic overall demand of ethanol in the United States.[1]

The supply elasticity of ethanol is also relatively elastic when there is a substantial capacity of ethanol plants because ethanol uses less than one-third of the limiting input, corn. In the early part of this decade, ethanol production used a very small share of corn production, but plant capacity was more limited. We expect the marginal cost of producing ethanol to rise with quantity produced and a 50% increase in cost of production with a doubling of ethanol production is consistent

---

[1] Elobeid and Tokgoz used -0.43 as the total demand elasticity in their simulation. Their study was calibrated using 2005 data when U.S. produces only half of current U.S. capacity.

with cost data when ethanol uses a significant part of the corn crop. Overall, the supply of ethanol from US producers is likely to be somewhat elastic and we take a supply elasticity of 2.0 as representative.[2]

Using an import market share of 8% in the above formula and a supply elasticity of 2.0, the demand elasticity for imports is

$$(-5.0)/(0.08) - ((1/0.08) - 1)(2.0)) = -62.5 - 23.0 = -95.5$$

in the case when η_t is -5, and

$$(-0.1)/0.08) - ((1/0.08) - 1)(2.0)) = -1.25 - 23.0 = -24.5$$

in the case when η_t is -0.1. Even at very inelastic overall demand ($\eta_t = -0.1$), the import demand elasticity is very large so long as the share of imports is relatively small and supply is even moderately elastic. For many purposes, such a very elastic demand functions are functionally equivalent to horizontal, and we will treat them as such in the empirical estimation of the elasticity of import supply below.[3] Even a demand elasticity that was in the smaller range of our calculations would bias the econometric estimates of the import supply function only marginally.

Three changes in underlying conditions could cause the demand facing imported ethanol to be less elastic. First, if the share of imports were to rise such that imports were a major part of the domestic ethanol market, the demand facing imports would be closer to the overall demand elasticity for ethanol. Second, if the overall demand for ethanol were very small, the residual demand would be smaller. As noted above, an inelastic demand for ethanol in the US market could occur if a binding mandate or environmental requirement meant that blenders would be required to buy a set amount of ethanol to blend with gasoline even if the price of ethanol were to rise substantially above the price of gasoline (on an energy content basis). Such price inversions occurred in the summer of 2006 when the California use of ethanol jumped and late in 2008 when the federal mandates are beginning to be binding on ethanol supplies. Note, however, that even if the overall demand were inelastic, a small share of imports and a relatively elastic domestic supply would make the import demand quite elastic.

Finally, a less elastic domestic supply of ethanol production would contribute to a less elastic import demand. For example, in the short run before additional domestic ethanol production capacity could be brought on line and no excess capacity was available, the domestic supply elasticity would remain small and if at that time import shares rose, the demand for imports would be less elastic. Again, this was the short-run situation in the summer of 2006 when US ethanol production capacity

---

[2] Elobeid and Tokgoz used the supply elasticity of 0.65 in their simulation.

[3] Martinez- Gonzalez et al. (2007) estimated the price elasticity of U.S. import demand and the estimated elasticity was 3.84.

was insufficient to meet the new inelastic California demand and prices rose substantially, but only temporarily. Under conditions that held in the market in 2007–2008, demand elasticity facing imported ethanol can be assumed to be very large (in absolute value). As a small share of the total fuel market, ethanol competes with gasoline when the relative price of ethanol is relatively low and supplies exceed the amount needed to meet environmental regulatory requirements.

Let us now turn to considering the overall market for ethanol imported into the United States. Figure 19.1 shows the impact of the ethanol tariff under the conditions that the demand for ethanol imports into the United States is elastic. In this case, a lower tariff increases the price received for imported ethanol and increases import quantity because of a movement along the supply curve. Under the conditions of Fig. 19.1, removal of the import tariff and the consequent increase in import quantity has a minimal effect on the domestic price of ethanol (which is equivalent to the duty inclusive import price) and the domestic quantity.

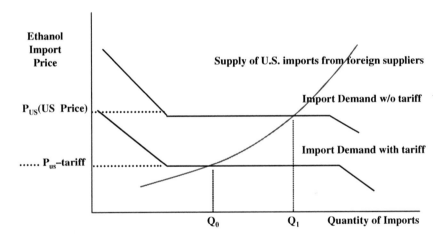

Fig. 19.1 Effects of tariff reduction on the quantity of imports when the US demand for imports is elastic and supply of imports to the United States is upward sloping

Movements in the position of the demand function holding import policy constant, say due to changes in the US price of gasoline or shifts in the cost of production of domestic ethanol, affect the price of ethanol and the quantity of imports. Such movements of the demand function may be used to trace out the import supply function. As shown in Fig. 19.1, the amount by which exogenous US ethanol price movements affects the quantity imported depends critically on the elasticity of supply of ethanol to the Unites States from export sources. We now turn to estimating that parameter econometrically.

19.4 Ethanol Import Patterns and the Elasticity of Supply of Imports in the United States

This section first describes the ethanol import patterns in some detail, clarifying alternative data sources and how shipments have varied over time. We then use these data sources and methods to estimate econometrically the elasticity of the supply of imported ethanol into the United States. We do not estimate a global model of ethanol supply nor do we focus on supply conditions in exporting countries. Our focus here is on the impact of the US domestic price on the supply of imports into the United States. If the United States is a significant share of the world market for ethanol imports and if ethanol supply globally is limited, then the import supply elasticity facing the United States will be less than perfectly elastic. We estimate that parameter. To provide the context, we first summarize the US import patterns in the recent decade. We, then, present our empirical model for the US ethanol import market and our results.

19.4.1 US Ethanol Imports and Data Sources

Here, we provide two data series of US monthly ethanol imports. The quantity of monthly ethanol imports from all sources and all ports are available from US import statistics as reported by the US International Trade Commission (ITC). An alternative source of data, from the US Department of Energy, Energy Information Agency (EIA), provides imports of "fuel ethanol." As shown in Table 19.1, according to the US ITC data, for the period January 1997–October 2008, the ethanol imports range from 1.4 million gallons per month to 108 million gallons per month. The average value for the EIA estimates is below the ITC estimates and the minimum is lower. In the earlier years, the fuel ethanol imports are generally below total imports suggesting that some ethanol imports are not used as fuel.

However, these sources are not fully consistent, indicated by the fact that for some months the EIA estimate of fuel ethanol imports has been larger than total imports as reported by the ITC. For example, in the August 2006 the total imports recorded by the ITC reached a maximum of 108 million gallons, which was exceeded by the fuel ethanol imports of 130 million gallons reported for that month by the EIA. Nonetheless, the two series are closely correlated with one another over the 2001–August 2008 period. They follow the same monthly pattern with similar quantities in most months, with a correlation coefficient of 0.96.

Figure 19.2 shows a scatter plot of the ITC import data indicating low imports through 2003 and then growing imports reaching peaks in the summers of 2006 and 2008. From the summer of 2005, we can also observe some cyclical pattern of summer spikes in ethanol imports, which corresponds with high demand for transportation fuel during the summer season. The US domestic price of ethanol also varied considerably during the data period. For the internal US ethanol price, we used

Table 19.1 Data and descriptive statistics

Variable	Data source	Mean	Standard deviation	Minimum	Maximum
Monthly ethanol imports (Mil. gallons)					
Jan 1997–Sept 2008	ITC[1]	21.3	20.6	1.4	108.5
Jan 2001–Sept 2008	ITC	26.2	23.6	4.4	108.5
Jan 2001–Aug 2008	EIA[2]	20.3	29.4	0.2	130.3
Ethanol price ($/gallon)					
Jan 1997–Sept 2008	NEB[3]	1.59	0.593	0.90	3.58
Jan 2001–Sept 2008	NEB	1.83	0.592	0.94	3.58
Monthly crude oil price ($/gallon)[4]					
Jan 1997–Sept 2008	EIA	0.87	0.605	0.21	3.12
Jan 2001–Sept 2008	EIA	1.10	0.628	0.38	3.12
Monthly corn price ($/bushel)					
Jan 1997–Sept 2008	NASS[5]	2.44	0.832	1.52	6.12
Jan 2001–Sept 2008	NASS	2.64	0.959	1.76	6.12

[1] United States International Trade Commission (Tariff and Trade DataWeb),
HTS Codes prior to June 2008: 2207-20-0000, 2207-10-6000
HTS Codes after June 2008: 2207-10-6010, 2207-10-6090, 2207-20-0090, 2207-20-0010
[2] Energy Information Agency, US Department of Energy (webpage for "fuel" ethanol imports).
[3] Nebraska Ethanol Board, Nebraska Energy Office.
[4] Monthly average derived from weekly United States spot price, FOB weighted by estimated import volume.
[5] National Agricultural Statistics Service, US Department of Agriculture.

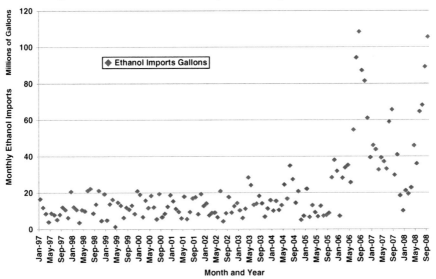

Fig. 19.2 The monthly pattern of US imports of ethanol, January 1997–September 2008, US ITC data

an average monthly domestic rack price made available by the Nebraska Ethanol Board. As summarized in Table 19.1, the domestic price of ethanol varied from less than $1.00 per gallon for several months in the sample period to more than $3.00 per gallon for several months during the period. We next turn to exploring the relationship between the US domestic price of ethanol and import supply of ethanol to the United States.

19.4.2 Statistical Specifications and Estimation Issues

We estimate the US import supply as a function of the US domestic ethanol price of ethanol and the U.S. price of crude oil. Import duties did not change over the period of the data used in the estimation. We interpret the oil price as a proxy for the price of a substitute for ethanol in the global market, and thus a potential shifter of the supply of ethanol into the US market. Other potential shifters, such as policies in other countries or the size of the sugar crop in Brazil, are unlikely to be correlated with the price of ethanol in the United States and thus would not bias the estimate of the price elasticity of supply.

In some specifications, we include the US price of corn, which is the major input into the production costs of domestic ethanol, which is a substitute for imported ethanol. If the specification is capturing demand-side impacts, we would expect the price of corn in the United States to exhibit a strong significant and positive effect on imports by shifting the downward sloping demand of imports, even holding constant the ethanol price. Including the corn price allows us to explore the robustness of the estimates and consider an informal check on the identification of the supply elasticity as opposed to demand-side bias.

These additional data used in the statistical analysis are obtained from the EIA for the US crude oil prices and from USDA for the US corn prices. The statistical summary for these additional data is also presented in Table 19.1.

For each of these three independent variables, the one-month lagged price is used given import decisions are likely made prior to the period when imports enter the country. We use a log–log specification and the coefficients may be interpreted as elasticities. In robustness check using linear specifications, we found very similar results to those reported in the tables. In all cases, we used time-series procedures to correct for significant first-order autoregression in the errors as indicated by low Durbin–Watson statistics in the uncorrected estimates.

Because of differences in import data and the fact that ethanol imports (and ethanol consumption in the United States) were very small in the 1990s, we examined the relationship between domestic price and imports with three separate data sets. We considered the ITC data on all ethanol imports for the period 1997–September 2008 and January 2001–September 2008. We used the EIA data for the period January 2001–August 2008 (the last month available, with June 2005 missing in these data.)

19.4.3 Econometric Estimates of the Import Supply Elasticity

Table 19.2 presents econometric results using the ITC data for the period from 1997 to September 2008. Note the coefficient of the log ethanol price is between 1.27 and 1.54 in all specifications and is between about four and nine times the size of its standard error. Thus the estimate is clearly larger than zero using a 0.01 significance level. None of the coefficients of the other right-side variables is even two times their standard error and the model would fail to reject the hypothesis of a zero impact for either of the other two explanatory variables or the trend.

Table 19.2 Estimate of ethanol import supply function to the United States, monthly data January 1997–September 2008

Dependent variable $=$ log(gallons of monthly imports of ethanol), data source: ITC

	Alternative regressions			
	1	2	3	4
Variable	Coefficient (S.E.)	Coefficient (S.E)	Coefficient (S.E)	Coefficient (S.E)
Intercept	15.94 (0.0889)	15.84 (0.122)	15.75 (0.362)	15.24 (0.432)
Lag log ethanol price	1.54 (0.167)	1.27 (0.294)	1.32 (0.346)	1.34 (0.333)
Trend		0.0028 (.0025)	0.0035 (0.0035)	0.0039 (0.0034)
Lag log crude oil price			−0.079 (0.268)	−0.231 (0.269)
Lag log corn price				0.489 (0.252)
R^2	0.50	0.510	0.51	0.52
Durban–Watson	2.04	2.04	2.03	2.03

As shown in Fig. 19.2, the period through 2000 had low levels of ethanol imports and relatively little variation. For that reason, we also examined the data over a more recent and relevant period from 2001 to 2008. These regressions using the ITC data are shown in Table 19.3 and the results from this truncated period reinforce the results in Table 19.2. Somewhat more of the sample variation in import quantities is accounted for by the ethanol price (R^2 is 0.59 compared to 0.50.) The coefficient now ranges from 1.17 to 1.70, but remains large relative to the estimated standard error in each specification. Again we reject the hypotheses of a zero impact at even a 0.01 level of significance. Over the shorter data period, the trend is more pronounced, between 1 and 2% per year, and is significantly different from zero statistically with the coefficient about two times the size of the standard error. The other explanatory variables are included to control for the price of alternative fuels in the global market and as a general control for commodity prices. As noted, if US

Table 19.3 Estimate of ethanol import supply function to the United States, monthly data January 2001–September 2008

Dependent variable = log(gallons of monthly imports of ethanol), data source: ITC

	Alternative regressions			
	1	2	3	4
Variable	Coefficient (S.E.)	Coefficient (SE)	Coefficient (SE)	Coefficient (S.E.)
Intercept	15.82	15.69	14.89	14.44
	(0.156)	(0.161)	(0.519)	(0.571)
Lag log ethanol price	1.70	1.17	1.48	1.64
	(0.241)	(0.356)	(0.394)	(0.383)
Trend		0.0089	0.021	0.018
		(.0045)	(0.0088)	(0.0085)
Lag log crude oil price			−0.808	−0.961
			(0.500)	(0.487)
Lag log corn price				0.578
				(0.353)
R^2	0.59	0.61	0.62	0.63
Durban–Watson	2.04	2.07	2.05	2.03

Table 19.4 Estimate of ethanol import supply function to the United States, monthly data January 2001–August 2008

Dependent variable = log(gallons of monthly imports of fuel ethanol), data source: EIA[1]

	Alternative regressions[2]			
	1	2	3	4
Variable	Coefficient (S.E.)	Coefficient (SE)	Coefficient (SE)	Coefficient (S.E.)
Intercept	13.06	12.47	11.31	11.60
	(0.395)	(0.296)	(1.053)	(1.116)
Lag log ethanol price	4.50	2.46	2.97	2.87
	(0.851)	(0.729)	(0.879)	(0.901)
Trend		0.0364	0.0535	0.0552
		(.0084)	(0.0162)	(0.0178)
Lag log crude oil price			−1.172	−1.068
			(0.967)	(0.972)
Lag log corn price				−0.360
				(0.920)
Log likelihood	−132.37	−122.17	−121.20	−121.02

[1] Import data have one missing value for June 2005.
[2] Estimation by an ARIMA (AR, 1) procedure to correct for autocorrelation in error terms.

demand were downward sloping, the corn price may represent demand shifts that should have significant impacts on quantity of imports, but it does not.

Table 19.4 reports on estimates of the four models using the EIA data that are specific to fuel ethanol imports over the period 2001–August 2008. These data are best suited to our question of fuel ethanol imports and thus we believe the estimates in Table 19.4 are the most relevant and reliable.

The coefficient on the domestic price of ethanol is larger using these data than those reported in Tables 19.2 and 19.3. The estimated supply elasticity now ranges from 4.5 when no other explanatory variables are included to about 2.5 when the trend term is the only other explanatory variable. In all cases, the coefficient is much larger than the standard error. Indeed, in all four specifications, we can reject the hypothesis that the coefficient is as low as 1.0 with degrees of significance from 0.05 to 0.01 or below. With these data there is a strong and highly significant trend of between 3.5 to 5.5% per year. The other two control variables, the lagged crude oil price and the lagged corn price, both have negative coefficients but neither coefficient is large relative to the standard errors. Thus, we reject the notion that a downward sloping demand function biases the estimates of export supply provided.

Given that the data is explicitly ethanol imports destined for the fuels ethanol market, we are inclined to put more weight on the estimates in Table 19.4. We interpret the estimates of the effect of domestic price of ethanol on the quantity of imports as the supply elasticity of imports into the US market. We also think that the estimates including the trend term are most appropriate with these data. The estimated supply elasticity is between about 2.5 and 3.0. The range of our estimated import supply elasticity is in line with the values previously estimated or implied. The import supply elasticity estimated by de Gorter and Just (2008) is 2.69 (used in their simulation), and a similar value, 2.22, was implied in the study by Elobeid and Tokgoz (2008).[4] However, these elasticities contrast with the one estimated at 0.024 by Martinez-Gonzalez et al. (2007). Given their estimation used the annual data of 1975–2006, and the majority of those years have almost no fuel ethanol imports by the United States. Thus, their estimates are not relevant to the current period and we believe our estimates are reliable indicators of the fuel ethanol supply elasticities facing the United States.

The facts that neither corn price nor the crude oil price has a significant effect on ethanol imports and that the coefficient of the ethanol price (and its estimated standard error) is only marginally affected suggests that the estimated relationship is indeed the supply function and not biased by demand-side concerns. As noted above, imports have been a small share of the market and are not significant enough to drive the domestic price of ethanol in the United States. Thus, we generally expect the demand function to have been approximately horizontal over almost all of the estimation period.

---

[4]Their simulation analysis indicates that the elimination of the U.S. tariff on ethanol imports increases the world price by 23.89% and Brazil net exports by 63.96%, which implies the import supply elasticity of 2.2 (note that some of Brazilian exports are rerouted through the Caribbean basin countries before it reached the U.S.).

19.4.4 Interpretation of the Import Data in the Postduty Drawback Period

Next, let us consider the pattern of imports during the period just before the duty drawback change on October 1, 2008. As noted above, the 2008 Farm Bill removed the eligibility of jet fuel as a like product that would qualify ethanol imports for a duty drawback. Figure 19.2 shows that just before the end of the drawback, imports of ethanol surged in August and September 2008 from 57 million gallons in July 2008 to 81 million gallons in August and 104 million gallons in September 2008. In October 2008, after the end of the duty drawback provision, US ethanol imports dropped to about 26 million gallons (not shown in Fig. 19.2). Imports directly from Brazil, which are not eligible for CBI duty provisions, surged in September 2008 to more than 60 million gallons only to collapse to about 3 million gallons in October 2008. Data through February 2009 show continued low ethanol imports relative to the period before the change in the duty drawback eligibility.

For imports that had previously used the drawback provision, their removal is equivalent to raising the import duty by $0.54. When the demand is horizontal this increase in the duty translates directly into a lower net price received by ethanol shippers to the US market. We can convert the impact of the lower price on export supply in elasticity terms. The base wholesale price of ethanol in late September was about $1.70 per gallon, so the $0.54 per gallon increase in the duty on imports that had not paid that duty tax previously is equivalent to a 32% reduction in the net price (54/170 = 0.32). An import supply elasticity of about 3.0 is just enough to eliminate almost all the ethanol imports that had previously received the import duty drawback (3.0 times a 32% reduction in price equals 96% reduction in quantity supplied). Some of the increase in imports in August and September and some of the collapse in October were due to anticipation of the elimination of the duty drawback, and hence partly a result of inventory management to minimize the tax outlays. Nonetheless, even the more than 50% decline from July to the period after October 1 indicates a substantial impact of the import duty on import supplies. This is especially true since a substantial proportion of remaining imports is shipped in under the CBI duty-free provisions. The decline in ethanol import quantity with removal of the drawback provides an early indication of large expected impact one may expect with a change in tariff policy. The bottom line from both the econometrics and the initial experience with removing duty drawback eligibility is that the demand elasticity for imports is almost completely elastic and the supply elasticity of imports of fuel ethanol into the United States is in the range of 3.0 or more.

19.5 Implications for Changes in Import Tariff Policy

Our estimates of the supply elasticity indicate that removal of the ethanol duty would imply a large percentage increase in ethanol imports. With a domestic ethanol price minus the import duty of approximately $1.20 per gallon, the removal of the $0.54

duty would cause a net price increase to importers of about 45%. Using a constant supply elasticity of 3.0, this implies a 135% increase in those imports subject to the duty. Assuming that after the duty drawback policy shift on October 1, 2008, about half of ethanol imports are subject to the duty, then total imports would increase by about 68% with removal of the tariff. This estimate would need to be adjusted if the demand curve for imports has become less elastic than previously. However, consider the extreme case in which the federal mandate is binding on ethanol capacity and so the price of ethanol is above the minimum set by the use of ethanol as a gasoline substitute and fuel extender. In that case, we might use the low end of our import demand elasticity range of −24.5, still a very elastic response. So long as the US ethanol production capacity remains high relative to demand, and imports have a low market share, the demand facing imports will be very elastic and almost all the effect of changing the tariff would be observed in the quantity of imports.

Imports would have a substantial effect on the domestic price of ethanol only if the mandates are binding, ethanol capacity is binding so that the supply function is quite inelastic in the short run, and the import share is large. This combination of circumstances is unlikely over the next several years. We note that our findings of a small impact of the import duty on the domestic price are consistent with the simulation results of de Gorter and Just (2008). Their simulation results indicate that under the nonbinding ethanol mandate, the elimination of the tariff decreases the US price by less than 1% and increases US imports by about 90%.[5]

Next, consider the extreme case, when the domestic demand for ethanol is perfectly vertical because the demand is fixed by a mandate that is determined by federal regulation. The elasticity of supply in the domestic market and the supply elasticity of imports into the United States determine the quantity response to the lower domestic price and determines how much the domestic price will decline if the import duty is removed. The effect of reducing the tariff on the domestic price of ethanol when the domestic demand is completely inelastic may be derived simply and summarized in a simple equation in log differential form as follows:

$$\mathrm{d}\ln P = [S_{\mathrm{it}}\varepsilon_{\mathrm{it}}(t/(P-t))\mathrm{d}\ln t]/[S_{\mathrm{it}}\varepsilon_{\mathrm{it}}(P/(P-t)) + S_{\mathrm{if}}\varepsilon_{\mathrm{if}} + S_{\mathrm{u}}\varepsilon_{\mathrm{u}}],$$

where S denotes market share, the subscript "it" denotes imports subject to special duty, the subscript "if" denotes imports not subject to the special duty (those under the CBI provisions), and the subscript u demotes domestic supply from the United States. As before, ε denotes a supply elasticity. The special duty per gallon rate is denoted as t and P is the domestic price. Assuming the import market share is 8% and half is subject to the special duty, then $S_{\mathrm{it}} = 0.04$, $S_{\mathrm{if}} = 0.04$, and therefore, $S_{\mathrm{u}} = 0.92$. The two supply elasticities of imports facing the US market are both 3.0 and the US supply elasticity is 2.0. The special duty, t, is $0.54 and P varies

---

[5]Elobeid and Tokgoz (2008) had a larger price impact. Their study indicates that the elimination of the U.S. tariff on ethanol decreases the U.S. ethanol price by 13.6% and increases imports by 200%.

from month to month, but a recent figure is \$1.74. Therefore, $t/(P-t) = 0.45$ and $P/(P-t) = 1.45$.

Using these data and parameters, consider the elimination of the tariff so $d\ln t = -100\%$. With the parameters outlined above the equation yields,

$$d\ln P = -0.12(0.45)/[0.12(1.45) + 0.12 + 1.84] = -0.025, \text{ or about } -2.5\%.$$

Even if the share of duty-paid imports in the domestic market were to more than double to 0.10 and the domestic supply elasticity were much smaller, say 1.0, the price effect of removing the import duty would still be less than 10% – given that the price of ethanol has often varied by 50% from month to month. The price impact of removing the export duty would be expected to be small in the context of this industry.

19.6 Further Consequences and Concluding Remarks

A number of issues surround international trade and the US biofuels industry. These include the position of the trade barriers and subsidy programs in the context of WTO dispute resolution. The environmental impacts of ethanol production and the impacts of rules defining "renewable" fuel and low carbon fuel on imports relative to domestic supply are also of global concern. Finally, the impacts of removing tariff protections have implications for the domestic industry, as well as importers and exporters. In this paper, we show that the elasticity of the supply of ethanol imports into the United States is likely to be relatively large, in the range of 3.0, but not infinitely elastic. The United States is not a small country in the market for ethanol. Removing trade restrictions would likely increase imports by between 50 and 100%, but this would have modest impacts on the price of ethanol because demand for imports is quite elastic. In response to a slightly lower price, the domestic production of ethanol would also decline modestly, probably by between 2 and 10%.

References

de Gorter H and Just DR (2008) The Economics of the U.S. Ethanol Import Tariff with a Blend Mandate and Tax Credit. *Journal of Agricultural & Food Industrial Organization*, Vol. 6, Iss. 2, Article 6

Economic Research Service, U.S. Department of Agriculture (2008) Farm Bill Side By Side. (http://www.ers.usda.gov/FarmBill/2008/Titles/TitleIXEnergy.htm)

Elobeid A and Tokgoz S (2008) Removal of Distortions in the U.S. Ethanol Market: What Does it Imply for the United States and Brazil? *American Journal of Agricultural Economics*, Vol.90, Iss. 4, 918–932

Martinez-Gonzalez A, Sheldon, I, Thompson, S (2007) Estimating the Welfare Effects of U.S. Distortions in the Ethanol Market Using a Partial Equilibrium Trade Model. *Journal of Agricultural & Food Industrial Organization*, Vol. 5, Iss. 2, Article 6, Available at http://www.bepress.com/jafio/vol5/iss2/art5/

Perloff, J. (2008) *Microeconomics: Theory & Applications with Calculus* Boston, MA: Pearson Addison Wesley

Slaski A (2008) The Economic and Environmental Effects of Eliminating the U.S. Tariff on Ethanol. Reg-Market Center, AEI Center for Regulatory and Market Studies

Yacobucci BD (2007) *Fuel Ethanol: Background and Public Policy Issues*, Congressional Research Service Report RL33290 (July)

Yacobucci BD (2008a) *Ethanol Imports and the Caribbean Basin Initiative* Congressional Research Service Report RS21930 (Updated March 18)

Yacobucci BD (2008b) *Biofuels Incentives: A Summary of Federal Programs*, Congressional Research Service Report RL33572 (July)

Chapter 20
The Welfare Economics of Biofuel Tax Credits and Mandates

Harry de Gorter and David R. Just

Abstract An ethanol consumption mandate is a tax on fuel consumers with a fixed oil price, but consumer fuel prices may decline with endogenous oil prices, putting the burden on oil producers. The ethanol consumption mandate has an ambiguous effect on total fuel consumption, CO_2 emissions, and miles traveled. A tax credit increases fuel consumption and miles traveled. But a tax credit subsidizes gasoline consumption in lieu of a binding mandate, contradicting energy and environmental policy goals while providing no extra support to farmers. A mandate of 36 bil. gallons by 2022 will cost taxpayers $28.7 bil., potentially generating up to $37 bil. in annual social deadweight costs of increased CO_2 emissions, pollution, miles traveled, and dependence on foreign oil. The intercept of the ethanol supply curve is above the gasoline price, implying part of the price premium due to ethanol policy is redundant and represents "rectangular" deadweight costs that dwarf standard measures. Historically, corn subsidies were required for any ethanol production to occur; ethanol import tariffs, mandates, production subsidies, and tax credits were not enough. The claim by proponents that ethanol policy reduces tax costs of farm subsidy programs is therefore in doubt as farm subsidies make ethanol policy more inefficient and vice-versa.

20.1 Introduction

Important political, economic, and environmental issues account for the increased focus worldwide on biofuel policies (Jank et al. 2007; Kojima et al. 2007). There are high expectations that biofuels can be a solution to a host of policy problems, ranging from reducing dependency on oil and tax costs of farm programs to improving farm incomes and environmental quality (Rajagopal and Zilberman 2007; Miranowski 2007). Hence, ambitious goals on the use of biofuels are being

D.R. Just (✉)
Department of Applied Economics and Management, Cornell University, Ithaca, CA, USA
e-mail: drj3@cornell.edu

M. Khanna et al. (eds.), *Handbook of Bioenergy Economics and Policy*,
Natural Resource Management and Policy 33, DOI 10.1007/978-1-4419-0369-3_20,
© Springer Science+Business Media, LLC 2010

set in many countries, including developing countries. Meanwhile, the most salient set of recent criticisms of biofuels relate to their impact on food prices and the environment. Rapidly escalating food prices have stressed many developing countries and poor households while recent studies argue that indirect land use changes due to biofuels may enhance greenhouse gas emissions (Runge and Senauer 2007; Searchinger et al. 2008). The potential misalignment of policy effects and stated objectives means it is very important to understand the social costs and benefits of government biofuel policies on agricultural, biofuel, and gasoline markets.

This chapter outlines a general framework to analyze the welfare economics of biofuel policies. To do so, we first enumerate the various public policy goals and categorize the concomitant policies adopted. We then develop our general analytical framework for determining the social costs and benefits of alternative biofuel policies. We use this framework to determine who benefits and who loses from each policy and suggest how policy reforms can better achieve policy goals. We show that policies have been counterproductive in several instances and so can be much improved. Before such analysis can be properly undertaken, it is important to understand how ethanol prices are determined in relation to corn and oil prices in the presence of alternative policies including a tax credit or a mandate.

In evaluating the social costs and benefits of each policy, we highlight the interaction effects between policies. For example, the sole cause of biofuel production in the United States historically for the most part was biofuel and feedstock production subsidies. Tax credits by themselves would have generated little if any ethanol production. Oil prices were so low that the intercept of the ethanol supply curve has been well above oil prices historically. Tax credits therefore had minimal impacts on corn prices relative to free market levels when oil prices are low. But at higher oil prices, the gap between oil prices and the intercept of the ethanol supply curve narrowed. The tax credit then had a larger impact on corn prices. If the oil prices are above the ethanol supply curve intercept, then the tax credit is shown to increase the corn price by a hefty $2.08 per bushel.

We also determine that mandates are more efficient than tax credits for the same level of ethanol production because mandates result in higher gasoline prices and lower CO_2 emissions and miles traveled. When tax credits are used in conjunction with mandates, the effects of biofuel tax credits are reversed. By themselves, tax credits subsidize biofuel consumption, but with mandates the same tax credit subsidizes gasoline consumption. This has major implications for countries worldwide that also use both tax credits and mandates.

This chapter is outlined as follows. We first define policy categories and policy objectives. We then begin by outlining the economics of a tax credit and derive the complex relationship linking the ethanol price to oil-based gasoline prices. If actual ethanol prices are higher than this predicted price, then the mandate was binding. A brief look at historical prices in the US ethanol and gasoline markets shows how the tax credit did not determine ethanol prices; de facto mandates due to environmental regulations and the value of ethanol as an additive to gasoline was the primary determinant of ethanol prices. We then explain how mandates work and compare their effects to tax credits. Because mandates result in higher consumer gasoline prices, they are preferred to tax credits in the face of suboptimal gasoline taxes. We then compare the effects of the tax credit to a mandate on fuel consumption, CO_2 emissions, and miles traveled.

Ethanol price premiums are very high compared to the corn price itself. The explanation requires the introduction of the concept of "water" in the tax credit to depict the cases where the intercept of the supply curve is above oil prices. This generates substantial "rectangular" deadweight costs such that corn producers often did not gain from tax credits in the past. Furthermore, we show that the corn price is below the free market corn price (with no ethanol tax credit or corn production subsidies) because of both corn subsidies and the ethanol tax credit. This means corn subsidies, interacting with the tax credit, were the sole cause of ethanol production for many years.

20.2 Policy Objectives and Instruments

Before we proceed, it is useful to summarize biofuel policies and the policy objectives. It is generally agreed that there are three broad categories of policy objectives, many of which are mutually reinforcing:

(1) To reduce dependence on fossil fuels due to rising oil prices, dwindling oil supplies (concerns over "peak oil" and energy shortages), instability in both oil prices and sources of supply (political instability in the Middle East and among developing country exporters), and the desire to diversify both energy use and energy sources, particularly reduced dependence on imported oil
(2) To improve the environment (reduce local air pollution and mitigate global climate change)
(3) To improve farm incomes, reduce tax costs of farm subsidy programs, and stimulate rural development

Given the plethora of policy objectives, governments have implemented a myriad of policies. Biofuel policies generally promote biofuel production and substitution for petroleum fuels in consumption. These policies can be classified as follows:

(1) Consumer excise-tax exemptions at the gasoline pump (blender's tax credit in the United States)[1]
(2) Mandatory blending or biofuel consumption requirements (from domestic and import supplies)
(3) Import tariffs and quotas on biofuels
(4) Production subsidies for biofuel feedstocks (e.g., corn) and biofuels themselves (grants, loan guarantees, tax incentives, etc.)

---

[1] Fuel tax reductions depend on presence and magnitude of excise taxes levied on petroleum fuels (some developing countries subsidize petroleum consumption).

(5) Subsidies for R&D of new technologies
(6) Policies that shift the:

    a. demand curve for nonbiofuel feedstocks to the right (e.g., US import quota on sugar increases the demand for corn used as a sweetener product), or
    b. supply curve for biofuel feedstocks to the left (e.g., US subsidies for other crops)

All policies enhance biofuel prices and production except category (6) where sugar policies divert corn from ethanol to corn syrup (while reducing the world price of sugar and hence reducing import prices of ethanol from Brazil) while subsidies to other field crops divert land from corn production.

Policies in categories (1) and (2) by themselves do not discriminate against trade. It is very difficult to determine a priori which of the tax credits (currently total-ing over 52 cents per gallon if we include both state and federal credits) or the mandates (several state and federal mandates exist either explicitly or de facto via environmental regulations) are more important. According to Kojima et al. (2007), "Among various support measures, fuel tax exemptions are most widely used" (p. 54). Meanwhile, de Gorter and Just (2009a) report that over 65% of total fuel consumption is affected by tax exemptions for biofuels. On the other hand, a recent FAO report concluded that "virtually all existing laws to promote...biofuels set blending requirements, meaning the percentages of biofuels that should be mixed with conventional fuels" (Jull et al. 2007, p. 21). The evidence is overwhelming that countries use both tax credits and mandates. We will not only compare a tax credit to a mandate but also show that if the mandates are binding, then the tax credits act as a petroleum-based gasoline subsidy rather than a subsidy on ethanol consumption as intended.

Most countries have significant import tariffs on biofuels while production sub-sidies for biofuel feedstocks and biofuels themselves are also important (Steenblik 2007). We will, therefore, also touch upon the effects of production subsidies for biofuels and biofuel feedstocks and of biofuel import tariffs.[2] Policies in category (5) are also important but may have a longer run impact.

20.3 How Tax Credits Affect the Ethanol Market

Consider a competitive market with biofuel and gasoline assumed to be perfect sub-stitutes in consumption with a fuel tax t. With no biofuel policies in place, the price ethanol consumers are willing to pay is given by

$$P_E^* = \lambda(P_G + t) \tag{20.1}$$

---

[2]For a comprehensive documentation of all types of U.S. ethanol policies including import tariffs and ethanol production subsidies, see Koplow (2007). For an analysis of the economic effects of the ethanol import tariff, see de Gorter and Just (2008c).

where P_G is the price of gasoline and λ is the ratio of miles per gallon of ethanol relative to gasoline and is approximately 0.70 when adjusted to an E-100 (100% ethanol) basis. Notice that the willingness to pay for ethanol by consumers P_E^* can be greater or less than P_G, depending on the relative values of P_G, t, and λ. In the United Kingdom, where taxes are over \$3.40 per gallon, the ethanol price consumers are willing to pay exceeds P_G, but in the United States, the opposite occurs.

Now introduce a tax credit for ethanol t_c. To take advantage of the government subsidy offered them, blenders of ethanol and gasoline will bid up the price of ethanol until it is above consumers' willingness to pay for ethanol by the amount of the tax credit (about 52 cents per gallon in the United States if we include state tax credits). Otherwise, blenders would be foregoing money represented in the tax credit. The tax credit is an ethanol consumption subsidy but because ethanol is a perfect substitute for gasoline and current ethanol production is too low to have a significant impact on gasoline prices, the incidence of the subsidy is such that ethanol producers get most of the benefit.

In reality, the tax credit for ethanol, t_c, is higher than the fuel tax in the United States. The net revenues to refiners for ethanol sales are given by

$$P_E + t - t_c \tag{20.2}$$

Setting Equation (20.2) equal to Equation (20.1) and solving for P_E gives the market price

$$P_E^\wedge = \lambda P_G - (1 - \lambda)t + t_c \tag{20.3}$$

It is possible for the producer price of ethanol to be either above or below the consumer price of ethanol P_E^*, depending on whether t_c is greater than or less than t. If the tax credit is eliminated, then the market price is equal to $\lambda P_G - (1 - \lambda)t$. It is interesting to note in this situation that the tax t is a disproportionate tax on ethanol because it is levied on a volume basis rather than based on ethanol's contribution to miles traveled. Increasing the fuel tax reduces the market price for ethanol because consumers are only willing to pay λP_G while suppliers have to pay for the full tax t. Hence, $(1 - \lambda)t$ becomes a net cost to suppliers. One implication is that if ethanol is to be encouraged, there should always be a tax credit of $(1 - \lambda)t$ to ensure ethanol does not get "over-taxed." Note also that domestic and foreign producers of ethanol benefit alike from this tax credit.

To assess the impact of ethanol policies in the past, Fig. 20.1 presents three price series: the actual ethanol price, the ethanol price if there was only a tax credit (given by Equation (20.3)), and the price of ethanol if there was no policy nor additive value for ethanol. There are several important conclusions when analyzing this historical experience in the United States. First, the price premium for ethanol over gasoline has exceeded the tax credit in the past 25 years. This is shown in Fig. 20.1 where the actual ethanol price is higher than the price that otherwise would be if only

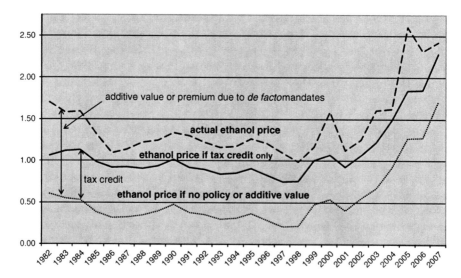

Fig 20.1 Ethanol prices: Actual; tax credit only; or no policy

a tax credit affected ethanol prices and consumers purchased ethanol only for its contribution to mileage. This means that because the blue line in Fig. 20.1 is above the pink line, the tax credit was dormant.[3] How can one explain the fact that the ethanol price premium was above the tax credit in these years? Mandates at local, state, and federal levels always existed but were never binding. Two explanations are plausible (Tyner 2007). First, there were de facto mandates due to environmental regulations (the Clean Air Act in the 1990s and the implicit ban on MTBE in this decade). Second, blenders purchased ethanol for its additive value as an octane enhancer or oxygenate. This means ethanol was purchased in fixed proportions to gasoline, thus demand for ethanol would function as if a blend consumption mandate were in place.

20.4 The Economics of Biofuel Mandates

Understanding the effects of mandates is very important. First, many countries have mandates. Second, historical price premiums for ethanol above the tax credit in the United States as shown in Fig. 20.1 suggests a mandate existed (de facto due to environmental regulations or due to ethanol purchased for its additive value). Third, the new renewable fuel standard (RFS) in the recently passed Energy Independence and Security Act (EISA) mandates the use of 36 billion gallons of renewable fuel

---

[3]Not exactly "dormant" because as we show in a moment, when the ethanol premium exceeds the tax credit, a de facto mandate or ethanol purchased on the basis of its additive value necessarily implies the effects of the tax credit are reversed: it subsidizes oil consumption!

by 2022 in the United States. Consider a biofuel consumption mandate of the level Q_E. Because no tax costs are involved with a mandate, the consumer has to pay the weighted average price of the biofuel and gasoline where the weights are formed by the required consumption of biofuels:

$$P_F = P'_E Q_E + P_G(C_F - Q_E) \qquad (20.4)$$

where P_F is the weighted average fuel price for consumers, C_F is the consumption of fuel, P'_E is the market price of ethanol, and Q_E is the mandated level of ethanol consumption. The consumer must pay this weighted average price if the blending market is competitive by the zero profit condition. The market equilibrium is depicted in Fig. 20.2 where the ethanol price premium due to the mandate is given by γ_M. The marginal cost, and hence supply curve of the mandated mixture, is S_F as P_E is the marginal cost of biofuel and P_G is the marginal cost of gasoline.

The market equilibrium is determined by the intersection of S_F with the demand for fuel D_F that determines the weighted average price P_F and total fuel consumption C'_F. Due to the weighted price formula, the supply curve for fuel is flat at P_E until Q_E is achieved; it is convex for all fuel consumption beyond Q_E and is asymptotic to the perfectly elastic gasoline supply curve (as the quantity of gas goes to infinity while the quantity of ethanol remains at the mandated level).

Because we assume the supply curve for oil is flat in Fig. 20.2, the transfer to ethanol producers is completely financed by an implicit consumer tax on gasoline.

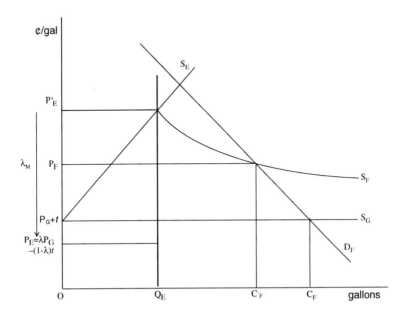

Fig 20.2 The economics of a biofuel consumption mandate

For the same level of ethanol production, this necessarily implies gasoline consumption is lower with a mandate compared to tax credit. Recall in the analysis of a tax credit there is no effect on total fuel consumption with a tax credit and fixed oil price. Total fuel consumption remains at C_F and gasoline consumption declines by Q_E. But with a mandate, total fuel consumption declines, necessarily resulting in a lower level of gasoline consumption by the amount of $C_F - C'_F$ compared to a tax credit.

Now, consider the case where the supply curve for gasoline is upward sloping. With a tax credit, total fuel consumption increases and world oil price declines. But in the case of a mandate, an upward sloping supply curve for oil will now result in the mandate acting as a tax on oil producers, but not always a tax on consumers, depending on market parameters. Sometimes a mandate will increase the price for consumers but in other cases it will reduce the price of fuel for consumers even. This even though there are no taxpayer costs. In this case, oil producers are transferring income to both ethanol production and fuel consumption through the reduction in the marginal cost of oil production.

This ambiguity is depicted in Fig. 20.3. Consider first one specific case where the price of fuel P_F is unaffected by the mandate. This occurs where the demand for fuel D_F intersects the supply curve for fuel S_F and the supply curve for gasoline S_G all at the same point, as depicted in Fig. 20.2. The price of gasoline falls to P'_G, but consumer fuel prices remain at the free market price $P_G + t$. The higher price for ethanol is financed completely by a lower price to gasoline producers. The mandate in this specific case acts as a monopsonist consumer for gasoline

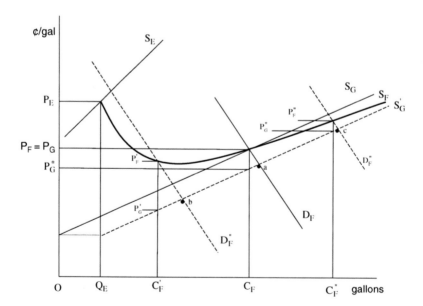

Fig 20.3 Consumption mandate with endogenous oil prices

where the excess revenues are transferred to ethanol producers. Fuel consumers are unaffected.

If instead the demand curve for fuel D'_F in Fig. 20.3 intersects S_F above S_G, then the mandate increases prices for both gasoline consumers and producers simultaneously. This means the higher ethanol price is financed by an implicit tax on gasoline by both fuel consumers and gasoline producers. A third possible equilibrium is when the demand for fuel D'_F in Fig. 20.3 intersects S_F below S_G, in which case the high ethanol price is again entirely financed by gasoline producers, but fuel consumers also are subsidized with the resulting lower gasoline price.

Under what conditions does a mandate increase fuel consumption? The outcome depends on the relative value of the elasticity of supply between biofuel and gasoline. But gasoline producers always lose while fuel consumers can gain or lose. See de Gorter and Just (2008b, 2009b) for full details and the formal conditions under which fuel prices increase or decrease with a mandate. Nevertheless, regardless of market conditions, compared to tax credits that achieve the same level of ethanol consumption, a mandate results in higher fuel prices and lower fuel consumption (even though a mandate can generate an increase in fuel consumption). This means a mandate is preferred to a tax credit when there is a suboptimal gasoline tax like in the United States. A mandate also saves taxpayer costs and does not incur the deadweight costs of taxation.

20.4.1 The Effects of a Tax Credit on Gasoline Consumption, CO_2 Emissions, and Miles Traveled

If the supply curve for oil is upward sloping so that oil prices are affected by ethanol production, then the effects of the tax credit in the absence of a binding mandate will be to increase the fuel supply such that the price of gasoline falls. This means less ethanol production and more fuel consumption. But the tax credit always increases fuel consumption (while lowering gasoline consumption). This means the effect of the tax credit on miles traveled is always positive because consumers buy ethanol on the basis of its contribution to mileage. With a mandate that achieves the same level of ethanol consumption, the impact on fuel consumption, miles traveled, and CO_2 emissions are all ambiguous, but are always lower than under an equivalent tax credit.[4] This result is important for at least two reasons. First, it is widely believed the United States has a suboptimal gasoline tax (Parry and Small 2005). Second, by far the highest cost externality due to fuel consumption is traffic congestion and traffic-related accidents (Parry and Small 2005). As a result, a mandate will reduce externality costs of transportation more than a tax credit.[5] Furthermore, a mandate

---

[4] For full explanation of these results, see the discussion in de Gorter and Just (2008b).

[5] It should be noted that the reduction in miles traveled with the implicit tax on gasoline with an ethanol mandate is less than the reduction in gasoline consumption because consumers invest in cars with higher miles per gallon when gasoline prices increase (Parry and Small 2005).

may save taxpayer costs and the deadweight costs or marginal excess burden of taxation in raising tax revenues (the "revenue recycling effect" as described by Goulder and Williams 2003) provided the tax costs of a tax credit in achieving the same level of ethanol consumption is greater than the lost tax revenues due to lower fuel consumption.[6]

20.4.2 The Economics of a Biofuel Mandate and Tax Credit Combined

So far, we have determined the equilibrium with a blend mandate and compared the efficiency of a mandate to that of taxes and subsidies under different policy goals. But policy makers seem intent on using mandates and tax credits in concert. We now evaluate these policies jointly. In the shadow of a public outcry over the role of biofuels in rapidly escalating world food prices and declining rainforests, President Bush signed into law the EISA on 19 December 2007, which established the largest increase in a biofuels mandate in history. The new mandate requires the use of at least 36 billion gallons of biofuels in 2022, a fivefold increase over current RFS levels. By 2022, biofuels could represent over 20% of US automobile fuel consumption.

Meanwhile, the new legislation calls for the continuation of the federal biofuel tax credit of 45¢ per gallon which, when combined with state tax credits, will potentially cost taxpayers over $28 bil. by 2022.[7] Tax credits by themselves encourage ethanol production as a replacement for oil-based gasoline consumption. But with mandates in place, the tax credits will unintentionally subsidize gasoline consumption instead. This contradicts the new energy bill's stated objectives of reducing dependency on oil, improving the environment, and enhancing rural prosperity. This result is independent of the issues related to indirect land use and CO_2 life-cycle analysis that is currently in the forefront of the public debate over biofuels.

The unintended result of a tax credit switching to a gasoline subsidy in the presence of a government mandate is easily explained. Consider first how the tax credit would work by itself. To take advantage of the government subsidy offered to them, blenders of ethanol and gasoline will bid up the price of ethanol until it is above the market price of gasoline by the amount of the tax credit. If the price premium over gasoline is less than the tax credit, then blenders will be making windfall profits from the government subsidy by pocketing the difference. But competition among blenders will ensure that there will be no "free money left on the table," and the

---

[6]Notice that we do not attribute any gains or losses associated with air pollutants between ethanol and gasoline. Jacobsen (2009) argues it is about the same although Hahn and Cecot (2008) attribute higher costs to ethanol. We keep ethanol consumption fixed in comparing tax credits to mandates.

[7]The federal tax credit is 45¢/gal and national average state tax credit is over 7¢/gal in 2007 (Koplow, personal communication). This means projected tax costs of biofuels for 2022 is $28.5 bil.

price of ethanol will therefore exceed that of gasoline by the full 52 cent per gallon tax credit.

Now consider the case where the ethanol price is determined by the binding mandate − 36 billion gallons by 2022 − with no tax credit. The consumer "fuel" price is a weighted average of the ethanol and gasoline prices. Implicitly, consumers pay a higher price for gasoline to finance the same ethanol production as before, when only the tax credit was in place. Now introduce a tax credit alongside the mandate. Because the ethanol price premium due to the mandate exceeds the tax credit, there is no incentive for blenders to bid up the price of ethanol as before. Instead, blenders will offer a lower fuel price to consumers to take advantage of the tax credit offered to them by the government. Because market prices of ethanol cannot decline due to the mandate, blenders will compete for the government subsidy by reducing the implicit price paid by consumers for gasoline in their fuel price. This increases gasoline consumption and thus increases the market price of gasoline and oil. The price of gasoline paid by consumers declines until the per unit subsidy on ethanol is exactly exhausted on an adjusted per-unit basis of gasoline consumption – hence the reversal of the intended policy effects.

An estimate of the social costs of having a tax credit when a mandate could have generated the same level of ethanol consumption is summarized for United States in Table 20.1. The first row gives the increase in gasoline consumption due to tax credits with a binding mandate in place. The long run (short run) refers to a price elasticity of demand for gasoline of −0.55 (−0.15) and an OPEC oil supply elasticity of 2.25 (0.25).[8] The bottom half of Table 20.1 provides the various deadweight costs. For example, total deadweight costs for 2022 are expected to range from $24.21 bil. to $37.2 bil in the long run. The low (high) estimate assumes a marginal excess burden of taxation of 35 (75) cents to the dollar and a price of carbon of $8.2 ($81.7) per ton. These deadweight costs are static annual deadweight costs and are all rectangular. The estimates of deadweight costs in Table 20.1 do not include the standard Harberger triangular deadweight costs of the mandate itself. We assume ethanol production to be the same with a mandate alone versus a mandate and tax credit together. So the data in Table 20.1 are only for the additional effects of a tax credit, fixing ethanol production with the mandate.

Due to the unique way in which mandates reverse the market effects of a tax credit, the intentions of policy makers cannot necessarily be faulted. There is no other example in the economics literature of the interaction between a price-based and quantity-based policy measure that generates such a unique result as that of a biofuel tax credit and mandate. Furthermore, this policy mistake is not unique to the United States but is a worldwide error of judgment as most countries use both mandates and tax credits simultaneously. The policy implication is clear: allow the mandate to work by itself, eliminate the tax credit, and save billions in taxpayer

---

[8]The source for the long run demand elasticity is Parry and Small (2005) while de Gorter and Just (2007) provides a range for the OPEC oil supply elasticity of 0.25–5.

Table 20.1 The social costs of adding a tax credit to a mandate*

Changes in...	2007	2015	2022
Gasoline price (¢gal.)	0.00228	0.00552	0.00989
Ethanol price (¢/gal.)	0	0	0
Ethanol consumption (bil. gals)	0	0	0
Fuel consumption (bil. gals)	1.05	2.23	4.40
Oil imports (bil. gals)	0.9	2.0	3.9
Taxpayer costs ($ bil.)	5.7	14.2	28.7
CO_2 (bil. tons)	0.0095	0.0203	0.0404
Miles traveled (bil.)	26.2	55.8	110.0
Social deadweight costs ($ bil.)			
International terms of trade	0.4	1.0	1.9
CO_2 emissions			
$8.17/ton (Nordhaus 2007)	0.077	0.166	0.330
$81.74/ton (Stern Report 2007)	0.773	1.661	3.303
Marginal excess burden of taxation			
35¢ (Browning 1987)	2.0	5.0	10.0
70¢ (Feldstein 1999)	4.0	9.9	20.1
Local air pollution	0.60	1.28	2.53
Traffic congestion	1.31	2.79	5.50
Traffic accidents	0.79	1.67	3.30
Oil dependency	0.157	0.335	0.660
Total Social Deadweight Costs ($ bil.)			
Low	5.3	12.2	24.2
High	8.0	18.7	37.2

*Ethanol prices and consumption are held equal to that with a tax credit.
Source: de Gorter and Just (2008b).

monies. This involves only a modest change in biofuel policy while dramatically improving policy achievements.

What if the tax credit is binding and the mandate is dormant? In other words, what if the ethanol price premium with a mandate and no tax credit would be lower than the current tax credit? In this case, the hypothetical ethanol price premium with a mandate only would represent the implicit subsidy on gasoline consumption due to the tax credit because the tax credit is not allowing the mandate to bind. Furthermore, one can easily increase the mandate to generate the same desired level of ethanol consumption as the tax credit is currently generating. In this way, one obtains the same level of ethanol consumption, but with no taxpayer costs. Using a mandate alone has an additional advantage: for the same level of ethanol consumption, a mandate generates a higher consumer price paid for gasoline and so further enhances energy and environmental policy goals.

20.5 The Link Between the Corn and Ethanol Markets

So far, we have established the link between ethanol and gasoline prices. Now the task is to link the corn to the ethanol price. Denote β as the gallons of ethanol produced from one bushel of corn and denote δ as the proportion of the value of corn returned to the market in the form of by-products. Then, the price of corn P_C (equal to the price of ethanol P_{Eb} in \$/bu) is given by

$$P_{Eb} = \left(\frac{\beta}{1-\delta} \right) (\lambda P_G - (1-\lambda)t + t_c) - c_0 \qquad (20.5)$$

where c_0 is the processing costs. For the purposes of this paper, we use estimates from Eidman (2007) such that β equals 2.8 and δ equals 0.31. The resulting value of $\beta/(1-\delta)$ is 4.06. Thus, a binding federal tax credit of 45¢/gal plus state tax credits of 7¢/gal translates into approximately a \$2.08 per bushel subsidy to corn farmers.[9] This means the corn price is very sensitive to a change in the price of ethanol (induced by either a change in the tax credit or world oil price). For every 1¢/gal change in the ethanol price, the corn price changes by \$0.0406/bu.

An important finding is that the actual observed corn price is always below the ethanol price premium until 2007/2008 (see Fig. 20.4). In fact, the corn price is observed to be lower than the tax credit itself in 9 of the past 25 years! This is at first glance puzzling – how can the implied subsidy of the tax credit be greater than the corn price itself? We explained earlier that the corn price is increased by the amount of the tax credit or ethanol price premium due to its additive value or de facto mandates.

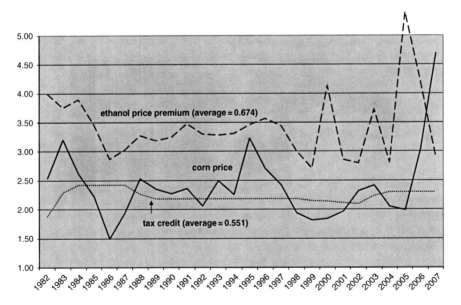

Fig 20.4 Ethanol price premium, corn price and tax credit in \$/bushel

---

[9]The 7 cents per gallon tax credits given by the states is estimated by Koplow (2007).

There are two keys to understanding this phenomenon (see de Gorter and Just 2008a, 2009a for complete details). First, one has to recognize that the intercept of the ethanol supply curve was above the gasoline price. In other words, if there were no ethanol price premium due to either its additive value or tax credits, there would be no ethanol production. Costs of production would exceed the price of gasoline. Because the intercept of the ethanol supply curve was above the oil price, farmers historically have not been able to take advantage of such a large subsidy. This means a significant part of the tax credit has been redundant. We call this "water" in the tax credit or ethanol price premium due to mandates or additive value. This means part of the ethanol price premium is what we call rectangular deadweight costs. A portion of the tax credit goes to cover the difference between the intercept of the ethanol supply curve and the oil price. This difference represents the rise in the ethanol price necessary to overcome the opportunity cost of corn in other uses. Consumers pay this difference, but nobody benefits as corn producers were already receiving this level of compensation in other uses of corn.

Second, not only was the intercept of ethanol supply below the price of gasoline, but it was also above the price of corn. The only way this can happen is with production subsidies for corn and/or ethanol. These subsidies are the only reason for ethanol production in these cases. In other words, even with the tax credit and premiums due to additive value, there would be no ethanol production unless there were production subsidies for corn and/or ethanol as well. Because the intercept of the ethanol supply curve in the United States has been far above the price of oil, the resulting "water" in the ethanol price premium generates "rectangular" deadweight costs. Rectangular deadweight costs are defined as that part of the price premium (assume here to be due to the tax credit) that is not a transfer to domestic producers or any other domestic or foreign interest group. This exacerbates the social costs of ethanol policies compared to standard analysis.

20.5.1 How the Tax Credit Affects the Taxpayer Costs of Farm Subsidies

Proponents of US ethanol policy argue that the tax credit reduces the tax costs of farm subsidy programs. There are two particularly important issues to analyze: the impact of the tax credit on both the tax costs and deadweight costs of the loan rate program, and vice-versa, how farm subsidies increase the tax costs of the tax credit and increase rectangular deadweight costs.

A summary of the costs of the tax credit with and without deficiency payments associated with the loan rate is given in de Gorter and Just (2009a). Taxpayer costs due to the tax credit with the loan rate averaged $1,914 mil. But the taxpayer costs of the tax credit would be $526 mil. less if there were no loan rate. Hence, the loan rate increases the taxpayer costs of the tax credit. On the other hand, the tax credit reduces the taxpayer costs of the loan rate by $2,092 mil. But rectangular deadweight costs due to the tax credit averaged $1,520, dwarfing the standard

Harberger triangles traditionally calculated for farm subsidy programs. The loan rate contributed $417 mil. to rectangular deadweight costs. The tax credit doubles the average annual deadweight costs from $613 to $1,291 mil. Furthermore, given the tax credit, the loan rate increases deadweight costs by $378 mil. Hence, one does not want to introduce the tax credit to mitigate the effects of the loan rate.[10]

One forgotten cost of the tax credit in reducing tax costs of farm subsidies is the economic inefficiency caused by the relative spike in food prices. This is called the "tax interaction effect" (Goulder and Williams 2003). The food price spike magnifies the inefficiency of a pre-existing distortion (the wage tax) because the price spike reduces real wages and so discourages work. This results in a leftward shift in the labor supply curve and generates further rectangular deadweight costs (the tax base erodes as consumers substitute away from the taxed good). This means a higher tax on wages is required to maintain tax revenues.

Finally, the assumption that the loan rate and target prices set by politicians would not be affected by the tax credit is erroneous. Higher crop prices due to biofuel policies give politicians an incentive to increase support prices (Swinnen and de Gorter 1998). Estimates of the welfare effects of one policy assuming the level of the other policy is unaffected can be seriously biased. For example, the recent Farm Bill includes an increase in 14 of 28 support prices for crops. If it were not for biofuel policies, Congress may otherwise have been proposing lower price supports so caution should be exercised in attributing any social benefits of biofuel policy in reducing the tax costs of price supports.

20.6 Concluding Remarks

Current policies involve several layers of incentives and regulations, requiring clear theoretical and empirical modeling for effective policy analysis. By specifying the proper economic relationships determining ethanol price linkages to oil and corn, our chapter reveals policy interaction effects that are often contradictory or self-defeating. These contradictory effects include potentially important costs and benefits of biofuel policy on the environment. There are several other reasons why ethanol policy has been more costly than the literature recognizes including an overemphasis on GHG emissions. The single largest external cost associated with transportation fuels is vehicle miles traveled and the resulting costs of traffic congestion, traffic accidents, and local air pollutants. Thus, when assessing biofuel policy, it is important to simultaneously account for GHG emissions as well as the induced

---

[10]The conditions under which the loan rate can eliminate, create, have no effect or have an ambiguous effect on rectangular deadweight costs is given in de Gorter and Just (2009a). The outcome depends on whether the market price of corn is above or below the price that would prevail without ethanol production, and on whether there is water in the tax credit with the loan rate. There are situations where the loan rate has been the sole cause of ethanol production (and therefore of rectangular deadweight costs), even with the tax credit. Corn producers do not benefit from an ethanol tax credit when the loan rate is in effect.

changes in local air pollutants and traffic congestion and traffic-related accidents. As a result, the benefits to US agriculture of several ethanol policies may be overstated, raising serious issues in regards to policy reform and risks to the competitiveness of US agriculture.

We determine that total fuel consumption and miles traveled always increase with a tax credit, but the impact on CO_2 emissions is ambiguous. With a mandate that achieves the same level of ethanol consumption, the impact on fuel consumption, miles traveled, and CO_2 emissions are all ambiguous but are always lower than under an equivalent tax credit. This result is important for at least two reasons. First, it is widely believed the United States has a suboptimal gasoline tax and second, a mandate may also save taxpayer costs and so incur lower deadweight costs or marginal excess burden of taxation in raising tax revenues.

On the other hand, if there is a suboptimal carbon tax on fossil fuels, then it may be possible that a mandate for ethanol consumption improves social welfare because, in order to pay for the higher ethanol price, a mandate may force consumers to pay a higher price for gasoline. This higher price of gasoline could compensate for a suboptimal carbon tax. The benefits of an ethanol mandate alone can contribute to reducing the external costs of miles traveled by many billions of dollars. Hence, it is possible under some circumstances that ethanol policy simultaneously contributes to energy, environment, and farm policy goals because of a reduction in gasoline use while enhancing ethanol and hence crop prices.

The increased resolve of the current administration and Congress to promote alternative energy sources and combat climate change has led to a bevy of new policy proposals including "cap and trade," carbon offsets, rebates to producers, and "green" tariffs. Many of these proposed, along with current policies, have no match in the lexicon of economic policy analysis. Further, these policies have been combined without being coordinated. These combinations of policies are not well understood to date. The current set of economic analysis has largely avoided this issue by examining only one of the policy incentives at a time. The results we present in this chapter show how absolutely necessary it is to understand the impacts of the various agro-energy-enviro policies together. Further work is necessary to determine how these policies can be adjusted to meet the stated environmental goals while maintaining a competitive agricultural sector.

Hence, complex policy designs currently in play will become even more complex with proposed legislation on renewable energy and global climate change. The proposed cap and trade policy will provide additional challenges for US agriculture with higher energy prices but will also provide opportunities in the sale of carbon "offsets." One key issue is compatibility with existing biofuel and agricultural price, trade, and environmental policies. The issue of leakage and loss of competitiveness due to higher energy prices will complicate the analysis further. Proposed legislation includes a proviso for rebates and "green" tariffs on products from countries with less stringent GHG emissions regulations. This means there will be even more complex interactions between agricultural, environmental, and energy policies.

Meanwhile, the global climate regime may be on a potential collision course with the global trading regime. National efforts to reduce GHGs instill fear of leakage and lost competitiveness. Leakage of emissions could come from energy-intensive industries relocating to countries without emissions commitments; and via a reduction in world energy prices (McKibbin and Wilcoxen 2008; Hauser et al. 2008). Even more salient politically is the related issue of competitiveness: because US agriculture is energy intensive, agriculture may be at a competitive disadvantage. Further, some policies (at home or abroad) may violate WTO agreements, leading to challenges or retaliation.

References

Browning EK (1987) "On the Marginal Welfare Cost of Taxation." American Economic Review 77 March: 11–23.

de Gorter H and Just DR (2007) "The Law of Unintended Consequences: How the U.S. Biofuel Tax Credit with a Mandate Subsidizes Oil Consumption and Has No Impact on Ethanol Consumption." Department of Applied Economics and Management Working Paper # 2007-20, Cornell University, 23 October. http://papers.ssrn.com/sol3/papers.cfm?abstract_id=1024525

de Gorter H and Just DR (2008a) "Water' in the U.S. Ethanol Tax Credit and Mandate: Implications for Rectangular Deadweight Costs and the Corn-Oil Price Relationship." Rev Agri Econ 30(3):397–410.

de Gorter H and Just DR (2008b) "The Welfare Economics of the U.S. Ethanol Consumption Mandate and Tax Credit." Department of Applied Economics and Management Working Paper unpublished, Cornell University, 17 April (available upon request).

de Gorter H and Just DR (2008c) "The Economics of the U.S. Ethanol Import Tariff with a Blend Mandate and Tax Credit." J Agri and Food Ind Org 6(2): Article 6.

de Gorter H and Just DR (2009a) "The Welfare Economics of a Biofuel Tax Credit and the Interaction Effects with Price Contingent Farm Subsidies" Am J Agri Econ 91(2):477–488.

de Gorter H and Just DR (2009b) "The Economics of a Blend Mandate for Biofuels." Am J Agri Econ 91(3):738–750.

Eidman VR (2007) "Ethanol Economics of Dry Mill Plants." Chapter 3 in Corn-Based Ethanol in Illinois and the U.S.: A Report, Dept. of Agr. and Consumer Econ., University of Illinois, November.

Feldstein M (1999) "Tax Avoidance and the Deadweight Loss of the income Tax" The Review of Economics and Statistics November 81(4): 674–680.

Goulder LH and Williams RC III (2003) "The Substantial Bias from Ignoring General Equilibrium Effects in Estimating Excess Burden, and a Practical Solution." J Polit Econ 111: 898–927.

Hahn R and Cecot C (2008) "The Benefits and Costs of Ethanol." Working Paper 07-17, AEI Center for Regulatory and Market Studies, August.

Hauser T, Bradley R, Childs B, Werksman J and Heilmayr R (2008) "Leveling the Carbon Playing Field: International Competition and U.S. Climate Policy Design." Peterson Institute for International Economics, Washington DC.

Jank MJ, Kutas G, Fernando do Amaral L and Nassar AM (2007) EU and U.S. Policies on Biofuels: Potential Impact on Developing Countries, The German Marshall Fund of the United States, Washington DC.

Jacobson MZ (2009) "Review of solutions to global warming, air pollution and energy Security." Energy Environ Sci 2(2):148–173.

Jull C, Redondo PC, Mosoti V and Vapnek J (2007) Recent Trends in the Law and Policy of Bioenergy Production, Promotion and Use. FAO Legal Papers Online #68, September, Rome.

Kojima M, Mitchell D and Ward W (2007) Considering Trade Policies for Liquid Biofuels. Energy Sector Management Assistance Programme (ESMAP) World Bank, Washington D.C., June.

Koplow D (2007) Ethanol–At What Cost? Government Support for Ethanol and Biodiesel in the United States. 2007 Update. Geneva, Switzerland: Global Subsidies Initiative of the International Institute for Sustainable Development, October.

McKibbin W and Wilcoxen P (2008) "The Economic and Environmental Effects of Border Adjustments for Climate Policy," Brookings Conference on Climate Change, Trade and Competitiveness: Is a Collision Inevitable?, Washington, DC.

Miranowski JA (2007) "Biofuel Incentives and the Energy Title of the 2007 Farm Bill", in The 2007 Farm Bill & Beyond, American Enterprise Institute, Washington D.C.

Nordhaus WD (2007) "A Review of the Stern Review on the Economics of Climate Change." J Econ Lit 45: 686–702.

Parry I and Small K (2005) "Does Britain or the United States Have the Right Gasoline Tax?," Am Econ Rev 95(4) (September): 1276–1289.

Rajagopal D and Zilberman D (2007) "Review of Environmental, Economic and Policy Aspects of Biofuels", Policy Research Working Paper WPS4341. The World Bank Development Research Group, September.

Runge CF and Senauer B (2007) "How Biofuels Could Starve the Poor." Foreign Aff 86(3):41–53.

Searchinger T, Heimlich R, Houghton RA, et al. (2008) "Use of U.S. Croplands for Biofuels Increases Greenhouse Gases Through Emissions from Land Use Change." Science Express 7 February 2008 pages 1–6.

Steenblik R (2007) "Biofuels–At What Cost? Government support for ethanol and biodiesel in selected OECD countries", The Global Subsidies Initiative (GSI) of the International Institute for Sustainable Development (IISD) Geneva, Switzerland, September.

Stern N (2007) The Economics of Climate Change: The Stern Review. Cambridge and New York: Cambridge University Press.

Swinnen J and de Gorter H (1998) "Endogenous Commodity Policy and the Social Benefits from Public Research Expenditures." Am J Agri Econ 80: 107–115.

Tyner WE (2007) "U.S. Ethanol Policy – Possibilities for the Future" Purdue University Working Paper ID-342-W, West Lafayette, Indiana.

Chapter 21
Biofuels, Policy Options, and Their Implications: Analyses Using Partial and General Equilibrium Approaches

Farzad Taheripour and Wallace E. Tyner

Abstract In this chapter, we highlight some important aspects of biofuels development and policies from partial and general equilibrium perspectives. We first examine US biofuel policy backgrounds to determine factors which caused the boom in the ethanol industry in recent years. Then, we use a partial equilibrium model to investigate the economic consequences of further expansion in the ethanol industry for the key economic variables of the US agricultural and energy markets. This analysis is done for a range of alternative policy options and crude oil prices. Finally, we extend our analyses to examine consequences of further biofuel production at a global scale.

This chapter shows that biofuels policies such as the blending subsidies and the effective ban on use of MTBE have had significant impacts on US biofuel expansion in recent years. In addition, the increased price of crude oil has significantly contributed to biofuel expansion. Our analyses indicate that corn ethanol is economic without a subsidy at crude oil prices more than \$60 per barrel (even though subsidies persist in the United States), but biodiesel would not exist even at high crude oil prices without government subsidies. Both ethanol and biodiesel production involve potentially large land use changes that take place all over the world. The model and data used for land use change analysis illustrated here need improvement. However, the preliminary results illustrate the kinds of estimated changes that can be produced from the global analysis.

21.1 Introduction

The biofuel industry has been experiencing a period of extraordinary growth, fueled by a combination of high oil prices, ambitious renewable fuel standards, subsidies,

F. Taheripour (✉)
Purdue University, West Lafayette, IN, USA
e-mail: tfarzad@purdue.edu

This chapter is adapted from the authors' paper that originally appeared in the *Journal of Agricultural and Food Industrial Organization* with the following complete reference: Tyner and Taheripour (2008).

and import protection. This rapid growth has important consequences for the United States and global economies. Research studies have examined expansion of biofuels from different perspectives. Some papers examined price and welfare impacts of these policies for the United States (Rajagopal et al. 2007; de Gorter and Just 2008; Khanna et al. 2008). Several papers have investigated interactions between energy and agricultural markets (Elobeid et al. 2007; Tokgoz et al. 2007; McPhail and Babcock 2008a, b; Tyner and Taheripour 2008b, c). A group of studies have conducted life-cycle analyses of biofuels (Wang 1999, 2005; Farrell et al. 2006; Argonne 2007). In another line of research on biofuels, several papers have studied global economic and land use implications of biofuel policies (Banse et al. 2007; Birur et al. 2007; Taheripour et al. 2008; and Hertel et al. 2008).

In this chapter, we highlight some important aspects of biofuel and biofuel policies from partial and general equilibrium perspectives. We first examine US biofuel policy backgrounds to determine factors which caused the boom in the ethanol industry in recent years. Then, we use a partial equilibrium model to investigate the economic consequences of further expansion in the ethanol industry for the key economic variables of the US agricultural and energy markets under alternative policy options which might be used to promote ethanol production in the future. Finally, we extend our analyses to examine consequences of further biofuel production at a global scale.

Each of these perspectives provides valuable insight into the functioning and impacts of ethanol policy alternatives, and each approach illustrates quite well the newly emerging integration of energy and agricultural markets. Historically, there has been almost no link between energy and agricultural commodity prices[1] (Tyner and Taheripour, 2008b, c). An extensive production of biofuels from agricultural resources links energy and agricultural markets tightly. This chapter demonstrates how the US ethanol policies affect the integrated agricultural and energy markets from partial and general equilibrium angles.

We develop a partial equilibrium analysis at the firm level to review factors which have contributed to the recent boon in ethanol production. This partial equilibrium analysis is built upon the firm level analysis that has been extensively reported in our previous work (Tyner and Taheripour 2007, 2008a). Unlike our earlier work, here we introduce a new break-even comparison for combinations of corn and ethanol prices, which keep a representative dry milling ethanol plant at the zero profit condition. We use this instrument to examine the profitability of the ethanol industry in 2000–2008.

Then, we extend our partial equilibrium analyses to examine the link between agricultural and energy markets. The partial equilibrium analysis at the market level extends our earlier work in this context (Tyner and Taheripour 2008b, c) and examines economic consequences of policies which are designed (or can be used) to

---

[1] Several articles have addressed the impacts of higher energy prices on the agricultural costs of production (examples are Dvoskin and Heady 1976; Christensen et al. 1981). These papers do not address the link between these markets on the demand side. Even with energy being an influence on cost of production, there has been very low correlation historically between energy and agricultural commodity prices.

promote ethanol production in the United States. In this chapter, we review impacts of fixed and variable subsidies, mandates, and their combinations.

Finally, we use a general equilibrium framework to examine the global economic and land use implications of international biofuel mandate policies. The global general equilibrium analysis is built upon Taheripour et al. (2008) and Hertel et al. (2008) and concentrates on the impacts of the US and EU biofuel mandate policies.

The chapter indicates that, from each of the perspectives, we have entered a new era in which crude oil prices will have a major impact on corn and other agricultural commodity prices, and that the policy alternative we choose will have a major influence on what happens in these markets. At relatively low oil prices, it is the US domestic biofuels policies that dominate, but as oil prices get very high, they become the major driver of the energy–agriculture interactions.

This chapter not only provides useful analysis of different biofuels policy alternatives, but also highlights the strengths and weaknesses of each analytical approach to the problem. The general equilibrium analysis, for example, is very good at capturing economy wide linkages and global impacts but is too aggregated to fully capture what happens in domestic markets and at the firm level. Each of the analytical tools provides useful information in understanding the impacts of US biofuels policy options.

21.2 Policy Background and the Ethanol Boom

In this section, we first develop a break-even analysis which measures profitability of a representative ethanol producer at all combinations of corn and ethanol prices. Then, we compare the break-even line with the actual observations to examine factors which caused the surge in ethanol industry in recent years. The break-even line (Fig. 21.1) presents combinations of all corn and ethanol prices,[2] which keep a new dry milling ethanol plant at the zero profit condition.[3] By zero profit, we mean zero economic profit. The break-even includes a 12% return on equity, so zero profit means no profit in excess of the 12% equity return.[4]

Figure 21.1 shows that the representative producer can operate at $2.50 per bushel of corn and $1.55 per gallon of ethanol with no added profit. At relatively higher prices of corn, this combination changes significantly. For example, at $6.50 per bushel of corn the producer must receive $2.80 per gallon of ethanol to operate at the zero profit condition. This indicates that the corn price increases faster than the ethanol price along the break-even line. This is due to the fact that ethanol from corn

---

[2] These prices are corn price paid and ethanol price received by the ethanol producer.

[3] We have modified the Tiffany–Eidman spreadsheet developed at the University of Minnesota (Tiffany and Eidman, 2003) to generate the break-even ethanol and corn prices.

[4] We also assumed that the price of DDGS is a function of corn and soybean prices. We used historical data to establish the link between these variables. To define a break-even line between corn and crude oil prices, one needs to establish a link between ethanol and crude oil prices. There is no need to establish such a link for the break-even line between corn and ethanol prices.

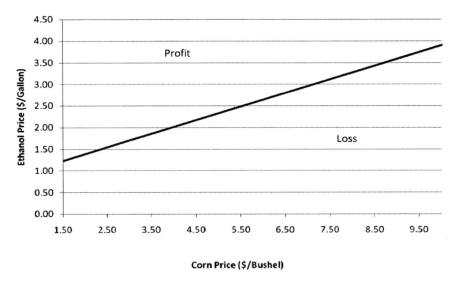

Fig. 21.1 Break-even ethanol and corn prices

is produced in conjunction with distiller dried grains with solubles (DDGS), which is a valuable feed used in livestock industry. Each bushel of corn can be converted to about 2.79 gallons of denatured ethanol and 18 pounds of DDGS by-product. This product is a good substitute for corn, and its price closely follows the corn price. With this introduction, we now examine factors which caused the boom in the ethanol industry. To facilitate our analysis, we add actual combinations of the corn and ethanol prices to the break-even graph. Figure 21.2 compares the break-even lines with the actual observations[5] from the period 2000–2008.

Figure 21.2 reveals that ethanol producers were close to the lower tail of the break-even line at the beginning of 2000s. Given the fact that a portion of the ethanol price received by the producers of this commodity is paid by the federal government (51 cents prior to 2009 and 45 cents starting in 2009),[6] we can conclude that at the beginning of 2000s the ethanol industry was not profitable without government subsidies. The tax credit helped the industry to operate with low profit margins at the beginning the century.

Subsidization of ethanol in the United States began with the Energy Policy Act of 1978. At the time, the main arguments that were used to justify the subsidy

---

[5] Actual observations are annual average prices of corn and ethanol in each year. The 2008 observation represents the average of the first 6 months. The monthly market corn (yellow number 2) prices are obtained from the United States Department of Agriculture website available at: http://www.ers.usda.gov/data/. The monthly rack ethanol prices are obtained from the Nebraska Government website available at: http://www.neo.ne.gov/statshtml/66.html.

[6] The ethanol industry share of the blending subsidy is a controversial issue. For a detailed discussion see Taheripour and Tyner (2008).

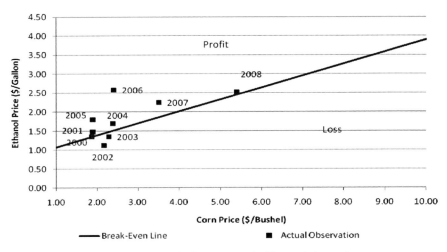

Fig. 21.2 Break-even corn and ethanol prices compared with actual observations

were enhanced farm income and, to a lesser extent, energy security. In 1990, the Clean Air Act was passed, which required vendors of gasoline to have a minimum oxygen percentage in their product. Adding oxygen enables the fuel to burn cleaner, so a cleaner environment became another important justification for ethanol subsidies.

By requiring the oil industry to meet an oxygen percentage standard instead of a direct clean air standard, the policy favored additives like ethanol that contain a high percentage of oxygen by weight. However, methyl tertiary butyl ether (MTBE), a competitor for oxygenation, was generally cheaper than ethanol, so it continued to be the favored way of meeting the oxygen requirements throughout the 1990s. The growth in MTBE use was short lived, as it began to crop up in water supplies in several regions in the country, and it is highly toxic. MTBE was gradually banned on a state-by-state basis.

In 2004, the crude oil price began its steep climb to over $100/bbl. The combination of MTBE ban, low prices of corn, higher crude oil price, and the ethanol tax credit raised profitability of ethanol industry beginning in 2004 and 2005. However, the big increase in ethanol profitability was in 2006 when use of MTBE was effectively banned leaving ethanol as the additive of choice. Indeed, ethanol prices peaked at $3.58/gallon in June 2006, shortly after the MTBE ban was complete. Since that time, the price of ethanol has been falling, as the demand for ethanol as an additive has become satiated. In 2007 and the first three quarters of 2008, it appeared that ethanol was increasingly being priced for its energy content – which is only about 70% of that provided by an equivalent volume of gasoline. While in 2006 each gallon of ethanol was making about $1 in profits, the industry moved toward the middle of break-even line in the first 6 months of 2008 with almost no above normal profit. In this time period, the ethanol price has weakly responded to the higher prices of crude oil due to the massive supply of ethanol and its weak demand. As the result, the profitability of the industry has declined significantly in 2008.

From Fig. 21.2 one can conclude that high profit margins from ethanol production in 2004–2007 encouraged a rapid investment in ethanol industry in these years. The contribution of the 51 cents per gallon of ethanol subsidy to the profit margin of the industry depends on the distribution of the subsidy among active agents in the corn, ethanol, and gasoline markets.

21.3 Future Ethanol Expansion and Alternative Policy Options

In this section, we extend our analyses to examine impacts of alternative policy options which can be used to promote the ethanol industry on the key economic variables of energy and agricultural markets. We use a partial equilibrium model, which we have developed earlier (Tyner and Taheripour 2008b, c). The partial equilibrium model links agricultural and energy markets and provides a consistent framework to evaluate economic impacts of ethanol supporting policies for the US economy.[7]

In this chapter, we design several prospective scenarios which depict the US fuel and corn markets in 2015 for crude oil prices ranging from $40 to $160 per barrel in $20 increments. All of the simulations were done with a 5% fuel demand shock, which basically means that higher income and population by 2015 would stimulate an increase in demand at any given oil price. The simulations also were done with a 40% corn export demand shock to account for the fall of the US dollar. Neither of these shocks was applied to the $40 or $60 oil price cases shown in the figures below, as those cases are included basically for calibration and validation of the model for data in the 2004–2006 period. To focus on impacts of crude price on corn and ethanol markets, we assume no technological improvement in corn production. Of course a major improvement in corn yield could increase corn supply. In designing all scenarios, we assume physical constraints such as inadequate infrastructure and the blending wall[8] will not restrict the market. While we recognize that the blending wall may be a major impediment for further growth in the ethanol industry, this analysis abstracts from that issue. The following scenarios are evaluated in this chapter:

---

[7]This model assumes that gasoline and ethanol are perfect substitutes and that the price received by ethanol producers is equal to its energy equivalent gasoline price plus the ethanol subsidy, if there is any.

[8]The blending wall refers to the physical limit on ethanol use with a 10% blend known as E10. We consume about 140 billion gallons of gasoline annually, so 10% would be 14 billion gallons. Due to lack of blending facilities, inability to blend ethanol in southern warm weather states in summer due to its high evaporative emissions, and other infrastructure issues, the effective limit is more like 12–12.5 billion gallons (Tyner et al. 2008). In addition, most vehicles are not flex-fuel and hence they cannot use gasoline blended with ethanol at a rate higher than 10%.

- A fixed ethanol subsidy of 45 cents per gallon, which took effect in January 2009
- No ethanol subsidy
- A variable ethanol subsidy beginning at $70 crude oil and increasing $0.0175 for each $ crude oil falls below $70
- A renewable fuel standard (RFS) of 15 billion gallons per year from corn ethanol

In what follows, we present impacts of alternative scenarios on the key economic variables of the fuel and corn markets.

21.3.1 Ethanol Production

The simulation results for ethanol production are shown in Fig. 21.3. Several important conclusions emerge. First, the second bar for each oil price is no subsidy, and it is clear that there is no ethanol production without subsidy unless oil is $60 or higher. This conforms to our earlier firm level analysis. Second, the level of ethanol production with 45 cents fixed subsidy at $40 oil is only 1.8 billion gallons, yet 2004 production was about 3.4 billion. In our previous work with a 51-cent subsidy, we actually got a level close to 3.4 billion, so at low oil prices results are sensitive to the level of the subsidy. Third, at $40 oil, the amount of ethanol production under the variable subsidy is considerably higher than under the fixed subsidy illustrating the potential of the variable subsidy to provide a safety net for low oil prices. Fourth, at oil prices less than $100, ethanol production is higher under the RFS than with the subsidy. However, at any oil price of $100 or higher, production is higher under the

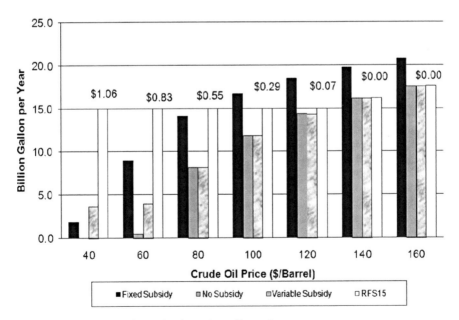

Fig. 21.3 Ethanol production under alternative policy options

subsidy policy. This is because the RFS mandate is no longer binding at $120 oil or higher. Fifth, the numbers above the RFS bar represent the implicit ethanol subsidy under the RFS. The binding RFS imposes an implicit tax on gasoline consumption and provides an implicit subsidy for ethanol producers to cover their production costs to satisfy the mandate. Consumers pay the implicit tax at the pump when they buy gasoline blended with ethanol at a price higher than what it would be in the absence of mandate. The implicit tax/subsidy approaches zero when the crude oil price increases. The implicit subsidy at $40 crude oil price is about $1.06 per gallon, and it goes to zero at $120 crude oil price. Sixth, at oil prices of $140 or higher, there is no difference between the no subsidy, variable subsidy, and RFS cases. The reason again is that at high oil prices, the RFS is not binding. The market alone produces more than the 15 billion gallon mandate.

Finally, it is important to note that as the crude oil price goes up, the ethanol production goes up at a decreasing rate. This is due to the fact that the opportunity costs of producing ethanol from corn goes up with an increasing rate when the crude oil price goes up. As one can see from Fig. 21.3, ethanol production jumps significantly when the crude oil goes from $40 to $60, but when the crude oil goes up from $140 to $160 the ethanol production goes up moderately. This indicates that market forces put automatically a cap on the ethanol production. The increase in corn price is enough to squeeze out profits, so the corn price increase slows substantially ethanol growth. Of course, any technological progress or yield improvement can push up the cap.

21.3.2 Corn Production

Figure 21.4 displays the amount of corn production for each of the policy options and oil prices. The level of corn production under $40 oil and fixed subsidy, 10.25 billion bushel, is about the actual 2004 value. Corn production must be substantially higher to meet the 15 billion gallon RFS. Again note, that above $120 oil, the RFS is not binding and corn production is market driven. With oil at $100 or higher, the subsidy induces considerably higher corn production. It is important to note that corn production is not only changing due to ethanol production. Corn production responds to other domestic and export demands as well. So corn production could change when the RFS is binding. The small reduction in corn production when no subsidy is available and the price of crude oil increases from $40 to $60 is due to reduction in corn demand for nonethanol uses (food, feed, and exports).

The fraction of corn used for ethanol increases with crude oil price under all alternative policies. For example, with the fixed subsidy in place, only 6.6% of production of corn is expected to be used in ethanol industry at $40 crude oil price. This figure increases to 57.8% at the $160 crude oil price.

21.3.3 Corn price

Figure 21.5 illustrates the corn price under these same policies and oil prices. At $40 oil, the corn price is a bit under $2 with the 45 cents fixed subsidy policy. In our

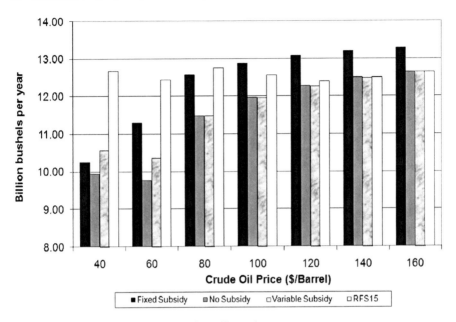

Fig. 21.4 Corn production under alternative policy options

previous work with the 51-cent subsidy, the corn price was a bit over $2, which was the case in 2004 with $40 oil. At $140 oil we get a corn price of about $6 under no subsidy, variable subsidy, and RFS. It is about $1 higher with the fixed subsidy. It is striking to see the progression of corn price along with crude oil price. Again, this result illustrates the tight link we will have in the future between crude oil and corn. Of course, in the short run, there will be perturbations due to supply shocks like the 2008 floods and other adjustments, but the long-run relationship should hold.

Figure 21.5 presents the break-even corn–crude oil prices under alternative policies at the market level. One can use this figure to decompose impacts of US ethanol subsidy and crude oil price on the corn price.[9] There is no doubt that ethanol production in the United States has contributed to higher corn prices. A large portion of the growth in corn demand is associated with growth in ethanol production. Between 2004 and earlier in 2008, crude oil went from $40 to $120. Over that same time period, corn went from about $2 to about $6. We can partition the $4 corn price increase into two parts: price increase due to the US ethanol subsidy and price increase due to the demand pull of higher crude oil price. Figure 21.5 shows that the corn price under the fixed subsidy is higher than the corn price under the no subsidy case by about $1 for all crude oil prices above $40. Therefore, about $1 of the corn price increase is due to the US subsidy and $3 to the crude oil price increase. The crude oil price increased due to many factors such as higher demand for crude oil,

---

[9]This analysis does not consider short-run adjustments in market prices and ignores short-term speculative demand for commodities.

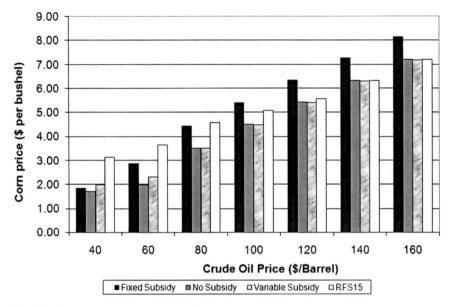

Fig. 21.5 Corn price under alternative policy options

devaluation of the US dollar, political instability in the Middle East, and many other factors. So the crude oil price is the major driver in corn price increase, and the US ethanol subsidy less so. Of course, that was not the case before the surge in crude oil prices. Prior to 2005, the ethanol industry would not have existed without the subsidy.

21.3.4 Corn Exports

Figure 21.6 shows the projected corn exports under alternative policies. Even though we have not yet seen exports fall due to higher corn prices, we would expect to see that after adjustment in global markets. The fall in corn exports shown here is much less that in our previous work, which did not include the 40% export demand shock to account for the falling dollar. Note that in every case, exports are less with the fixed subsidy. Also, as the crude oil price goes up, corn exports decrease as more corn is demanded for ethanol.

21.3.5 Policy Costs

Figure 21.7 shows the cost of each of the policy instruments for each oil price. Of course, the no subsidy case always has a cost of zero. The RFS cost is paid by the consumer at the pump, whereas the fixed and variable subsidy costs are financed through the government budget. Note that the RFS cost is high at low oil prices and low or zero at high oil prices. The fixed subsidy cost rises linearly with oil price as

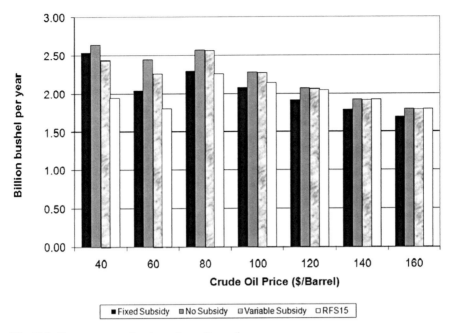

Fig. 21.6 Corn exports under alternative policy options

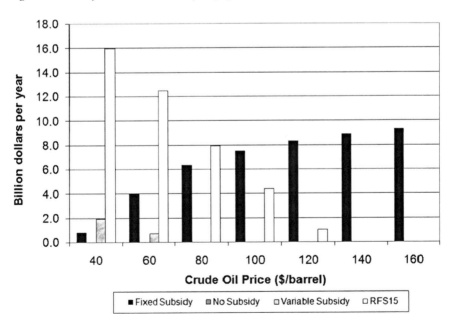

Fig. 21.7 Costs of alternative policy options

the subsidy supports more ethanol production demanded by the market at higher oil prices. The variable subsidy modeled here has very low costs, and these manifest only at low oil prices. Figure 21.7 indicates that at low crude oil prices, the RFS imposes significant costs on consumers. Alternatively, at oil prices above $80, the cost of the RFS is always lower than the fixed subsidy cost.

21.4 Global Biofuels Impacts

Many countries have announced and implemented plans and programs to increase production and use of biofuels renewable energy. In both the United States and the EU, programs are already in effect that either require or provide incentives for significant production of bioenergy. China, India, Indonesia, and Malaysia, among others, also have announced and implemented biofuels initiatives. More than 13 billion gallons of bioethanol and about 2 billion gallons of biodiesel were produced globally in 2007. This large-scale global implementation of bioenergy production causes global economic, environmental, and social consequences. It can affect the global economy in several ways. In addition, it induces major land use changes across the whole globe, which may lead to significant environmental impacts.

To assess the global impacts of biofuel production, we use a computational general equilibrium (CGE) model which is built upon the standard Global Trade Analysis Project (GTAP) modeling framework (Hertel 1997). GTAP is a CGE model which considers production, consumption, and trade of goods and services by regions and at a global scale. In this model, consumers maximize their utilities according to their budget constraints and producers minimize their production costs subject to resource constraints. The model determines demands for and supplies of goods and services according to consumer and producer behaviors. Resources are labor, capital, land, and natural resources, and they are owned by consumers.

A special version of GTAP, developed by Burniaux and Truong (2002) and modified by McDougall and Golub (2007), incorporates energy into the GTAP framework. Birur et al. (2007) have introduced biofuels into this model. They augment the model by adding the possibility for substitutability between biofuels and petroleum products. Taheripour et al. (2008) have extended the GTAP modeling framework to handle production, consumption, and trade of biofuel by-products. In particular, they introduced distillers dried grains with solubles (DDGS) and oilseed meals as by-products of ethanol and biodiesel into the model.[10] In a recent work Hertel, Tyner, and Birur (2008) have augmented this model with a land use module to accurately depict the global competition for land among land use sectors. The

---

[10]In the extended GTAP model, ethanol and DDGS are jointly produced by the ethanol industry, and biodiesel and oilseed meals are jointly produced by biodiesel industry. So ethanol and DDGS are joint products and biodiesel and oilseed are also joint products. So defining DDGS and oilseed meals as the by-products does not undermine their importance.

land use module disaggregates land into 18 Agro-Ecological Zones (AEZ). These AEZs share common climate, precipitation and moisture conditions, and thereby capture the potential for real competition between alternative land uses. In this chapter we use Hertel et al. (2008) and Taheripour et al. (2008) to address two key issues regarding the global impacts of biofuel production. We first highlight the implications of a multinational biofuel mandate for land use change across the world, and then we discuss the importance of incorporating biofuel by-products into the global assessment of biofuel production.

Hertel et al. (2008) have examined the implications of US and EU biofuel mandate policies for the world economy during the time period of 2006–2015.[11] The United States and EU are major producers of agricultural and livestock commodities. They are also big players in international agricultural markets. Therefore, their biofuel mandate policies are expected to affect agricultural markets and increase competition for land across the world. In addition, their biofuel policies could interact and affect direction of changes in resource allocation and land use changes across the world. So it is important to examine consequences of their joint biofuel mandate policies for the global economy. According to Hertel et al. (2008), these mandate policies are expected to affect land use changes across the world and in particular in the United States, EU, and Brazil. To jointly meet the biofuel mandate policies of the United States and EU (ceteris paribus), the cropland areas in the United States, EU, and Brazil would be increased by 0.8, 1.92, and 1.98%, respectively (Table 21.1).

Table 21.1 Percentage change in land cover due to the US–EU biofuel mandate policies (2006–2015)

Area	Cropland	Forest	Pasture
US	0.8	−3.14	−4.93
EU-27	1.92	−8.34	−9.68
Brazil	1.98	−5.13	−6.3

Note: Figures are percentage changes in productivity adjusted hectares.

We now decompose the impacts of the US and EU policies on cropland. As shown in Table 21.2, about 29% of the change in the cropland areas in the United States occurs due to the EU mandate polices while only about 7% of the change in the EU cropland areas is linked to US mandate policies. Table 21.2 shows that about 27% of changes in the cropland areas of Brazil would be associated with US mandate policies, and the rest would be induced by EU mandate policies. As shown in Table 21.1, the expansion of cropland causes reductions in forest and pasture

---

[11] According to these mandate policies, the US corn ethanol production will be equal to 15 billion gallons in 2015. The European Union Biofuels Directive requires that 10% of liquid fuel should be biofuel in the EU region by 2020 (European Commission (2007)). To compare and contrast the EU biofuel directive with the US mandate, it is assumed that the EU will obtain 6.25% of its transportation fuels from biofuels in 2015.

Table 21.2 Decomposition of land cover impacts of the US and EU biofuel mandate policies (%)

Description	Cropland			Forest			Pasture		
	US–EU mandates	Only US mandates	Only EU mandates	US–EU mandates	Only US mandates	Only EU mandates	US–EU mandates	Only US mandates	Only EU mandates
US	100.0	71.3	28.8	100.0	72.9	27.1	100.0	68.0	32.0
EU-27	100.0	6.8	93.2	100.0	6.5	93.5	100.0	6.8	93.2
Brazil	100.0	27.3	72.7	100.0	29.4	70.6	100.0	25.4	74.6

lands. The US and EU mandate policies jointly reduce the forest and pasture land areas of the United States by 3.1 and 4.9%, respectively (Table 21.1). About 73 and 68% of reductions in the US forest and pasture land areas are respectively due to its own biofuel mandate policies, and the rest are due to the EU mandates (Table 21.2). US mandate policies have a small impact on forest and pasture land areas in the EU. While both the US and EU mandate policies induce land use changes in Brazil, the EU mandate policies induce more land use changes in this country compared to the United States (up to 3 times) mainly because the EU needs Brazilian soybeans and soybean oil to meet its biodiesel mandate.

These results clearly indicate that international biofuel policies can interact and induce economic and land use consequences across the world. In other words, consequences of producing biofuel in one region can be passed on to other regions as well.

Taheripour et al. (2008) have shown the importance of incorporating biofuel by-products into the economic analysis of biofuels policies. The model with by-products reveals that production of distillers dried grains with solubles (DDGS) and biodiesel by-products (BDBP) would grow sharply in the United States and EU. For example, the US production of DDGS would grow from 12.5 million metric tons in 2006 to 34 million metric tons if it reaches to its target for ethanol production in 2015. A major portion of this by-product would be used within the United States, and the rest would be exported to other regions such as Canada, EU members, Mexico, China, African and Asian countries. On the other hand, the EU production of BDBP would grow from about 6.1 million metric tons in 2006 to 32.5 million metric tons if it reaches to its target for biodiesel production in 2015. The EU production of BDBP would be mainly used within this region. The CGE models with and without by-products tell quite different stories regarding the economic impacts of the US and EU biofuel mandates policies. Impacts of these policies on the outputs and prices of agricultural commodities are shown in Tables 21.3 and 21.4 for the United States, EU, and Brazil.

Table 21.3 Percentage changes in the outputs of agricultural commodities due to the US–EU biofuel mandate policies (2006–2015)

Agricultural commodities	Without by-products			With by-products		
	US	EU	Brazil	US	EU	Brazil
Coarse grains	16.4	2.5	−0.3	10.8	−3.7	−2.8
Other grains	−7.5	−12.2	−8.7	−5.0	−12.2	−8.5
Oilseeds	6.8	51.9	21.1	8.6	53.1	19.0
Sugar crops	−1.8	−3.7	8.2	−0.9	−3.3	8.4
Livestock	−1.2	−1.7	−1.3	−0.7	−2.1	−2.1

While both models demonstrate significant changes in the agricultural production patterns around the world, the model with by-products shows smaller changes in the production of cereal grains and larger changes for oilseeds in the United States and EU, and the reverse for Brazil. For example, as shown in Table 21.3 and Fig. 21.8,

Table 21.4 Percentage changes in the supply prices of agricultural commodities due to the US–EU biofuel mandate policies (2006–2015)

Agricultural commodities	Without by-products			With by-products		
	US	EU	Brazil	US	EU	Brazil
Coarse grains	22.7	23.0	11.9	14.0	15.9	9.6
Other grains	7.7	13.7	8.8	6.0	11.5	7.8
Oilseeds	18.2	62.5	20.8	14.5	56.4	18.3
Sugar crops	12.6	16.2	18.6	9.4	14.0	17.5
Livestock	3.6	4.6	4.0	3.1	6.0	5.6

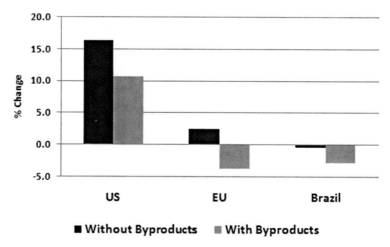

Fig. 21.8 Percentage change in coarse grain production due to the US and EU biofuel mandate policies 2006–2015

the US production of cereal grains increases by 10.8 and 16.4% with and without by-products, respectively. The difference between these two numbers corresponds to 646 million bushels of corn, which could be used to produce about 1.7 billion gallons of ethanol. This is really a big number to ignore and disregard in the economic analyses of biofuel production.

With by-products included in the model, prices change less due to the mandate policies. For example, as shown in Table 21.4, the model with no by-products predicts that the price of cereal grains grows 23% in the United States during the time period of 2006–2015. The corresponding number for the model with by-products is 14%.

Introducing by-products into the model alters the trade effects of the US–EU mandate policies as well. For example, as shown in Table 21.5, the model with no by-products estimates that the US exports of coarse grains to the EU, Brazil, and Latin American region would be sharply dropped by –4.8, –25.5, and –12.7%, respectively. The corresponding figures for the model with by-products are –2.1, –15.7, and –7.9%.

Table 21.5 Percentage changes in the quantities of US exports of grains and oilseeds to the selected regains due to the US–EU biofuel mandate policies (2006–2015)

Commodity	Without by-products			With by-products		
	EU	Brazil	LAEEX[a]	EU	Brazil	LAEEX[a]
Coarse grains	−4.8	−25.5	−12.7	−2.1	−15.7	−7.9
Other grains	32.0	−9.3	−8.7	31.4	−4.7	−5.9
Oilseeds	105.7	−4.1	−11.3	109.4	0.4	−8.7

[a]LAEEX is an abbreviation for Latin American Energy Exporting Countries.

21.5 Conclusions

Clearly, biofuels policies have major impacts on the level of production and the global distribution of impacts. In addition, other factors such as the effective ban on use of MTBE in the United States in 2006 also have had a significant impact. Clearly also the increased price of crude oil has a very important impact on ethanol production. It is important to note that ethanol production both from sugarcane and corn is largely driven by the oil price, whereas biodiesel is driven mainly by government policies. That is, ethanol is economic without a subsidy at crude oil prices more than $60 per barrel (even though subsidies persist in the United States), but biodiesel would not exist even at high oil prices without government subsidies. Both ethanol and biodiesel production involve potentially large land use changes that take place all over the world. The model and data used for land use change analysis illustrated here need improvement. However, the numbers shown here illustrate the kinds of results that can be produced from the global analysis.

References

Argonne National Laboratory (2007) Greenhouse gases, regulated emissions, and energy use in transportation (GREET) computer model. Center for Transportation Research, Energy Systems Division, Argonne National Laboratory, Argonne, IL.

Banse M, van Meijl H, Tabeau A, Woltjer G (2007) Impact of EU Biofuel Policies on World Agricultural and Food Markets. Presented at the 10th Annual Conference on Global Economic Analysis, Purdue University.

Birur D, Hertel T, Tyner W (2007) Impact of Biofuel Production on World Agricultural Markets: A Computable General Equilibrium Analysis. GTAP Working Paper No 53, Center for Global Trade Analysis, Purdue University.

Burniaux J, Truong T (2002) GTAP-E: An Energy-Environmental Version of the GTAP Model. GTAP Technical Paper No. 16, Center for Global Trade Analysis, Purdue University.

Christensen DA, Schatzer RJ, Heady EO, English BC (1981) The Effects of Increased Energy Prices on U.S. Agriculture: An Econometric Approach, Card Report 104 Center for Agricultural and Rural Development, Ames, IA.

de Gorter H, Just DR (2008) "Water" in the U.S. Ethanol Tax Credit and Mandate: Implications for Rectangular Deadweight Costs and the Corn-Oil Price Relationship Review of Agricultural Economics Fall, 30(3), 397–410.

Dvoskin D, Heady EO (1976) U.S. Agricultural Production under Limited Energy Supplies, High Energy Prices, and Expanding Agricultural Exports. Card Report 69 Center for Agricultural and Rural Development, Ames, IA.

Elobeid A, Tokgoz S, Hayes DJ, Babcock BA, Hart CE (2007) The Long-Run Impact of Corn-Based Ethanol on the Grain, Oilseed, and Livestock Sectors with Implications for Biotech Crops. AgBioForum 10(1):11–18.

European Commission (2007) Impact Assessment of the Renewable Energy Roadmap-March 2007, Directorate-General for Agriculture and Rural Development, European Commission, AGRI G-2/WM D.

Farrell AE, Plevin RJ, Turner BT et al. (2006) Ethanol Can Contribute to Energy and Environmental Goals. Science 311, 5760, 506–508.

Hertel T, Tyner W, Birur, D (2008) Biofuels for all? Understanding the Global Impacts of Multinational Mandates. GTAP Working Paper No. 51, Center for Global Trade Analysis, Department of Agricultural Economics, Purdue University.

Hertel T (1997) Global Trade Analysis, Modeling and Applications. Cambridge University Press, Cambridge.

Khanna M, Ando A, Taheripour F (2008) Welfare Effects and Unintended Consequences of Ethanol Subsidies. Review of Agricultural Economics Fall 30(3):411–421.

McDougall R, Golub A (2007) GTAP-E Release 6: A Revised Energy-Environmental Version of the GTAP Model. GTAP Research Memorandum No. 15, Center for Global Trade Analysis, Department of Agricultural Economics, Purdue University.

McPhail LL, Babcock BA (2008a) Short-Run Price and Welfare Impacts of Federal Ethanol Policies. Staff General Research Papers 12943, Iowa State University Department of Economics.

McPhail LL, Babcock BA (2008b) Ethanol, Mandates, and Drought: Insights from a Stochastic Equilibrium Model of the U.S. Corn Market. Staff General Research Papers 12878, Iowa State University, Department of Economics.

Rajagopal D, Sexton SE, Roland-Holst D, Zilberman D (2007) Challenge of Biofuel: Filling the Tank Without Emptying the Stomach? Environmental Research Letters 2:1–9.

Taheripour F, Tyner W (2008) Ethanol Subsidies, Who Gets the Benefits? In Joe Outlaw, James Duffield, and Ernstes (eds), Biofuel, Food & Feed Tradeoffs, Proceeding of a conference held by the Farm Foundation/USDA, at St. Louis, Missouri, April 12–13 2007, Farm Foundation, Pak Brook, IL, 91–98.

Taheripour F, Hertel T, Tyner W, Beckman J, Dileep K (2008) Biofuels and their By-Products: Global Economic and Environmental Implications. Presented at the 11th GTAP Conference, June 12–14 2008, Helsinki, Finland and at the 2008 American Agricultural Economics Association meeting in Orlando Florida.

Tiffany D, Eidman V (2003) Factors Associated with Success of Fuel Ethanol Producers. Staff Paper P03-7. St. Paul, MN: University of Minnesota, Department of Applied Economics.

Tokgoz S, Elobeid A, Fabiosa J, et al. (2007) Emerging Biofuels: Outlook of Effects on U.S. Grain, Oilseed and Livestock Markets. Staff report 07-SR 101 Iowa State University (www.card.iastate.edu)

Tyner W, Taheripour F (2007) Renewable Energy Policy Alternatives for the Future, AJAE, 89(5):1303–1310.

Tyner W, Taheripour F (2008a) Future Biofuels Policy Alternatives, in Joe Outlaw, James Duffield, and Ernstes (eds), Biofuel, Food & Feed Tradeoffs, Proceeding of a conference held by the Farm Foundation/USDA, at St. Louis, Missouri, April 12–13 2007, Farm Foundation, Pak Brook, IL, 2008, 10–18.

Tyner W, Taheripour F (2008b) Policy Options for Integrated Energy and Agricultural Markets. Presented at the Allied Social Science Association meeting in New Orleans, January 2007, and published in the Review of Agricultural Economics, 30(3):387–396

Tyner W, Taheripour F (2008c) Policy Analysis for Integrated Energy and Agricultural Markets in a Partial Equilibrium Framework. Paper Presented at the Transition to a Bio-Economy: Integration of Agricultural and Energy Systems conference on February 12–13, 2008 at the Westin Atlanta Airport planned by the Farm Foundation.

Tyner W, Taheripour F (2008D) Biofuels, Policy Options, and Their Implications: Analyses Using Partial and General Equilibrium Approaches, Journal of Agricultural & Food Industrial Organization, 6(2): Article 9. Available at: http://www.bepress.com/jafio/vol6/iss2/art9

Tyner W, Dooley F, Hurt CS, Quear J (2008) Ethanol Pricing Issues for 2008. Industrial Fuels and Power, February, pp.50–57.

Wang M (2005) Updated Energy and Greenhouse Gas Emission Results of Fuel Ethanol. Presented on the 15th International Symposium on Alcohol Fuels, Sept. 26–28, 2005, San Diego, CA, USA.

Wang M (1999) GREET 1.5 – Transportation Fuel-Cycle Model Volume 1: Methodology, Development, Use, and Results. Center for Transportation Research, Energy Systems Division, Argonne National Laboratory, Argonne, Illinois.

Chapter 22
Welfare and Equity Implications of Commercial Biofuel

Fredrich Kahrl and David Roland-Holst

Abstract New sources and uses of biofuel energy offer the prospect of climate change mitigation and less reliance on fossil fuels. At the same time, biofuel represents an unusual precedent in economics, the possibility of substitution between two essential but very different commodities, food and energy. The first characteristic has dramatically heightened interest in biofuel production around the world, but particularly in high-income economies, whose expenditure patterns are most energy-intensive. Rising concerns about the need for climate stabilization and rapid innovation to reduce greenhouse gas (GHG) emission have stimulated visions of a booming new agribusiness energy industry, with pervasive induced effects on transportation and related sectors. At the same time, diversion of agricultural resources to energy production has implications for food markets that are only beginning to be fully understood, but are of special concern to poor countries whose expenditure patterns are mostly food-intensive. Both commodities are essential to human well-being, and their prices are important determinants of real living standards.

In this paper, we assess the consequences of energy and food price uncertainty for the poor, using a variety of empirical assessment techniques. Beginning with traditional poverty indicators, we appraise the effect of energy and food price vulnerability on the existing burden of poverty, providing concrete indications for policy makers about distributional incidence. A second approach decomposes energy and food price effects across the economy, showing how the embodied costs of these basic commodities affect overall household purchasing power. Finally, we use a general equilibrium framework to elucidate the complex pathways through which energy and food prices interact to affect both household incomes and the cost of living. Each approach offers different insights that can support better foresight and more effective policy responses, but a few general patterns emerge. It is clear from this analysis, for example, that a North−South dichotomy between relative

F. Kahrl (✉)
Department of Agricultural and Resource Economics, University of California, Berkeley, CA, USA
e-mail: fkahrl@berkeley.edu

M. Khanna et al. (eds.), *Handbook of Bioenergy Economics and Policy*,
Natural Resource Management and Policy 33, DOI 10.1007/978-1-4419-0369-3_22,
© Springer Science+Business Media, LLC 2010

energy and food dependence, respectively, may lead to persistent contention regarding the use of agricultural resources for energy production. In particular, food price vulnerability is strongly regressive across income distributions, while energy price vulnerability is strongly progressive. This conclusion reaffirms productivity oriented, and intensive rather than extensive approaches to biofuel development.

22.1 Introduction

In modern times, food prices have been kept low by productivity gains and policy intervention. While energy prices have risen gradually in high-income countries, they were often buffered in low-income countries by subsidies. The result of these trends (apart from individual crises) has been relatively stable food and energy living standards within and between economies both North and South. However, sustained recent escalation of global fuel prices threatens to disrupt this status quo, and biofuels might compound this problem. Direct energy price effects will reduce real incomes and divert expenditure from other necessities, while indirect stimulus to biofuel development will increase pressure on food prices, eroding living standards from a different direction. In this chapter, we use a variety of empirical assessment techniques and data from a diverse set of lower and middle income countries to assess the significance of these price effects for living standards, and the implications of this for biofuel policy.

Of course, economic sustainability depends on different rules today than it did in previous centuries. Resources are not infinite, and their costs will continue to rise with increasing and prolonged exploitation. Higher prices for resources will have two economic impacts, incentives for efficiency and rationing. Economic efficiency incentives can be expected to trigger innovations that make more productive use of existing resources, like the Green Revolution, which dramatically increased agricultural yields and food security in developing countries. Fuel efficiency, renewable energy innovation, recycling – all these are rational and technologically progressive responses to rising energy prices. The innovation process has not only helped us overcome scarcity in the past, but triggered new waves of prosperity in knowledge-intensive industries.

Rationing is the second effect of scarcity-induced price increases and is more ominous. As something becomes more expensive, those without the right combination of willingness and purchasing power will be driven out of the market. In the case of luxury goods, like elephant ivory, this may be a personal tragedy but not a social one. When the goods and/or resources in question are essential (food, water, clean air, etc.), however, this is a much more serious matter. During 2007–2008, a number of factors contributed to rapid escalation of prices for basic cereals, including rice, wheat, and corn. These staples still provide half the protein for the world's poor majority, 58% of humanity who live on $2.50 or less per day.[1]

---

[1] CEpsteinhen and Ravallion (2008).

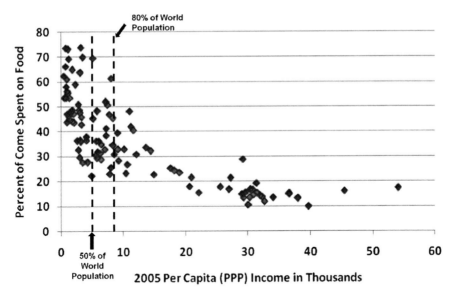

Fig. 22.1 Food and poverty (Source: Author estimates from World Bank and USDA data)

Figure 22.1 indicates how profound the social implications of these price increases can be. For a sample of 114 countries, we see expenditure on food as a share of total income (vertical), plotted against GDPPC.

Now contrast this with Fig. 22.2, which shows patterns of national energy dependence by income and population. Clearly, expenditure patterns in higher income countries are much more energy intensive on a per capita basis. These two figures reveal a basic fundamental dichotomy in North–South energy–food dependence, with important implications for biofuel policy. Food is of course essential to everyone, but while Northern countries will be relatively more sensitive to energy scarcity, Southern countries will be relatively much more sensitive to food scarcity. In terms of vulnerability to the price effects of such scarcity, however, there is an important asymmetry in the North–South food–energy dichotomy – ability to pay. Whichever way scarcity might drive a global food-fuel auction, the poor would be at a severe disadvantage.

For these reasons, food scarcity and price effects require special policy consideration in contexts like biofuel, where food–fuel tradeoffs may emerge. This is so regardless of whether biofuels enter markets in response to spontaneous economic forces or policy interventions. In the first case, one might argue that entry confers a lower priced alternative on global energy consumers, yet this corresponds to an indeterminate externality for food consumers. When policy makers influence the economics of biofuels, they may be responding to interests of some stakeholders with incomplete information regarding spillover effects that will play out in the marketplace.

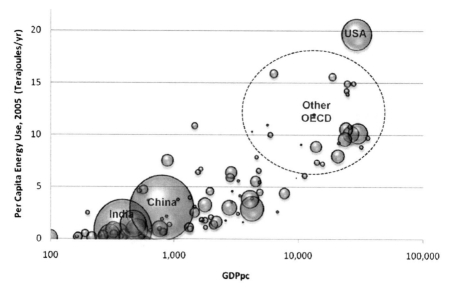

Fig. 22.2 Energy and prosperity (Source: Author estimates from OECD (2005) and World Bank (2005) data. Bubble diameter is proportional to population)

To more fully assess the complex domestic and international welfare effects of the modern biofuel economy, a new generation of assessment tools is needed. This paper proposes three families of metrics based on established tools for analyzing poverty and inequality. Beginning with traditional poverty indicators, we appraise the effect of energy and food price vulnerability on the existing burden of poverty, providing concrete indications for policy makers about distributional incidence. A second approach decomposes energy and food price effects across the economy, showing how the embodied costs of these basic commodities affects overall household purchasing power. Finally, we use a general equilibrium framework to elucidate the complex pathways through which energy and food prices interact to affect both household incomes and the cost of living. These are then estimated with observed data from a representative set of countries to illustrate the generality of the measures and the importance of heterogeneity of initial conditions. These results indicate the complexities of issues that biofuel potential could create between energy and fuel, and how domestic and international disparities in real incomes and welfare may be affected by these interactions. For example, energy price escalation could become contagious for food prices if it leads to diversion of agricultural resources. This could reverse an important and virtuous 50-year trend of convergence between high- and low-income country living standards.

22.2 Price Effects on Poverty and Inequality

Consumers in OECD economies have some flexibility when responding to rising energy or food prices. Both goods are small shares of total expenditure, and

accumulated savings enable these household to consider efficient technology investments as a buffer against higher energy and agricultural price trends. In developing
countries, energy price rises have been amplified by the collapse of domestic subsidy schemes and food security is low and in many cases aggravated by external,
cheap food dependence. In both cases, the scope for price hedging through domestic substitution or more efficient resource use is limited, and to a significant extent
these countries must take energy and food prices as given to them by international
markets.

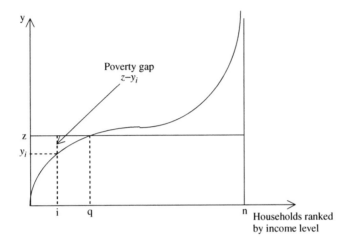

Fig. 22.3 Poverty profile

To assess the distributional incidence of energy and food prices, we exploit
traditional poverty and inequality measures. In particular, we consider how decomposable poverty indices, decile income status, and Gini coefficients respond to price
variation in these essential commodities. To begin, consider the traditional family
of additively decomposable (sometimes called FGT for Foster-Greer-Thorbecke:
1992) indexes:

$$P^{\alpha} = \frac{1}{n} \sum_{i=1}^{q} \left(\frac{z - y_i}{z} \right)^{\alpha} \tag{22.1}$$

where n denotes a population of individuals (i) with incomes y_i, and we identify
a threshold level below which ($y_i < z$) there are $q < n$ people in poverty. Finally
the parameter α defines a category of poverty measurement. For example, when
$\alpha = 0$, $P^0 = q/n$, the headcount ratio or proportion of the population in poverty.
Figure 22.3 above illustrates a schematic poverty profile, or poverty calibrated
income distribution. Here we can think of $\frac{z-y_i}{z}$ as representing the normalized
"poverty gap" or the amount individual i would need to receive (as a percent of z)
to reach the threshold. Adding these over the population (q) of eligible poor yields
a Poverty Burden, equal to the total cost of poverty alleviation given by

$$B = nzP^1 = \sum_{i=1}^{q} (z - y_i) \tag{22.2}$$

We will also measure poverty by expenditure rather than income.[2] In this case, we can express y_i as a simple linear expenditure system

$$y_i = p_0 + \sum_{k=1}^{K} p_k X_{ik} \tag{22.3}$$

for savings p_0 and K commodities with prices P_k and i th individual consumption of good k given by X_{ik}. Now assume the price of one commodity (k) changes. This will change the *severity* of poverty or inequality among the poor

$$P^2 = \frac{1}{n} \sum_{i=1}^{q} \left(\frac{z - y_i}{z} \right)^2 = \frac{1}{nz^2} \sum_{i=1}^{q} (z - y_i)^2$$

as follows

$$\frac{dP^2}{dp_k} = -\frac{2}{nz^2} \sum_{i=1}^{q} (z - y_i) \frac{dy_i}{dp_k} \tag{22.4}$$

which will be positive for rising prices and any good with negative own-price elasticity.

To illustrate the importance of energy and food prices to domestic poverty agendas, we examine evidence from Living Standards Measurement Surveys (LSMS) for Thailand and Vietnam. These samples are nationally representative and consist 44,000 and 65,000 observations, respectively. Schematically, the poverty profiles for this middle- and low-income country are compared in Fig. 22.4. After normalizing to equalize top decile income in each country, national poverty lines (z) are depicted by the gray band. A relatively larger middle class in Thailand explains the gap between the two profiles.

Table 22.1 presents the poverty-impact assessments, expressed as percent changes for 1% variation in prices for food and two types of energy. These measures are expressed first in terms of national and international ($2/day) poverty thresholds, followed by elasticities of decile expenditure averages with respect to both prices. The most arresting finding here is the elasticity of the overall poverty gap (Burden of Poverty) with respect to food prices. In both countries, the poor have very high food

---

[2]Expenditure-based welfare analysis is well established in developing countries, where income reporting is very unreliable.

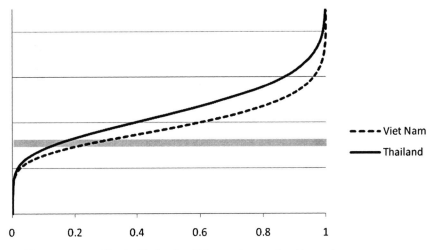

Fig. 22.4 Poverty profiles for Thailand and Vietnam (poverty band in gray)

Table 22. 1 Poverty and expenditure elasticities with respect to food and energy prices (all figures expressed as percent change with respect to a 1% change in price)

| | | Food | | Electricity | | Transportation fuels | |
		Vietnam	Thailand	Vietnam	Thailand	Vietnam	Thailand
Poverty burden	National line	2.00	2.17	0.050	0.118	NA	0.189
Average	Mean	0.47	0.30	0.024	0.026		0.063
expenditure	Decile 1	0.70	0.53	0.021	0.030		0.029
	Decile 2	0.64	0.48	0.021	0.032		0.046
	Decile 3	0.60	0.45	0.022	0.030		0.050
	Decile 4	0.58	0.43	0.022	0.029		0.054
	Decile 5	0.55	0.40	0.022	0.028		0.056
	Decile 6	0.52	0.38	0.022	0.028		0.059
	Decile 7	0.50	0.35	0.022	0.027		0.064
	Decile 8	0.46	0.32	0.022	0.027		0.070
	Decile 9	0.42	0.28	0.023	0.026		0.075
	Decile 10	0.35	0.18	0.027	0.022		0.065

Sources and Notes: Thailand data are from the 2006 Thailand Socioeconomic Survey; Vietnam data are from the 2002 Vietnam Living Standards Survey. Poverty burden calculations for Thailand are based on a 2006 poverty line of 1,386 baht/person-month ($34/person-month); for Vietnam these are based on a 2002 national poverty line for Vietnam of 1,915 thousand dong/person-year ($10/person-month). Electricity expenditure elasticities are based on household, rather than per capita, expenditures.

expenditure shares, and thus their real incomes would be significantly undermined by food price inflation. Here we see that the compensation needed to raise the poor to the poverty line would increase at twice the rate of food price increases. Seen as a targeted subsidy, this finding is significant for both welfare and fiscal policy.

The poverty impact of energy prices is much less significant, reinforcing the impression gained from cross-country comparison in the previous section. The poorest among the population spend only a small fraction of their income on energy, and would thus be much less adversely affected by equal percent changes in prices of this commodity.

The basic conclusions regarding those below the poverty line are sustained when we look across the rest of the income distribution. In particular, compensated income elasticities with respect to food prices decline monotonically with income decile in both countries (shown graphically in Fig. 22.5), much as we saw between countries in Table 22.1 above.

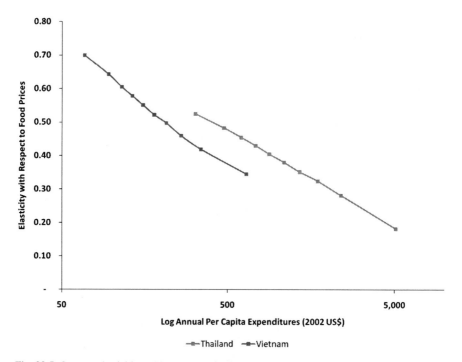

Fig. 22.5 Income elasticities with respect to food prices

The situation for energy prices is more complex, however. Figure 22.6 illustrates two important findings for these countries. First, energy price vulnerability is relatively low in both countries, an order of magnitude less than food price dependence. Second, price vulnerability rises from the lowest incomes but then falls sharply in Thailand while continuing its ascent in Vietnam. The most likely explanation of this difference has to do with household technology adoption. Recall that Fig. 22.2 represents total domestic energy use per capita (i.e., including industry), while LSMS data account only for household energy use. As inhabitants of a newly emerging economy, Vietnamese households are still in the early stages of electrification generally and appliance acquisition in particular. For these reasons, electricity

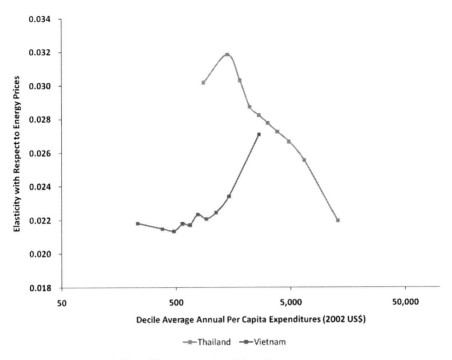

Fig. 22.6 Income elasticities with respect to electricity prices

is still likely to be a luxury good, rising with incomes as a share of expenditure. In Thailand, by contrast, electricity is now a necessity for most households, falling with income as a share of expenditure. Given the absolute magnitudes of all these compensated income effects, however, electricity remains significantly less important than food for majorities of both countries.

22.3 Multiplier Estimation of Price Vulnerability

In this section, we examine how food and energy price changes are transmitted through the economy, ultimately reaching households through extensive price pass-through effects. To elucidate these pathways, we use the Social Accounting Matrix (SAM), an economywide accounting device that offers a disaggregate view of value flows, detailing the direct linkages among its component sectors and institutions and pointing out the scope of the underlying indirect interactions.[3] Inflows from exogenous sectors that stimulate the level of activity of a production sector, for

---

[3] For more background on this approach, see e.g. Pyatt and Round (1985) or Stone (1981) who won the Nobel Prize for his work on social accounting. This section applies a dual multiplier approach developed by Roland-Holst and Sancho (1995).

instance, will also induce change in factor incomes that, once distributed among households, affect final demand for producer goods and services. Like its ancestor, input−output accounting, columns of the SAM indicate payments and rows tally receipts.[4] For each group, total spending is necessarily equal to total receipts, i.e., column and row totals of the matrix are equal. Although detailed path analysis is outside the scope of this example, it can be quite instructive to decompose multiplier linkages for this purpose.[5]

An SAM-based quantity model is derived from the SAM-table by distinguishing endogenous and exogenous groups and assuming activity levels may vary while prices are fixed. This assumption is justified in the presence of excess capacity and unused resources in production activities. Suppose group 1 is chosen as endogenous and 2, 3, and 4 are exogenous. Let A_{ij} denote the matrix of normalized column coefficients obtained from the underlying SAM transactions matrix Tij and let v_i denote that the incomes of groups $i=2,3,4$ are taken as given exogenously. Then the income level of group 1 can be expressed by

$$Y_1 = A_{11}Y_1 + A_{13}v_3 + A_{14}v_4$$

$$= (I-A_{11})^{-1}(A_{13}v_3 + A_{14}v_4) = M_{11}\mathbf{x} \qquad (22.5)$$

where $M_{11} = (I-A_{11})^{-1}$ is the usual inter-industry Leontief inverse and $\mathbf{x}$ is a vector of exogenous income levels. Since (22.5) implies $\Delta Y_1 = M_{11}\Delta\mathbf{x}$, matrix M_{11} is also termed the multiplier matrix. Column i of M_{11} shows the global effects on all endogenous activity levels induced by an exogenous unit inflow accruing to i, after allowing for all interdependent feedbacks to run their course.

Consider now the dual case, where prices are responsive to costs but not to activity levels. The justifying assumption here, in addition to the usual excess capacity condition, is generalized homogeneity and fixed coefficients in activities. This is a situation where the classical dichotomy between prices and quantities holds true and prices can be computed independently of activity levels. Let now p_i denote a price index for the activity of group i.[6] With the same classification of endogenous and exogenous accounts and identical notational conventions, column 1 of the SAM yields

$$p_1 = p_1A_{11} + \pi_2A_{21} + \pi_4A_{41}$$

$$= (\pi_2A_{21} + \pi_4A_{41})(I-A_{11})^{-1} = \mathbf{v}_1M_{11} \qquad (22.6)$$

---

[4]Note that no flows are associated with cells (1,2), (2,2), (2,3), and (3,1).

[5]See Defourny and Thorbecke (1984) for more on this approach.

[6]The notion of price should be taken in the same broad sense as the notion of income of a sector or institution has in a SAM framework.

where v_1 In this section, we examine (i.e., factor payments, taxes, import costs) and M_{11} is the same multiplier matrix as in (22.5). Notice that from (22.6) we have $\Delta p_1 = \Delta v_1 M_{11}$ so we can reinterpret the Leontief inverse by reading across rows. Row j of M_{11} displays the effects on prices triggered by a unitary exogenous change in sector j costs. This is a straightforward but seldom used interpretation of the Leontief multiplier matrix.

Starting from Equation (22.5) of the basic linear model, SAM-based quantity models yield extensions to encompass a larger and more complete view of the income-generating process. In the same way, we believe SAM-based price models departing from expression (22.6) may prove to be useful generalizations for evaluating the extensive cost linkages that pervade the relationships among households, factors, and producers.

To give content to this approach, consider each one of these groups as undertaking an economic activity. Producers pay for raw materials (T_{11}) and factors (T_{21}) which are combined to generate output; factors make use of household endowments (T_{32}) to provide firms with labor and capital services. Finally, households purchase output (T_{13}) from production to obtain consumption.[7] Additionally, each group is liable to pay taxes or import costs to the consolidated group 4. In terms of taxes, the government collects indirect production taxes from firms, taxes on the use of labor and capital from factors, and indirect consumption taxes and income taxes from households. Thus, each of these activities has an implicit cost or price index, which is linked to the rest of the price indices through the coefficient submatrices of the SAM. However, as it stands, price expression (22.6) omits these linkages and falls short of a satisfactory representation of interdependencies in the economy.

These links can be coherently integrated into a model by considering the three sets of accounts comprising producers, factors, and households as endogenous and taking the consolidated account as exogenous. Using the column normalized expenditure coefficients and reading down the SAM columns for endogenous accounts yields

$$p_1 = p_1 A_{11} + p_2 A_{21} + \pi_4 A_{41}$$

$$p_2 = p_3 A_{32} + \pi_4 A_{42}$$

$$p_3 = p_1 A_{13} + p_3 A_{33} + \pi_4 A_{43} \tag{22.7}$$

Define a matrix A of normalized coefficients

$$A = \begin{bmatrix} A_{11} & 0 & A_{13} \\ A_{21} & 0 & 0 \\ 0 & A_{32} & A_{33} \end{bmatrix} \tag{22.8}$$

---

[7] Transfers among households T_{33} can be thought of as distribution costs linked to consumption.

let $\mathbf{p} = (p_1, p_2, p_3)$ be the vector of prices for the endogenous sectors of the SAM, and set the vector of exogenous costs (taxes, import costs) as $\mathbf{v} = \pi_4 A$ (4), where A (4) is the submatrix of the SAM composed by column adjoining A_{41}, A_{42}, and A_{43}. In matrix notation

$$\mathbf{p} = \mathbf{p}A + \mathbf{v} = \mathbf{v}(I-A)^{-1} = \mathbf{v}M \qquad (22.9)$$

where M is the multiplier matrix. For the same classification of endogenous and exogenous accounts, M is also the multiplier matrix of the endogenous income determination model

$$Y = (I - A)^{-1}\mathbf{x} = M\mathbf{x} \qquad (22.10)$$

The interpretation of M is different, however, depending on whether we read its entries across the rows or down the columns. To clarify this distinction, M will be referred as the (standard) multiplier matrix whereas its transpose M' will be termed the price-transmission matrix.

The approach outlined here can be applied to any country with a SAM, a group that now includes a significant majority of LDCs.[8] To illustrate pass-through price effects for food and energy, we applied these methods to a Social Accounting Matrix for Morocco, a lower middle-income country in Africa. Table 22.2 indicates the total and direct household CPI effects of price changes in food and energy, by decile household group. For the Moroccan case, first note the relatively small expenditure shares for food among the poor, the reason for this is that this SAM accounts only for marketed food, and most of this country's lower deciles are in

Table 22.2 Total and direct CPI impacts of food and energy prices, by Moroccan household income level

	Food			Energy		
	Total	Direct	Ratio	Total	Direct	Ratio
Decile 1	0.62	0.26	2.37	0.23	0.04	5.23
Decile 2	0.80	0.39	2.06	0.21	0.02	8.89
Decile 3	0.77	0.36	2.10	0.22	0.03	7.27
Decile 4	0.74	0.35	2.13	0.21	0.03	7.79
Decile 5	0.71	0.33	2.17	0.21	0.03	6.97
Decile 6	0.66	0.29	2.26	0.21	0.03	7.09
Decile 7	0.60	0.25	2.35	0.20	0.03	6.95
Decile 8	0.55	0.22	2.52	0.19	0.03	6.90
Decile 9	0.48	0.18	2.72	0.18	0.03	6.91
Decile 10	0.37	0.12	3.06	0.16	0.03	5.95

[8]For example, SAMs are maintained for every country in the GTAP data set, although detail on household income distribution varies (see www.gtap.org).

the subsistence-oriented rural sector. This food self-sufficiency buffers them against direct and indirect food price shocks, yet we see that they remain vulnerable.

Even when subsistence agriculture meets an important part of food requirements for the poor, we again see that food expenditure shares (direct impact column) are an order of magnitude higher than energy shares. Of more interest in the present analysis, however, are the cumulative indirect effects of food and energy price pass-through. When the economy-wide impact (total) of food prices is incorporated into all consumption goods, the household CPI rises by over twice as much as does the CPI (direct) component of food prices alone. Both direct and total effects are relatively higher for the poor (outside subsistence), although the difference between the two rises as food consumption for higher income groups includes more processing value added.

The situation for energy is more dramatic. Even though it remains a small percent of (direct) expenditure, energy so pervades the economy that the (total) pass-though effect far outweighs this. Across deciles, total energy price impacts are 5–7 times higher than direct effects. Overall, our general finding regarding aggregate CPI importance is robust (i.e., food matters more than energy), but we see that, within the country, total CPI effects of energy are greater for the poor than for higher income groups. Having said this, total effects are more homogeneous than direct effects, which is logical in light of the economy-wide averaging effect of accounting for embodied energy.

From a policy perspective, these findings sustain the argument that food price vulnerability is greater than energy price vulnerability for a lower income country, and greater for lower income groups within that country. The pass-through analysis further suggests that targeting relief policies can be more effective for food, where direct price effects are more important, than for energy, where CPI impacts are much more diffuse.

22.4 General Equilibrium Estimation of Price Vulnerability

The assessment methods used above offer important insights into the incidence of food and energy price volatility, but restrict most of their attention to the demand side. This may be appropriate for energy, where in most cases prices are exogenous and households are strictly consumers. In developing countries, however, the majority of households (and a much larger majority of the poor) are tied to the rural sector and agricultural production. For these households, variations in food prices have important implications for both sides of the balance sheet, and a more comprehensive assessment of food price effects should account for income as well as consumption effects. We achieve this in the present context by using a computable general equilibrium model, applied here to the economy of Senegal.[9]

---

[9]See Roland-Holst and Otte (2006) for details. Drèze and Sen (1989) examine market price vulnerability in a broader context, while FAO (2005) provides extensive data on direct food price vulnerability.

Table 22.3 Real household income by location and quintile (percentage change for a 1% change in food or energy prices)

	Food	Energy
Rural 1	−.285	.055
Rural 2	−.144	.028
Rural 3	.162	−.005
Rural 4	.184	−.003
Rural 5	.609	.020
Urban 1	−.192	−.013
Urban 2	−.422	.054
Urban 3	.176	−.008
Urban 4	−.029	.019
Urban 5	.236	−.045

Table 22.3 presents the results of two comparative static simulations, varying exogenous (international) prices of food and energy, respectively. Since domestic prices are endogenous, there is less than complete first-order pass-through in these experiments. For food, domestic prices rose by 0.9% for 1% increase in external prices, while for energy the pass-through was 0.4%. Using these values, we calculated approximate elasticities of real household income by rural and urban quintiles. The most arresting feature of this table is the mix of positive and negative effects for both commodities. Energy price effects are again an order of magnitude smaller, even after accounting for general equilibrium linkages, and are in any case quite negligible.

Food effects are varied in more important and revealing ways. While the lowest rural deciles are mostly landless and adversely affected, farmers with capacity to market output experience income gains. Among urban households, the poor are losers but quintiles 3 and 5, attached to downstream food processing and export, gain. A better sense of the relative magnitudes can be gained from Fig. 22.7.

Clearly, supply side income effects can be as important as expenditure impacts to the incidence story. Even if the nation as a whole is adversely affected by higher food prices, some will benefit and among these beneficiaries could be large poor populations. Whether the implied redistribution is consistent with public policy objectives or not will depend on the country in question, its state of development, and the relative political importance of different stakeholder groups. Two examples offer relatively stark contrast. During recent global food price escalation, low-income rice exporters closed their doors to external markets. This represents a striking political calculus, choosing income security of urban poor minorities against income improvement for rural poor majorities. The former are politically more dangerous, and in any case linked to low-wage export industrial competitiveness, but this decision limits the upside potential of rising food prices seen in the present example. In contrast to this, China effected the greatest wealth distribution in its history in the early 1990s by liberalizing domestic agricultural markets. Poor farmers, allowed to

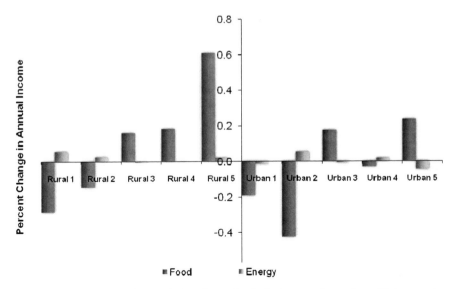

Fig. 22.7 Real household income by location and quintile (percent change for a 1% increase in food or energy prices)

sell their own produce to urban populations, quickly shifted from staples to more expensive, higher value products.

22.5 Conclusions and Extensions

The advent of modern biofuels has animated an intense debate about food and fuel tradeoffs. To the extent that these controversies focus on living standards, there is a pressing need for better evidence on the real global incidence of food and energy price volatility. This chapter presents a set of empirical approaches that can elucidate these issues, including examples from three low- and middle-incomes countries.

Our results show that heterogeneity matters in this context, both between and within countries. Generally speaking, a North–South dichotomy is strongly apparent, where high-income countries are relatively more energy dependent and low-income countries relatively more food dependent in terms of price effects on per capital real incomes. This dichotomy will have lasting implications for the biofuel debate and is likely to drive policy toward more intensive and productivity centered, rather than extensive approaches to biofuel development. In other words, as the food–fuel debate that flared in 2007 reminds us that, rather than simply expanding the scope of exploitation, we must do more with limited resources. In the interest of future social stability, we must also consider ability to pay, i.e., higher income groups can outbid the poor in both categories when scarcity emerges.

A second important set of insights concerns the effects of prices within low- and middle-income economies. Again, food prices are generally much more welfare-influential than energy prices, but vulnerability to both generally declines with income. More significantly, some groups in these economies can experience important gains from food price increases. These benefits must be tallied against vulnerabilities elsewhere, and can seriously complicate the political economy of state responses to both external price shocks and domestic food and energy policy.

This research sets out a variety of different assessment methods, with a few country examples, but it could be extended in many directions. Within the universe of poverty research, a more diverse set of welfare measures exists that could be applied in like manner. Social accounting approaches have been a fertile field of development studies, yet energy has received little attention with this tool. CGE modeling has been applied in both food and energy studies, but is only now entering the biofuel policy research arena. Given our present findings on the mixed domestic welfare effects of food price escalation, GE approaches will probably be important contributors to better understanding and guidance for North−South biofuel issues.

References

Chen, S. and M. Ravallion (2008). "The Developing World Is Poorer Than We Thought, But No Less Successful in the Fight against Poverty," Policy Research Working Paper 4703, World Bank, August.

Defourny, J. and E. Thorbecke (1984). "Structural Path Analysis and Multiplier Decomposition within a Social Accounting Matrix." *Economic Journal* 94, no. 373: 111–136.

Drèze, J. and A.K. Sen (1989). *Hunger and public action*. Oxford England New York: Oxford University Press.

FAO (2005). *The State of Food Insecurity in the World*, Food and Agriculture Organization, Rome.

OECD (2005). *International Development Statistics*. Organisation for Economic Co-operation and Development/Development Assistance Committee (www.oecd.org/dac).

Pyatt, G. and J. Round (eds.) (1985). *Social Accounting Matrices: A Basis for Planning*. Washington, DC: The World Bank.

Roland-Holst, D. and F. Sancho (1995) "Modeling Prices in a SAM Structure," *Review of Economics and Statistics*, 77: 361–371.

Roland-Holst, D. and J. Otte (2006). "Livestock Development Goals: Definitions and Measurement." Internal PPLPI Working Document, FAO, Rome.

Stone, J.R.N. (1981) *Aspects of Economic and Social Modelling*. Lectures delivered at the University of Geneva. Droz, Geneva.

World Bank (2005). *World Development Indicators*. World Bank (www.worldbank.org/data).

Chapter 23
European Biofuel Policy: How Far Will Public Support Go?

Jean-Christophe Bureau, Hervé Guyomard, Florence Jacquet, and David Tréguer

Abstract The strong biofuel expansion experienced in the European Union (EU) originates in the incentives set up by Member States (MSs) within a global framework provided by the EU. A significant part of the EU rapeseed production (more than half) is now channeled into the energy market. MSs support the development of biofuels through subsidies/tax exemptions, mandatory blending, and import barriers (at least for ethanol). Several motivations for supporting biofuels have been put forward. For some MSs, the motivation was clearly to increase agricultural income (e.g., in France and Germany). In other cases, biofuels (produced domestically or imported) were mainly seen as a means to abate GHG emissions. Public support for biofuels has recently been questioned. Some potentially negative effects of biofuels (most notably the indirect land use change) have fueled the debates in the EU Parliament and Commission over the adoption of the 2008 Renewable Energy Directive. These discussions have led to the adoption of less stringent mandatory incorporation targets for 2020 (with respect to the initial 2007 proposal), as the objectives are now set in terms of "renewable fuels" (i.e., biofuels, hydrogen, and green electricity). In spite of an agreement on targets for 2020, some important questions still need to be addressed. The indirect land use change triggered off by biofuel production is the most critical. The ability of the Commission to come up with a clear methodology to address this issue is a necessary condition for the enforcement of the biofuel targets.

23.1 Introduction

The development of biofuels in the European Union (EU) has largely been driven by incentives set up by European and national public authorities in both the agricultural and energy sectors. Without the current set of subsidies, tax reductions, and

J.-C. Bureau (✉)
AgroParisTech, UMR Economie Publique, F-75005 Paris , France
e-mail: bureau@grignon.inra.fr

M. Khanna et al. (eds.), *Handbook of Bioenergy Economics and Policy*,
Natural Resource Management and Policy 33, DOI 10.1007/978-1-4419-0369-3_23,
© Springer Science+Business Media, LLC 2010

exemptions, as well as mandatory blending levels, production increases would certainly have been much more limited. At the EU level, the Common Agricultural Policy (CAP) provided incentives to produce crops for energy use, and framework directives set targets on incorporation rates of biofuels in road transport fuel. The main driving force has nevertheless been the measures taken at the Member State (MS) level aiming at increasing the use of biofuels, through tax exemptions, subsidies, and mandatory incorporation in transport fuel. This, together with significant import barriers, at least for ethanol, has led to a considerable increase in production over the recent period. Meanwhile, concerns relative to the overall environmental effect of biofuels and potential competition for land with food production have triggered erosion in the public image of biofuels. This has led some MSs to review their initial ambitions, as well as intense debates within the European Commission and the European Parliament. The November 2008 reform of the CAP ended the subsidies for the production of energy crops, and the new energy policy directives now focus on renewable energy mandates rather than biofuel blending mandates which leaves some doors open for lower biofuel targets. Several MSs also reduced their tax exemptions and credits on first-generation biofuels. Even though a deal was struck in December 2008 for the development of biofuels until 2020, some important questions still linger. The most critical issue deals with the indirect land use change triggered off by biofuel production, a fact that had been overlooked until the beginning of 2008 and which now jeopardizes the credibility of the whole framework.

23.2 Biofuel Development in the EU

23.2.1 The Role of Public Policies

EU authorities invoke several motives to legitimize public support to biofuels. Climate change is the first one. The EU has been more active than many other developed countries in implementing the constraining provisions of the Kyoto Protocol. In a context where the transport sector accounts for around 30% of greenhouse gases (GHG) emissions, biofuels are presented as a significant instrument of the EU strategy for lowering emissions in the transport sector. Biofuels are also part of a strategy to meet air quality targets.[1] The second motivation is to reduce dependence on foreign oil supply. Around 80% of EU oil consumption is covered by imports, and reducing dependence on foreign suppliers is a strong motivation given the recurrent threats of gas export restrictions by Russia and the fears of major crises that would limit oil supply by OPEC countries. However, the objective of providing a larger outlet to the farm sector and by this way, increasing farm income, has certainly been a major determinant of the support to biofuels, particularly in "agricultural" MSs

---

[1] Air quality objectives are set out in the Air Quality Directive 2008/50/EC (European Parliament and Council, 2008).

such as France and Germany. The main decisions regarding biofuels were made at a time when the farm sector was hit by low prices and painful reforms, when the need to comply to World Trade Organization panels limited subsidized exports (sugar) and with the prospect of further trade liberalization under the Doha Round of multilateral negotiations.

Biofuel development in the EU has been shaped by three key political decisions, each marking a new stage in biofuels expansion: the 1992 CAP reform, the 2003 biofuel directives, and the 2008 Climate and Energy Package. We shall now review each of these steps in the build-up of biofuel policy.

The first impulse for biofuel production development in Europe was the 1992 CAP reform. In 1992, mandatory land set-aside was introduced to adjust supply and demand for grains by reducing the area under cultivation. Farmers were allowed to grow nonfood crops on set-aside land (see, e.g., Sourie et al. 2005). By 2002, this policy had led to a small development of energy crops on idled hectares. In addition, during this period, some MSs had implemented on their own initiatives incentives to encourage the blend of biofuels in conventional fuel through tax reductions. France was one of the MSs that pushed harder for such tax cuts.

A more ambitious biofuel policy was launched in 2003 at the EU level. On the production side, a new premium for energy crops grown outside set-aside land was implemented under the CAP (these payments were phased out at the end of 2008 under the "CAP Health Check" process). The major event was the adoption of two directives aimed at promoting the use of biofuels. The biofuels use directive (Council and Parliament Directive 2003/30/EC) set short- and medium-run targets for the percentage of biofuels to be incorporated into conventional fuel (2% in 2005 and 5.75% in 2010). It is noteworthy though that these targets were not mandatory and there was no penalty for noncompliance; the only constraint was an annual report to the European Commission by the MS indicating its progress in achieving European targets (see European Commission 2006a, b). Furthermore, a companion directive, i.e., the energy taxation directive (Council and Parliament Directive 2003/96/EC), has allowed MSs to grant tax reductions and exemptions on biofuels.[2] This framework gives MSs considerable leeway in terms of instruments that can be implemented. These directives bestow on MSs many degrees of freedom. As a result, biofuel development shows a great level of heterogeneity among MSs both in their biofuel blending targets and in the development level they have reached, as may be observed in Table 23.1 (which shows the blending levels reached by MSs in 2005 and their expected target for 2010). In 2005, only two MSs had reached significant levels of biofuel consumption, that is Germany (with 3.7% of biofuels incorporated in conventional fuel) and Sweden (2.2%). Globally, biofuels only accounted for 1% of the EU transport fuel market in 2005, which is half the target

---

[2]The blending of biofuels in fossil fuels is also constrained by the Fuel Quality Directive 98/70/EC (European Parliament and Council, 1998). A proposal for its revision has been formulated by the EC in January 2007 and adopted in December 2008 as part of the "Climate Change and Energy Package". The final version mainly aims at tightening environmental quality standards for a number of fuel parameters and enabling higher blending levels for ethanol in gasoline.

Table 23.1 Biofuels blending targets in the different member states

Member state	2005 target (%)	2010 target (%)
Austria	2,5	5.75
Cyprus	1	5.75
Czech Republic	3,7 (2006)	5.55
Denmark	0	na
Estonia	2	na
Finland	0.1	na
France	5,75 (2008)	7
Germany		
Ethanol	1.2 mandate (2007)	3.6 mandate
Biodiesel	4.4 mandate (2006)	6.25 mandate
Greece	0.7	5.75
Hungary	0.4–0.6	Na
Ireland	0.6	Na
Italy	1	2.5
Latvia	2	5.75
Lithuania	2	5.75
Luxembourg	na	5.75
Malta	0.3	Na
Netherlands	2 mandate (2007)	5.75 mandate
Poland	0.5	5.75
Portugal	2	na
Slovakia	2	5.75
Slovenia	0.65	5
Spain	2	na
Sweden	3	5.75
United Kingdom	0.3	3.5
EU target	2	5.75

Source: Paul Hodson (DG –Tren), European Commission (2006) Presentation.

of 2%. Several MSs have strengthened their policies since 2005 and the EU rate of incorporation was 2.7% by mid-2008. Many MSs have also amended their policy toward mandatory incorporation besides or instead of tax exemption. France has set more ambitious targets than the EU recommendation, with a 7% incorporation rate target for 2010, and has combined fiscal incentives with penalties for not complying with the national target (the French scheme is very close to a mandatory blending policy). Sweden is one of the few countries where government support to biofuels is not motivated primarily by farm support objectives, which explains that Sweden largely relies on imports to achieve its incorporation target. Germany was one of the few countries which met the 2005 target (2%) with a biofuel market share of 3.7%. This has been the result of an ambitious tax exemption plan initially implemented without quantitative limits. However, from August 2006, the German government went back to a limited exemption tax. Currently, the tax level amounts to 0.09€/l compared to 0.47€/l for diesel. By 2012, both fuels will face the same taxation level (a mandate has been introduced to replace the tax cuts, see Table 23.1). Finally, note that Belgium has introduced an original tax scheme to develop biofuels: the government has increased taxes on fossil fuels (+0.013€/l on diesel and +0.037€/l on gasoline) and compensates oil companies for blending biofuels (0.0102€/l and 0.0305€/l, respectively). For a more extensive presentation of MS policies, see Box 23.2 in the annex section.

The last step in the unfolding of the EU biofuel policy was reached with the adoption of the Renewable Energy Directive (RED) in December 2008 (European Commission, 2008) as a piece of the so-called Climate and Energy Package. Note that the provision for renewable fuels is only part of the Renewable Energy Directive, which is itself only an element of the broader "Energy and Climate Change package"; the latter inter alia includes a directive on the EU Emissions Trading System, a decision on the effort sharing between MSs for CO_2 reduction, a directive on the geological storage of carbon dioxide, and a revision of the fuel quality directive in order to account for the new blending rates of biofuels. All the components of the "Energy and Climate Change package" have been adopted by the European Parliament on 17 December 2008 (see European Parliament, 2008a, b, c, and d). We shall now discuss the details of the process that led to the adoption of the RED.

In March 2007, EU leaders committed to a binding minimum target of 10% of transport fuel in each MS to be provided by biofuels by 2020 (Council of the European Union, 2007). This declaration was followed by a project of a directive presented by the European Commission, the so-called Renewable Energy Directive, the scope of which is larger than biofuels since it sets a 20% target for renewable energy to be reached by 2020. However, during the year 2008, several reports produced by MS bodies (see, e.g., the Gallagher Report, 2008, in the UK) or by the Commission (De Santi et al., 2008) suggested that the overall impacts of a 10% target were questionable. These analyses followed the same trail as Searchinger et al. (2008) and Fargione et al. (2008): they show that the carbon balance of biofuels might indeed be negative, owing to a "carbon debt" incurred by biofuels when their production implies land use changes leading to a carbon release (e.g., primary forest or peatland transformed into crop fields to produce biofuels). In the wake of such path-breaking publications, the growing criticisms addressed to biofuels expanded beyond NGOs circles to reach international institutions: OECD (2008), FAO (2008), and the World Bank (2008). In July, the Environment Committee of the European Parliament voted to reduce the biofuel mandatory blending target to just 4% by 2015. Shortly after, EU environment and energy ministers meeting in Paris put forward the fact that the Directive proposal was not limited to biofuels and dealt with the wider notion of "renewable" fuels. In September, the Industry and Energy Committee of the European Parliament voted for a more lenient version of the biofuel target in the directive: while keeping the 10% objective for 2020 and setting an interim target of 5% for 2015, the Committee introduced subtargets to be met from second-generation biofuels or cars running on green electricity and hydrogen (at least 20% in 2015 and 40% in 2020). Hence, the final version formally adopted by the European Parliament in December 2008 expanded the target to "renewable fuels," i.e., including electricity and hydrogen, rather than a strictly biofuel target. Moreover, the final compromise introduces sustainability criteria: minimal GHG savings have to be achieved (biofuels must provide at least 35% carbon emission savings compared to fossil fuel in 2010, and this level will rise to 45% by 2013 and 50% by 2017), some types of land are unfit to grow biofuels crops (primary forests, protected areas, grassland with a rich biodiversity, wetlands, and peatlands), and social standards have to be met (domestic and foreign production should comply with 8 conventions from the International Labour Organization).

Besides, subtargets for first- and second-generation biofuels were finally not taken into account and no legally binding reference to "indirect land use" aspects was kept in the final compromise, a decision that upset NGOs (Birdlife International, European Environmental Bureau and Friends of the Earth Europe, 2008, and European Federation for Transport and Environment, 2008). The European Commission has been asked to come forward with proposals by the end of 2010 to limit indirect land use change. The Parliament and the Council will then have to make a decision based on these proposals before 2012.

23.2.2 Biofuels Use in the EU Is Strongly Oriented Toward Biodiesel

In 2007, 64% of EU fuel consumption used in road transport was diesel and 36% gasoline. The share of diesel has been steadily increasing and diesel engines still account for the larger part of new car registrations. Because of the joint production by refiners, the EU imports increasing quantities of diesel and exports gasoline. This emphasis on diesel explains why biodiesel accounts for a larger share than ethanol in the EU biofuel production, respectively 80 and 20%. In 2006, the share of biofuels in the total fuel used in transport was 1.8% on average, 2.3% for biodiesel but only 0.8% for ethanol. In 2007, it reached 3% for biodiesel and 1% for ethanol.

The EU biodiesel supply relies largely on rapeseed. The huge development of biodiesel production since 2002 has resulted in a considerable increase in domestic utilization of rapeseed oil for transformation into biodiesel (61% in 2006/2007). Hence, the EU, which had been a major exporter of rapeseed oil during the 1990s, became a net importer in 2006. The rapeseed market is the one where the externality of the biodiesel policy on food market seems larger. It is, however, difficult to isolate the "EU biofuel effect" from other forces driving the surge in agricultural market prices observed over the 2006–2008 period. Other factors, including supply and demand conditions worldwide, stock changes, fossil oil price and US dollar evolutions, climatic conditions, speculation and trade policies adopted by importers and exporters certainly play a role (see, e.g., World Bank, 2008 and Abbott et al., 2008).

EU biodiesel production is highly concentrated as three countries only (Germany, France, and Italy) account for more than 70% of quantities (Fig. 23.1).[3] Germany

---

[3] Note however that the market structure greatly varies across countries. In France, Diester Industrie controls 7 plants and enjoys a near monopoly position. In Germany, production is more evenly divided between 5 firms.

is by far the EU leading producer (50% of EU production). Biodiesel production growth has been particularly pronounced in this country thanks to a 100% tax exemption on pure biodiesel until 2006. However, the law that came into force in 2006 re-established a partial taxation associated with quotas, leading to a sharp decline in the growth rate in 2007 (+8.5% in 2006–2007, as opposed to an average annual growth of 55% in the period 2002–2006, see Fig. 23.1). In France, the growth in biodiesel production and consumption has been increasing after new regulations were set in 2005.

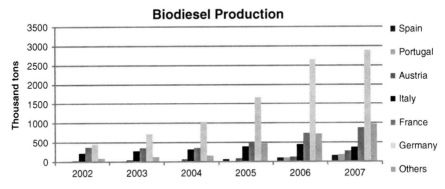

Fig. 23.1   Biodiesel production in the EU-25 (Source: EurObserv'ER (Biofuel Barometer), 2008)

Ethanol in the EU is essentially produced from wheat and to a lesser extent from sugar beet (production from corn is marginal). Germany, France, and Spain are the largest ethanol producers, accounting for two-thirds of production (Fig. 23.2). Ethanol is still a very minor outlet for EU cereals since it adds up to around 1% of end use of the latter. It is a relatively larger outlet for sugar beet (5% of EU production). Because of the combined effect of the 2006 reform of the EU sugar regime, the "Everything But Arms" initiative (that fully opened the EU market to least developed countries' exports), and the WTO panel that led to cut exports significantly, the EU sugar production had to be cut dramatically. In that perspective, the development of beet-based ethanol has been a welcomed outlet for producers, but production costs of beet-based ethanol have been high so far compared to wheat-based ethanol.

23.2.3 Trade in Biofuels

Until 2006, EU biodiesel consumption was entirely domestically produced while 10% of ethanol consumption was imported. This situation changed in 2007 with increases in imports of both ethanol and biodiesel.

Figures reported in Table 23.2 show a sharp increase in biodiesel consumption (+ 1,977,000 tons), which is not caught up with a similar growth in production

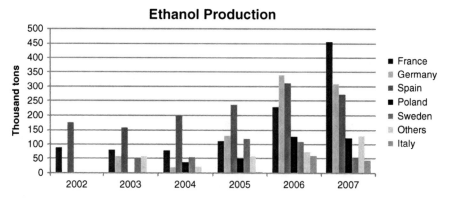

Fig. 23.2 Ethanol production in the EU-25 (Source: EurObserv'ER (Biofuel Barometer), 2008)

Table 23.2 Supply and demand of biofuels in the EU-27 (thousand metric tons)

	2006		2007*	
	Production	Consumption	Production	Consumption
Biodiesel	4,890	4,737	5,713	6,714
Ethanol	1,258	1,363	1,398	1,822

Source: EurObserv'ER (Biofuel Barometer), 2008; *estimated values.

(+ 823,000 tons). As for ethanol, the same dynamics are at play, although to a lesser extent: consumption increases by 459,000 tons while production only rises by 140,000 tons. One can explain, at least partially, these evolutions in the sharp increase in the prices of grains hurting the competitiveness of EU ethanol with respect to the Brazilian one, and competition from US biodiesel whose exports seem to be subsidized.[4]

According to the Comext database, the EU-27 imported 913,000 tons of undenatured ethanol (code 220710) and 76,700 tons of denatured ethanol (code 220720) in 2007. Roughly 10% of this volume was imported under the regime of inward processing, involving subsequent re-exports. As there is currently no specific customs classification for ethanol used as a biofuel, it proves impossible to establish from trade data whether or not imported ethanol is used in the fuel ethanol sector in the

---

[4]The European Union (EU) has launched anti-dumping investigations into imports of biodiesel from the US following complaints by the European Biodiesel Board (EBB) trade association. The EBB claims that the EU biodiesel industry has been damaged unfairly by imports from the US where biodiesel produced in that country or elsewhere can be blended with a very small amount of mineral diesel (to form B99 fuel) to receive a $1/gallon ($0.26/liter) blending subsidy, before being shipped to the EU. The US subsidy amounts to $264/m3 ($300/t or €200/t), said the EBB (FO Licht, 2008).

EU.[5] Despite this uncertainty, it can be reasonably assumed that the increase in EU ethanol imports (from 161,270 tons in 2002 to nearly 1 million tons in 2007; see Fig. 23.3) has largely been caused by biofuel demand. Brazil is by far the main origin for EU-27 ethanol imports (more than 80%).

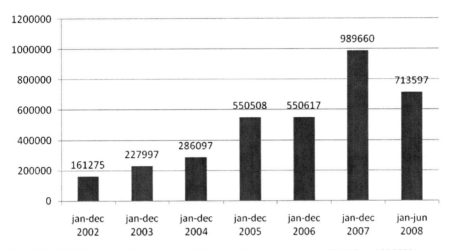

Fig. 23.3 EU-27 imports of ethanol (tons) (Source: Comext, categories 220710 and 220720)

While many countries face limited barriers to trade when exporting ethanol to the EU, the main supplier, Brazil, faces high tariffs. Indeed, imports from both Brazil and the United States face a Most Favored Nation (MFN) tariff that amounts to 19.2€/hl for undenatured alcohol and 10.2€/hl for denatured alcohol. This is not the case for many developing countries who, thanks to the various preferential agreements in force in the EU, can export ethanol duty free. This is particularly the case for the Least Developed Countries (eligible to the Everything But Arms regime) and countries that benefit from the Generalized System of Preferences (GSP) under the "GSP+". In practice, this grants duty-free access for ethanol to 14 countries including all Latin American countries except Argentina, Brazil, Chile, Paraguay, and Uruguay. African, Caribbean, and Pacific countries can also export ethanol at zero or reduced tariff. With the growing number of developing countries interested in accessing the EU market under the GSP+, it is expected that these preferential imports will keep growing.[6]

Until 2007, there were almost no imports of biodiesel mainly due to the fact that the EU was by far the world's largest producer. However, the increase in biodiesel use from 2002 to 2007 led to drastic changes in raw materials trade, that is rapeseed, palm oil, and soybean oil. In 2007, estimates suggest that around 1 million ton of

---

[5] Hence, the difference between consumption and production of ethanol in 2007 observed in table 2 (424,000 tons) is smaller than the total imports of ethanol in 2007 observed on figure 3 (989,000 tons), which entails other uses of ethanol, such as food.

[6] For an extensive discussion on the international trade of biofuels, see ESMAP (2007).

biodiesel was imported from the United States; the early figures for 2008 show that this amount has been increasing (see Table 23.3).[7] The 1 million ton figure can be drawn from consumption and production data (Table 23.2). However, the customs nomenclature does not allow determining the exact quantities of biodiesel imported. Imports were part of larger categories and took place under different CN codes. Despite efforts and progress from 2008 onwards, raw material including vegetable oil used in the production of biodiesel remains classified under other categories, and there is still a problem for identifying mixes of biodiesel with other types of diesel.

Table 23.3 EU-27 Biodiesel imports in tons per origin in 2008 (code 3824 90 91)

Partner/period	Jan. 2008	Feb. 2008	Mar. 2008	Apr. 2008	May. 2008	Jun. 2008	Jan–Jun 2008
United States	29,834	52,892	74,908	127,046	102,945	152,316	539,940
Indonesia						10,610	10,610
Argentina				4,294	41		4,334
Japan	2,479			997	0	0	3,476
United Arab Emirates						3,204	3,204
Canada		62	62	953	219		1,295
China	2	0	21	1	2	110	136
Others	163	1 018	50	64	51	71	1,417
Total	32,478	53,972	75,041	133,354	103,257	166,311	564,413

Source: Comext, category 38249091.

23.3 How Far Can the EU Public Support to Biofuels Go?

23.3.1 Questions on the Future of Biofuel Policy

The increase in the EU biofuel production can largely be explained by national public incentives which have resulted in large, although heterogeneous, degrees of subsidization. The costs for MS budgets have become significant (see, e.g., Kutas and Lindberg, 2007), up to the point that several countries are now moving toward decreased levels of tax exemptions and more constraining targets for mandatory blending of biofuels in fuels used for road transport. While this has been limiting the budget pressure, such a policy ends up passing significant costs to the final consumers, who have already expressed their discontent (UFC, 2007).

   More generally, the public controversies that have emerged, which have led to questioning the EU biofuel policy in bodies such as the European Parliament, have

---

[7]Note that Germany has introduced a provision to prevent biodiesel produced from soybean and palm oil to benefit from mandates and other tax cuts until sustainability criteria are decided.

made central the question of the overall environmental impact of biofuels. Clearly, the legitimacy of the public support to biofuel production is subject to the provision of more evidence of environmental benefits. The combination of budgetary pressures, as well as environmental concerns, is such that one can now consider that support to biofuels in the EU will have to pass both cost–benefit analysis and sustainability impact assessment to be continued. There will be a need to keep public support consistent with major market forces, or at least with the valuation of the actual positive externalities. More practically, either biofuels will have to compete with fossil fuels in terms of cost (either by reducing the production costs of biofuels or because oil prices will be higher), or the subsidies should be in line with what can be considered as a reasonable price of GHG emissions avoided. This raises several questions on which there is still a considerable degree of uncertainty in the EU. The first one is the extent of the actual positive externalities as far as GHG emissions are concerned. The second one is the actual degree of competitiveness of EU biofuels, compared to fossil fuel and biofuels produced in foreign countries. All these elements play a crucial role in the cost–benefit analysis of the EU program.

Sustainability criteria are, nevertheless, difficult to implement. Several research institutions have been commissioned for coming up with proposals for sustainability measurement. The intention is to account for the strong criticisms of environmentalists who claim that the EU biofuel policy ends up encouraging imports that result in deforestation, in particular for planting palm oil in tropical areas. Countries like Brazil, but also EU interest groups and in particular car makers (who now have a broader interest in promoting biofuels because their use would make less constraining the mandatory reduction in CO_2 emission standards) fear that sustainability standards would act as a trade barrier. On the other hand, EU farm groups consider that such standards are impossible to enforce, claiming in particular that imports of biodiesel from apparently righteous countries include some re-exportation of reprocessed palm oil from areas where rainforest was destroyed.

Hence, even if the enactment of the RED ought to provide a more stable framework for biofuel development (as the blending of "renewable energy" becomes legally binding), the enforcement of this policy now takes place in a context of growing skepticism among MSs as to the suitability of biofuel policy in the face of growing evidence of environmental adverse effects.

23.3.2 The Pressure on Land Use in the EU

The food crisis in early 2008 (FAO, 2008) has also led to question the EU biofuel policy on the basis of potential competition with food production and its impact through agricultural prices on consumers' welfare. If one attempts to estimate the energy crops acreage required to meet a 10% target, two majors issues are whether the EU target will be achieved through imports or not, and whether second-generation biofuel technology will make it possible to lift the constraint on land. As a benchmark, we assume here that the 10% target is met only by first-generation bio-fuels produced within the EU. These choices clearly lead to an upper bound estimate of the EU acreage that would be needed to reach the 10% blending target. We also assume that the EU-27 road transport in 2020 is around 346 million toe (European Commission, 2008), shared between diesel (55%) and gasoline (45%), and that the 10% objective is fulfilled by local production of biodiesel and ethanol in the respective proportion of 19.03 and 15.57 million toe. Assuming that biodiesel production is obtained from rapeseed (80%) and sunflower (20%), this would involve production of 41.88 million tons for rapeseed and 9.75 million tons for sunflower. Under quite optimistic assumptions on yields (see Table 23.5 for details), rapeseed production would require about 10.2 million hectares and sunflower production about 5 million hectares. In aggregate, the acreage in energy oilseeds would thus add up to 15.1 million hectares, which represents a very large increase in the total acreage currently devoted to oilseeds in the EU-27 (9.2 million hectares in 2007). If ethanol were obtained from wheat (2/3) and sugar beet (1/3), the required ethanol production would use roughly 8.47 million hectares of wheat and 1.27 million hectares of sugar beet (Table 23.4). These figures ought to be compared with areas of 22 million hectares for wheat and 1.7 million hectares for sugar beet in 2007.

Table 23.4 Acreage requirements for a local supply of biofuel in order to reach a 10% target (EU-27)

Biofuel production required to meet the target in 2020	
Total fuel consumption (million toe)	346
Target	10%
Biodiesel target equivalent (million toe)	19.03
Bioethanol target equivalent (million toe)	15.57
Biodiesel production (million tons)	20.25
Bioethanol production (million tons)	24.33
Acreage requirements	
Rapeseed production (million tons)	41.88
Rapeseed acreage requirement (million ha)	10.22
Sunflower production (million tons)	9.75
Sunflower acreage requirement (million ha)	4.89
Wheat production (million tons)	56.60
Wheat acreage requirement (million ha)	8.47
Sugar-beet production (million tons)	104.61
Sugar-beet acreage requirement (million ha)	1.27

Source: Authors' estimates.

If the EU biofuel production were to rely on its own domestic production only, meeting the 10% incorporation target would require a considerable amount of land, roughly 24.5 million hectares or approximately one fourth of the current arable land surface in the EU-25. This would have a major impact on market equilibriums and prices. As for biodiesel, relying on domestic raw material only would call for a doubling of the current area of oilseeds, which would be then totally dedicated to

Table 23.5 Coefficients used in Table 23.4 calculations

Assumptions Biofuel conversion coefficients (a)		
Biodiesel	ton/toe	0.940
Ethanol	ton/toe	0.640
Rapeseed/biodiesel	toe/ton	2.585
Sunflower/biodiesel	toe/ton	2.409
Sugar beet/ethanol	toe/ton	12.90
Wheat/ethanol	toe/ton	3.490
Crop yields (b)		
Rapeseed	t/ha	4.10
Sunflower	t/ha	2.00
Sugar beet	t/ha	81.98
Soft wheat	t/ha	6.68

Calculations on the basis of EU-25 2005–2007 yields (Eurostat) and assumptions on the yield growth trends +0.8% per year for wheat, +2.30% per year for sugar beet, +1.8% for rapeseed, and +1% for sunflower (according to European Commission, 2008 and the Gallagher Review, 2008).

energy uses. And devoting more land to cereals or beets for energy use would necessarily entail a reduction in EU grain exports. All these increases in area would not be possible without a surge in domestic prices[8] and very likely harsh environmental concerns due to more intensive practices.

The possibility of relying on second-generation technologies is often put forward as the solution to reach more ambitious targets. For example, in its estimates of the impact of the directive proposal, the European Commission finds that 17% of total arable land should be devoted to biofuel production in order to reach the 10% target (European Commission, 2007).[9] The advantages of second-generation biofuels are potentially numerous: they can be produced from a wide variety of crops, they can be developed with less agronomic constraints and on different soil types, and they are more efficient in terms of energy produced per hectare. This is particularly the case for the BtL (Biomass to Liquid) technology, which produces synthetic fuels that are perfect substitutes for diesel, using less land resources than the first generation.

However, as of today, second-generation biofuels are still not competitive with the first generation (Cormeau and Gosse, 2007). It is hard to make predictions regarding the development of cellulosic ethanol and BtL that are being experimented. Pilot plants are being constructed in several EU MSs, relying on a variety

---

[8]This would also require that a high degree of public support is maintained as well as the current (high) level of border protection as the wheat/sugar-beet ethanol would hardly compete with the sugarcane Brazilian ethanol.

[9]The scenario involves 20% of biodiesel consumption coming from imports. Besides, 30% of the biofuels are obtained from second-generation technologies.

of technologies (see Box 23.1). There is however little evidence of the economic viability of a large-scale process. The Commission recognizes the large uncertainty around the efficiency of second-generation technologies, including the promising BtL. A very recent study from the EC-JRC (De Santi et al., 2008) is much more skeptical than earlier assessments by other commission bodies. Taking into account not only the direct imports, but also the indirect ones (i.e., exports and other domestic uses diverted toward energy production), the authors estimated that the 10% target would not be feasible without imports of half of the biofuels consumed, as they anticipate that second-generation biofuels will not be available before 2020.

Box 23.1 Biomass for Biofuels and Bioenergy

In the so-called Biomass Action Plan, the European Commission sets targets for 2010 as regards the relative contribution of biomass (whatever form it might take) to the total energy consumption (European Commission, 2005). It is noteworthy that liquid biofuels only represent a small share of the total biomass energy consumption, the larger part being made up by solid biomass used for heat and electricity production (see Table 23.6).

Electricity production from solid biomass in the EU comes mainly from power plants in the timber and wood pulp industries. These industries primarily use their own waste (wood wastes and other by-products) as an energy source in large-scale power plants using combined heat and power (CHP) generation. These power plants use either biomass only or biomass mixed with other fuels. Heat production comes second. This category includes domestic heating and heat production in thermal and CHP plants. The countries heavily relying on biomass-based heat production are those which have developed district heating networks (the Scandinavian countries, Germany, and Austria). Public policies have spurred on the construction of power plants consuming very diversified biomass sources like straw, wood pellets, crop harvest residues, lumbering slash, and waste (for more information, see Faaij, 2006).

However, production growth has dramatically slowed down in both sectors. Many power plants projects have been delayed due to the financial crisis and the slump in oil price. Taking these new elements into account, current forecasts for 2010 show that the targets set in the 2005 EU Biomass Action Plan will not be met. A total consumption of 105.8 Mtoe of biomass energy is forecasted, whereas the European Commission established a target of 149 Mtoe (Eur'Observer no 188, 2008).

According to the European Biomass Association, reaching the European objective of 20% of renewable energies in total energy consumption by 2020 would call for a doubling of the current production level, which seems out of reach considering the current trend.

Several small-sized projects involve the use of used food oil (restaurants, food industry), of rendering fats (slaughterhouses), by-products of the refining industry, and the fat component of sewage residues. The economic prospects look promising, but the overall production potential remains limited. While a quick development of this processing seems likely, the impact on the overall EU target might be small.

Second-generation biofuels projects show a great heterogeneity across MSs. Compared to the United States, the development of second-generation biofuels in the EU is more careful, in the sense that investors mainly develop pilot plants and there is more caution before shifting to large-scale production. This can be seen either as a lack of confidence in the market, or as an option value-based strategy.

The main resources being tested are straws and other agricultural by-products. While competition is feared for alternative use (livestock) and there are concerns regarding the agronomic impact of exporting organic material rather than returning it to topsoil, the resource is potentially significant. Residues from the timber industry are seen as another major source.

Several MSs are largely involved in the development of lignocellulosic-dedicated energy crops. Sweden has taken an original path with dedicated short-rotation trees (willows), while Finland has chosen a dedicated grass (mainly *Phalaris arundinacea*). Germany, and to a lesser extent the United Kingdom, Austria, and France have initiated programs mainly based on miscanthus. Those crops are used to produce heat or electricity, but they are considered as good candidates for the second generation of biofuels. The economic aspects of lignocellulosic-based biofuels are however highly uncertain, given the alternative uses of the raw materials.

Several pilot plants are also being constructed in order to produce biodiesel from algae, using either ponds or confined bioreactors. The production has been very limited so far, but the technology looks promising (although production costs are still very high).

Table 23.6 Total biomass energy consumption in the EU-27, in million ton of oil equivalent (Mtoe)

	2006	2007	2010 Forecast
Biofuels	5.6	7.7	17.4
Biogas	4.9	5.9	7.8
Renewable municipal solid waste	5.8	6.1	7.4
Solid biomass	65.7	66.4	73.2
Total	82	86.1	105.8

Source: Eur'Observer no 188, 2008.

23.3.3 Environmental Impacts of Biofuel Production

The issue of the energy and GHG balance of EU biofuels is a matter of considerable controversy. For a long time, the debates have been confined to a rather academic and industry audience. However, over the past 2 years, many stakeholders, including environmental organizations, farmers' unions, and the media have shown a considerable interest in the matter, leading to a very lively debate in the EU. Indeed, many figures regarding the actual energy and GHG balance have been issued, ranging from very positive figures to slightly negative ones. Table 23.7 provides an estimate of the findings of the main studies on the EU regarding GHG. Without even mentioning the most extreme findings of a variety of life-cycle analyses, this group of studies shows the extent of the uncertainty and the controversies on the energy and, therefore, GHG emission impact of biofuels.

Table 23.7 suggests that most life-cycle analyses find that biofuels do reduce GHG emissions. Overall, the EU biofuel which displays the most favorable results in terms of GHG emission reduction is the use of pure vegetable oil, which raises technical problems in transportation fuel. Next come biodiesel and sugar-beet ethanol, and last wheat ethanol. Some studies conclude to large intervals regarding the results. This can be explained by the fact that a particular biofuel can be produced

Table 23.7 Reduction in GHG emissions compared with fossil fuel emissions (percentage reduction if positive)

Source of the study	Year of the study	Bioethanol from sugar beets	Bioethanol from grains	Biodiesel from rapeseed
RAC-F	2006	Positive (44%)	Positive (24–48%)	Positive (74%)
WTW	2005	Positive (37–44%)	From negative (minus 6%) to positive (+43%)	Positive (16–62%)
VIEWLS	2005	Positive (20–73%)	From negative (minus 21%) to positive (32%)	Positive (18–64%)
Imperial college	2004	*From negative (minus 11%) to Positive (63%)*	*Positive (5–68%)*	*Positive (48–80%)*
IEA	2004	Positive, 40%	Positive, (18–46%)	Positive (43–63%)
Mortimer et al. (Sheffield Hallam University)	2002	*Positive (47% to 54%)**	*Positive (62% ot 67%)**	Positive (54%)
ADEME/price waterhouse cooper	2002	Positive (75%)	Positive (75%)	Positive (74%)

Note: The figures in italics are quotations from the Table 5.2.1 of the impact assessment by the EU Commission (European Commission, 2006b); we did not access the primary source. The RAC-F and ADEME study refer to France. References of the studies are listed in the bibliography.

through different processes; in particular, some might use energy sources that are themselves renewable. For example, ethanol produced in a biorefinery which uses straw to produce electricity and heating will show much more favorable energy and GHG emission results than ethanol from a conventional unit. Compared to gasoline, the reduction in GHG emissions would be 60% in the first case, and only 15% in the second one (Concawe-WTW, 2007).

However, following studies by Fargione et al. (2008) and Searchinger et al. (2008), the traditional view of biofuels as a means to reduce GHG emissions has been challenged. Indeed, these articles have shown that the life-cycle analyses which had been undertaken until then omitted a crucial aspect of biofuel production: the use of land to produce the agricultural feedstock for biofuels, which leads to GHG emissions that can potentially negate the positive GHG externality of biofuels (calculated without considering these land use changes effects). The indirect effects from land use change or intensification tend to worsen the GHG balance of EU biofuels. A large part of the recent controversy on the blending targets comes from the fact that some studies (not specific to the EU) published in 2008 take into account the land use displacement at the world level and as a result, show less favorable environmental effects of biofuels.[10] For biofuels produced in the EU, there are only limited land substitution effects in terms of deforestation (for example, beets use for ethanol replace beets formerly used for sugar production). However, there are concerns that land which is today set-aside or used for permanent pasture be ploughed and used for energy crops. In addition, for large levels of biofuel use in the EU, the global consequences in terms of higher imports of biodiesel or vegetable oil, as well as reduction in EU exports of grains, need to be taken into account. If the EU target is achieved through large volume of imports, change in the global use of land, and therefore the deforestation and/or the ploughing of meadows, could result in an increase, rather than a decrease, in the EU indirect GHG emissions. For instance, De Santi et al. (2008) build on the analysis by Fargione et al. (2008) and show that the positive GHG effect of the whole EU biofuel program would be negated if only 2.4% of rapeseed oil was replaced with palm oil produced on peatlands as the emissions from the oxidizing peat cancels out the benefits from biodiesel consumption. These uncertainties regarding the global impact of the EU biofuel target are now fuelling a very controversial debate within the EU.

Few studies provide results on other environmental effects apart from GHG emissions. Some NGOs have expressed concerns regarding water resources, given that corn and to some extent wheat use irrigation in countries like France or Spain. Sugar beets, corn, and wheat also use a significant amount of pesticides. In a simulation of the effects of various economic scenarios on groundwater pollution on French and German regions, Graveline et al. (2006) find that the extension of biofuels as a

---

[10] As pointed out by Fargione et al. (2008), the sizeable production of biofuels on a global scale will lead to the conversion of land (mainly in the Americas and Southeast Asia) for biofuel production (or food/feed production when the existing agricultural land has switched to biofuel production). The conversion of land within previously undisturbed ecosystems will lead to large CO_2 emissions stemming from the burning and microbial decomposition of the organic carbon sequestered in the soil and plants.

way to cut GHG emission is actually the worst case among their scenarios regarding nitrate pollution of groundwater. A major explanation for this result is the extension of rapeseed production. They point out the trade-off between GHG emission and water pollution.

One must also account for the increase in the arable crops acreage on the conservation programs. From that point of view, ambitious targets on biofuels, in particular when reached by the production of biodiesel which is more land consuming for one ton of petrol equivalent, might be in contradiction with the present incentives to promote environmental set-aside and have an adverse impact on the efforts to promote biodiversity through agri-environmental measures. There is a risk of a serious contradiction between various CAP instruments, given that agri-environmental measures might no longer remain attractive if the production of energy crops becomes profitable enough.

The Commission points to the fact that there might be some positive externalities due to the production of energy crops themselves (European Commission, 2007). In some areas, maintaining agricultural production might prevent erosion, sometimes landslides. However, the direct positive impact would certainly be very limited, at least with the current generation of biofuels which are not particularly adapted to the regions where many of these problems occur. Energy crops might provide more incentive for crop rotations and have positive agronomic effects, but it is likely that overall, the non-GHG environmental balance of growing more energy crops is negative. Most of the environmental organizations, including many who supported biofuels as a way to reach Kyoto objectives a few years ago, are now expressing serious doubts regarding the environmental consequences of a large-scale program like the 10% target.

23.4 Conclusion

The EU biofuel policy has experienced an unprecedented growth over the past 5 years. This evolution is the result of two directives enacted in 2003, which gave MSs a target to reach (i.e., 5.75% biofuel blending) and the right to amend their fuel tax policies accordingly. It is also the result of policies developed at MSs level. Indeed, EU countries have implemented these directives setting their own national objectives. In some cases, the motivation was clearly to provide an outlet for farmers (e.g., in France). In other cases, biofuels, including imports of Brazilian ethanol, were seen mostly as a way to reduce GHG emissions.

During year 2008, the discussions on the future of biofuel development in the EU have been caught up with two major events: the recent price surges observed in agricultural markets and the drastic changes brought in the methodology for assessing the impacts of biofuels in terms of GHG emissions and more generally, in terms of environmental effects. Even if biofuel production is not the only factor triggering agricultural price hikes, its influence seems nevertheless sizeable (Dronne and Gohin, 2008). This is all the more true for vegetable oil markets in which the EU biodiesel demand could add up to 19% in 2020, should the 10% target be reached at this time. As regards GHG methodology, the recent studies that question the indirect consequences of biofuels through changes in global land use have fueled the controversy.

In spite of all the pitfalls on the road to the adoption of the mandatory blending target of biofuels in the EU, the final decision reached in December 2008 seems to be a victory (albeit partial) for biofuels backers. The adoption of the EU legislative framework is a necessary condition for the widespread development of biofuels; it is in no way sufficient, for two main reasons. First, the legislative framework is still incomplete as it does not address the problem of indirect land use change with enough clarity. As long as the Commission will not have come up with a satisfying framework to solve this central problem, the development of biofuels will be hampered by the growing evidence of their adverse environmental effects. Second, as this article has striven to show, much depends on the way the European directive is transposed into national law. Hence, some MSs may drag their feet to adopt the biofuel mandate, invoking the cost of the biofuel program in the context of a deep economic crisis in order to get rid of a very inconvenient policy.

Acknowledgments Financial support received by the "New Issues in Agricultural, Food and Bioenergy Trade (AGFOODTRADE)" (Small and Medium-scale Focused Research Project, Grant Agreement no. 212036) research project, funded by the European Commission, is gratefully acknowledged. The views expressed in this paper are the sole responsibility of the authors and do not necessarily reflect those of the Commission.

Appendix

Box 23.2 Summary of biofuel supporting policies within some MSs

Bulgaria: The country has introduced a dual support for biofuels with a 3% tax reduction for biofuels–fossil fuels blend and a 5% mandatory blending target which began on January 1st, 2008.

Czech Republic: The government has set targets on a volumetric basis: as of January 1st, 2009 the mandatory blending rates are 3.5% for gasoline and 4.5% for diesel.

Germany: The total tax exemption enjoyed by B100 came to an end on August 1, 2006. Currently, the tax level amounts to 0.09€/l compared to 0.47€/l for diesel. By 2012, both fuels will face the same taxation level. However, a mandate has been introduced, starting at 6.25% (energy content) in 2009 and reaching 8% by 2015.

Finland: It has introduced a mandatory blending scheme which came into force on January 1, 2008. The level gradually increases so as to reach 5.75% by 2010.

France: The French government combines 2 instruments to develop biofuels. The first instrument is a tax reduction of the excise tax on fuels (tax cuts are granted for limited quantities, after a tender at the EU-level). In addition, wholesalers selling petroleum products are subject to another tax (TGAP), which they can avoid paying by incorporating biofuels. Tax rates increase over time in line with the increase in the incorporation target up to 7% in 2010. This measure results in a high penalty for fuel distributors who do not respect the share of biofuels to be incorporated and thus could be considered similar to a biofuel mandate. Note that the former instrument will be definitively phased out by 2012.

Italy: As in the French case, a partial tax break (within quotas) and a mandatory blending rate (2% in 2008, 3% in 2009, energy content) have been applied.

The Netherlands: A mandate has been in place since January 1, 2007, which imposed a 2% blending rate in 2007 to be gradually increased to 5.75% by 2010.

Slovenia: A law has been passed in order to impose a mandatory blending of biofuels in fuels, the rate will reach 5% in energy content by 2010.

Spain: The government has forecasted to introduce a mandatory blending law in 2009.

Sweden: A 5% blending rate of ethanol in gasoline is already in place. Besides, ethanol and biodiesel are exempted from environmental taxes levied on fossil fuels. Other incentives like, e.g., subsidies to buy flex-fuel vehicles are also awarded.

The United Kingdom: A Renewable Transport Fuel Obligation (RTFO) has been introduced, which is a form of mandatory blending scheme. The blending rate of biofuels within fuels will reach 5% by year 2010.

Source: Reports of the different MSs to the European Commission, listed in the reference list.

References

Abbott PC, Hurt C and Tyner WE (2008) What's driving food prices? Farm Foundation Issue Report, July, 82 pp.

ADEME/PWC (2002) Bilans énergétiques et gaz à effet de serre des filières de production de biocarburant. Rapport technique, novembre, Agence française pour la Maitrise de l'Energie, Paris (Price Waterhouse Coopers).

Birdlife International, European Environmental Bureau and Friends of the Earth Europe (2008) First cars, now renewable – EU governments failing climate tests. http://www.foeeurope.org.

Concawe-WTW (2007) Well-to-Wheels analysis of future automotive fuels and powertrains in the European context 2007, Technical report, EUCAR / JRC / CONCAWE/

Cormeau J and Gosse G (2007) Les biocarburants de deuxième génération : semer aujourd'hui les carburants de demain, Demeter, 2008.

Council of the European Union (2007) Brussels European Council 8/9 March 2007 Presidency Conclusions, 7224/1/07 REV 1, Council of the European Union: Brussels.

Czech Republic (2007) Report of the Czech Republic for the European Commission for 2006 on implementation of Directive 2003/30/EC of the European Parliament and of the Council of 8 May 2003.

De Santi G, Edwards R, Szekeres S, Neuwahl F and Mahieu V (2008) Biofuels in the European Context: Facts and Uncertainties, Joint Research Centre (JRC), Petten, The Netherlands.

Dronne Y and Gohin A (2008) Les principaux déterminants de l'évolution des prix agricoles mondiaux. Paper presented at the Pluragri conference, December 14–15, Paris.

ESMAP (2007) Considering trade policies for liquid biofuels. Special report 004/07. Washington D.C.: World Bank.

Estonia (2007) Report on the promotion of the use of biofuels and other renewable fuels in transport.

EuObserv'ER (2008) Biofuels Barometer, no 185, June 2008.

European Commission (2005) Biomass Green energy for Europe, Brussels.

European Commission (2006a) Biofuels Progress Report – Report on the progress made in the use of biofuels and other renewable fuels in the Member States of the European Union, COM(2006) 845 final, January 2007.

European Commission (2006b) Commission Staff Working Document, Accompanying document to the Biofuels Progress Report – Report on the progress made in the use of biofuels and other renewable fuels in the Member States of the European Union, COM(2006) 1721 final, January 2007.

European Commission (2007) Proposal for a Directive of the European Parliament and of the Council amending Directive 98/70/EC as regards the specification of petrol, diesel and gas-oil and the introduction of a mechanism to monitor and reduce greenhouse gas emissions from the use of road transport fuels and amending Council Directive 1999/32/EC, as regards the specification of fuel used by inland waterway vessels and repealing Directive 93/12/EEC, COM(2008)18: CEC: Brussels.

European Commission (2008) Proposal for a Directive of the European Parliament and of the Council on the promotion of the use of energy from renewable sources, COM(2008)19, CEC: Brussels.

European Council (2003) Council Directive 2003/96/EC of 27 October 2003 restructuring the Community framework for the taxation of energy products and electricity, Brussels.

European Federation for Transport and Environment (2008) EU fails to guarantee emissions savings from biofuels, December 4th, http://www.transportenvironment.org.

European Parliament (2008a) Report on the proposal for a directive of the European Parliament and of the Council on the promotion of the use of energy from renewable source, Committee on Industry, Research and Energy, A6-0369/2008, European Parliament: Brussels.

European Parliament (2008b) European Parliament legislative resolution of 17 December 2008 on the proposal for a directive of the European Parliament and of the Council on the promotion of the use of energy from renewable sources (COM(2008)0019 – C6-0046/2008 – 2008/0016(COD)).

European Parliament (2008c) European Parliament legislative resolution of 17 December 2008 on the proposal for a directive of the European Parliament and of the Council amending Directive 2003/87/EC so as to improve and extend the greenhouse gas emission allowance trading system of the Community (COM(2008)0016 – C6-0041/2008 – 200/0014(COD)).

European Parliament (2008d) European Parliament legislative resolution of 17 December 2008 on the proposal for a decision of the European Parliament and of the Council on the effort of Member States to reduce their greenhouse gas emissions to meet the Community's greenhouse gas emission reduction commitments up to 2020.

European Parliament and Council (1998) Directive 98/70/EC of the European Parliament and of the Council of 13 October 1998 relating to the quality of petrol and diesel fuels and amending Council Directive 93/12/EEC, Brussels.

European Parliament and Council (2003a) Directive 2003/17/EC of the European Parliament and of the Council of 3 March 2003 amending Directive 98/70/EC relating to the quality of petrol and diesel fuels, Brussels.

European Parliament and Council (2003b) Directive 2003/30/EC of the European Parliament and of the Council of 8 May 2003 on the promotion of the use of biofuels or other renewable fuels for transport, Brussels.

European Parliament and Council (2008) Directive 2008/50/EC of the European Parliament and of the Council of 21 May 2008 on ambient air quality and cleaner air for Europe, Brussels.

Faaij APC (2006) Bio-energy in Europe: Changing technology choices. Energy Policy, Vol. 34, No. 3, pp. 322–342.

Fargione J, Hill J, Tilman D, Polasky S, and Hawthorne P (2008) Land clearing and the biofuel carbon debt. Science, 319, 1235–1238.

Finland (2007), 2007 Report of the Finnish Ministry of Trade and Industry pursuant to Directive 2003/30/EC on the promotion of the use of biofuels or other renewable fuels for transport in Finland.

FO Licht, http://www.agra-net.com/portal/, accessed January 2009.

Food and Agriculture Organization of United Nations (2008), The State of Food and Agriculture 2008, Rome.

Gallagher E, Berry A and Archer G (2008) The Gallagher Review of the Indirect Effects of Biofuel Production, Renewable Fuels Agency. http://www.dft.gov.uk/rfa.

Germany (2007), Fourth national report on the implementation of Directive 2003/30/EC of 8 May 2003 on the promotion of the use of biofuels or other renewable fuels for transport.

Gohin A (2008) Impacts of the European bio-fuel policy on the farm sector: a general equilibrium assessment. Review of Agricultural Economics, Vol. 30, No. 4, pp. 623–641.

Graveline N et al. (2006). Integrating economic with groundwater modelling for developing long term nitrate concentration scenarios in a large aquifer. Bureau de Recherches Géologiques et Minières, Montpellier. (International Symposium Darcy 2006 – Aquifer SysteMember States Management).

Hodson P (2006) European Commission, DG TREN, Presentation.

Italy (2007) National report on the implementation of Directive 2003/30/EC of 8 May 2003 on the promotion of the use of biofuels or other renewable fuels for transport for 2006.

Kutas G and Lindberg C (2007) Government support for ethanol and biodiesel in the European Union. Global Subsidies Initiative of the International Institute for Sustainable Development. Geneva: GSI.

Kingdom of Belgium (2007), Progress report on the promotion of biofuels in Belgium, Brussels.

Mortimer N, Elsayed M and Matthews R (2002) Carbon and energy balances for a range of biofuels options, Technical report, Sheffield Hallam University.

Netherlands (2007), Report from the Netherlands for 2006 pursuant to Article 4(1) of Directive 2003/30/EC on the promotion of the use of biofuels or other renewable fuels for transport.

OECD (2008) Economic assessment of biofuel support policies, Directorate for Trade and Agriculture, OECD Paris.

PriceWaterhouseCoopers (2005) Biofuels and other renewable fuels for transport. A study commissioned by the Federal Public Service of Public Health Food Chain Safety and Environment, Brussels, Belgium.

RAC-F (2006) Les biocarburants, quelle perspective? Réseau Action Climat-France, Etude réalisée par EDEN, Normandie.

Republic of Bulgaria (2007), The Republic of Bulgaria's 2006 report to the European Commission on the implementation of Directive 2003/30/EC of the European Parliament and of the Council of 8 May 2003, Sofia.

République Française (2007), Rapport (prévu par l'article 4-1 de la Directive 2003/30/CE) faisant état du bilan et des actions en faveur des biocarburants en France au cours de l'année 2006, Note des autorités françaises.

Searchinger T, Heimlich R, Houghton R, et al. (2008) Use of US croplands for bio-fuel increases greenhouse gases through emissions for land-use change, Science 319, 1238–1240.

Slovenia (2007), The use of biofuels in transport in the Republic of Slovenia in 2006, Ministry of the Environment and Spatial Planning, Ljubljana, June.

Sourie J-C, Tréguer D, Rozakis S (2005) L'ambivalence des filières biocarburants INRA Sciences Sociales No 2 Décembre 2005.

Spain (2007) National Report on the implementation of Directive 2003/30/EC of 8 May 2003 on the promotion of the use of biofuels or other renewable fuels for transport for 2006.

UFC (2007) Les biocarburants. Union Fédérale des Consommateurs-Que Choisir?, Paris.

United Kingdom (2007), Promotion and Use of Biofuels in the United Kingdom during 2006: UK Report to the European Commission under Article 4 of the Biofuels Directive (2003/30/EC).

VIEWLS (2005) Environmental and economic performance of biofuels.: VIEWLS (Clear Views on Clean Fuels, Data, Potentials, Scenarios, Markets and Trade of Biofuels), Directorate for Energy and Transport, European Commission.

World Bank (2008) Rising food prices: policy options and World Bank Responses, Washington D.C., 11 pp.

WTW (2005) Well-to-Wheel analysis of future automotive fuels and power trains in the European context. Concawe, Eucar, JRC Ispra (http://ies.jrc.cec.eu.int/WTW).

.

Chapter 24
Conclusions

Madhu Khanna

Renewable liquid fuels offer a pragmatic and low-carbon alternative to depletable crude oil with current vehicle technology and infrastructure, while serving a number of other domestic objectives as well, such as energy security and rural economic development. Biofuels are produced from plants that capture solar energy through photosynthesis and convert it to starch that can be converted to fuel and combusted. A dominant portion of the carbon emissions released from conversion of biomass to energy is carbon sequestered by plant growth in both above-ground and below-ground biomass, making this potentially a carbon neutral source of energy. The extent to which biofuels are actually carbon neutral depends on the amount of fossil fuels used in the production of the feedstock crop and in its conversion to usable fuel. While photovoltaic technology offers greater promise in its ability to provide usable solar energy per unit of land (Nelson), its high cost and the current status of battery technology limit its usefulness in the near future. Biofuels produced with first-generation technologies, primarily using food crops, that can convert sugar and starch-based crops to fuel are more economically viable in the interim.

However, concerns about first-generation biofuels have surfaced due to their low yields per hectare, the intensive use of chemical and energy inputs in their production, their direct impact on food prices due to the diversion of food for fuel production, and their indirect effects that cause leakage of GHG emissions. These concerns have led to a growing attention on second-generation biofuels, which are produced from lignocellulosic feedstocks. The main source of these feedstocks is perennial grasses, which are considered more promising due to their ability to supply greater energy per unit land, their high radiation use, water, and nitrogen efficiencies. A key bottleneck to the utilization of cellulosic feedstocks for biofuel production is the availability of chemical−physical technologies for cell wall deconstruction. Current techniques for deconstruction result in the formation of products that inhibit yeast and other fermentation microbes essential for conversion of cellulose to fuel. Ongoing research in the area of metabolic engineering and systems

M. Khanna (✉)
Department of Agricultural and Consumer Economics, University of Illinois, Urbana-Champaign, IL, USA
e-mail: khanna1@illinois.edu

M. Khanna et al. (eds.), *Handbook of Bioenergy Economics and Policy*,
Natural Resource Management and Policy 33, DOI 10.1007/978-1-4419-0369-3_24,
© Springer Science+Business Media, LLC 2010

biology has the potential to develop microbial strains that can grow in the presence of these inhibitors and pave the way for commercially viable second-generation lignocellulosic biofuels (Blaschek, Ezeji and Price). Current projections of costs of production of biofuels from various feedstocks suggest that the cost of producing cellulosic biofuels, even using a mature technology, will be considerably higher than first-generation technologies (Khanna et al.). To promote their production and consumption in the United States, the recently enacted energy bill, EISA, 2007, has set ambitious mandates for advanced (second generation) biofuel production in the next 15 years. Other policy supports for biofuels include a volumetric tax credit (that is higher for cellulosic biofuels than for corn ethanol), various subsidies to cover the costs of establishment and harvesting of perennial grasses, and a tariff on ethanol imported from Brazil. While these policies seek to level the playing field between gasoline and biofuels, they also have an impact on the price of the feedstock used to produce the biofuel.

Several papers in this handbook provide the conceptual framework and the empirical evidence of the linkages among the prices of oil, corn ethanol, and corn. De Gorter and Just show the linkage between gasoline and ethanol price, under the assumption that biofuel and gasoline are perfect substitutes in consumption, and the impact of the biofuel tax credit on the price for corn ethanol received by producers. They also show the relationship between the ethanol price and the price of corn taking into account the value of the coproduct (DDGS) credit, assuming it is a perfect substitute for corn as feed. These relationships will also be affected by biofuel mandates. If the mandate is binding, the price of corn will affect the price of ethanol, while if it is not binding then it will be affected by both the price of corn and gasoline. Taheripour and Tyner use similar relationships to obtain the breakeven price of corn at various crude oil prices that would keep a new dry milling ethanol plant at zero profits (in the absence of a mandate). Their representation of the long-run equilibrium relationship between corn price and crude oil price is based on the premise that demand for feedstocks will be perfectly elastic in the long run when entry and exit in the ethanol industry results in zero normal profits. Using data for the 2005−2007 period, Serra et al., Zilberman, Gil et al. find mixed econometric evidence to support this. While they find that the price of ethanol is positively related to the prices of corn and crude oil, they find that corn prices are not influenced by ethanol or crude oil prices. This apparent contradiction is explained by the difference in the order of magnitude of the ethanol and the food and fuel markets and the different type of data used in simulation vs. econometric studies. Serra et al., Zilberman, Gill et al. analyze changes in daily prices while de Gorter and Just and Taheripour and Tyner consider long-run equilibrium prices. Changes in daily prices of ethanol are not found to affect the price of corn, but seasonal changes in the quantity of ethanol sold could affect the seasonal price of corn. Furthermore, the ethanol market is much smaller than food and fuel markets and the large random shocks that affect these markets make it difficult to identify the impacts of changes in the ethanol markets. These studies indicate that ethanol is increasingly being considered a substitute for crude oil in the fuels market and that an increase in corn prices in the future could pose a threat to the competitiveness and expansion of the US ethanol industry. They also suggest that there may be short-run constraints on ethanol refining capacity, distribution infrastructure, and flex-fuel vehicle adoption that influence these relationships.

To understand the impact of crude oil (and thus ethanol price) variability on feedstock prices in the presence of various structural constraints, Meyer and Thompson distinguish between voluntary and involuntary demands for ethanol, which differ in their responsiveness to price and derive the corresponding demands for feedstock. Their analysis shows that the extent to which oil prices influence corn prices and the transmission of volatility in oil prices to corn prices depend on the presence of short-run constraints and is uncertain. However, it does appear likely that biofuel mandates increase volatility in the corn market in the short run.

Biofuels have not only created a linkage between energy and agricultural markets, they have also contributed to an increase in food prices; estimates of the extent to which they have done so in the recent past differ across studies. Fabiosa et al., Beghin, Dong et al. and Ferris and Joshi examine the effects of biofuel mandates for food prices in the future and the sensitivity of these impacts to crude oil prices. These studies show that the effect of the Renewable Fuels Standard (RFS) on food and fuel prices depends on the crude oil price. Ferris and Joshi show that high crude oil prices over the period 2010–2017 result in higher ethanol and biodiesel prices, higher feedstock prices, greater production of corn and soybeans, and higher land values, but lower livestock production. Results obtained by Fabiosa et al., Beghin, Dong et al. show that the RFS is expected to lead to higher imports of sugarcane ethanol from Brazil and higher ethanol prices. A higher oil price leads to a change in land use, with an increase in land devoted to corn and other coarse grains and a decrease in land devoted to wheat and oilseeds in the United States and in other countries, notably, Brazil and South Africa.

Msangi, Ewing, and Rosegrant discuss the importance of improving energy and crop technologies to prevent the food vs. fuel trade-off with limited land availability, growing demand for food, rising incomes, and urbanization. They show that increasing the rate at which yield improvements occur could lead to a decrease in cereal prices, a decrease in total cereal area, and an increase in per capita consumption of cereals in North America/Europe and Sub-Saharan Africa regions. Cellulosic biofuels from crop residues and high-yielding perennial grasses, such as miscanthus, offer another promising approach to mitigate the competition for land for food and fuel (Dohleman, Heaton and Long). But technological innovations that lower the cost of producing cellulosic biofuels are needed to make them economically viable. Hochman, Sexton, and Zilberman emphasize the need for technical innovations in the production of biofuels, such as development of high-yielding energy crops and low-cost conversion of cellulosic biomass to fuel, as well as for more widespread adoption of agricultural biotechnology that increases crop yield to lessen the severity of the land constraint. In the interim, there is also a need for policy to induce the development of second-generation biofuel technologies. With current estimates of the high costs of producing them, Khanna et al., Chen, Onal, et al. show that biofuel mandates are needed to induce their production and that crop residues are likely to be the lowest cost feedstock that will be used before perennial grasses.

A key justification for biofuels is the environmental benefits they promise, particularly, compared to gasoline; thus the food vs. fuel debate needs to be analyzed using a broader framework that recognizes that consumers obtain utility from food, fuel, and environmental amenities and that the provision of all three goods/services relies on land. Hochman, Sexton, and Zilberman discuss the trade-offs that arise when a given amount of land has to be allocated to meet these three sources of utility for households. These trade-offs are heightened in the presence of policies that are directed to achieve particular objectives. For example, they show that a carbon tax may reduce social welfare because it creates incentives to allocate more land to produce biofuels; if natural land is converted to biofuel production this may reduce other environmental services provided by that land, such as biodiversity. Similar findings are obtained by Khanna et al., Chen, Onal, et al. which show that expanding biofuel production to meet the RFS would reduce GHG emissions but increase nitrogen use (to produce the corn and corn stover needed as feedstocks). The adverse consequences of using corn stover for fertilizer runoff and water quality are also highlighted by the analysis by Kurkalova, Secchi, and Gassman. They show that the use of corn stover for biofuels has the potential to exacerbate soil erosion, nitrogen runoff, and phosphorus runoff. These findings suggest that if biodiversity loss and water quality degradation are a greater threat to social welfare than climate change, then policies such as the RFS and even a carbon tax could reduce social welfare. They point to the limitations of using a single-policy instrument to correct the multiple sources of market failure that exist and the need to recognize the multiple environmental impacts (some positive and some negative) associated with biofuel production.

Furthermore, even the focus on a single environmental impact, such as GHG mitigation, is served better by a broad evaluation of the mitigation strategies available rather than a narrow one that focuses exclusively on biofuels. With electric power generation being a bigger source of GHG emissions than the transportation sector, McCarl, Maung, and Szulczyk show that biomass from crop residues and perennial grasses have greater potential to offset GHG emissions from coal than from gasoline; therefore, the goal of GHG mitigation may be best met through reliance on bio-based electricity. The latter offers a low-cost strategy for GHG mitigation (together with soil carbon sequestration) and is likely to be adopted even at low gasoline and carbon prices. In contrast, high gasoline and high carbon prices will be needed to induce a shift toward cellulosic biofuels. They also show the need for concern about the indirect land use effects induced by diversion of land toward biofuels and the need to design policies that can prevent a leakage of GHG emissions. Rajagopal, Hochman, and Zilberman discuss the challenges of designing a regulatory framework that accounts for the direct and indirect life-cycle GHG emissions generated by biofuel production. While a carbon tax accompanied by a policy that also prices other environmental amenities would be first best, more politically feasible policies may need to be considered even though they are second best. These include thresholds or certification standards that ban the production of biofuels with large negative environmental impacts. They emphasize the need for coordinated policies across countries to address carbon emissions, which is a global public bad, as well as the need for international environmental agreements that recognize the multiple environmental services provided by land.

Several chapters in the handbook describe existing biofuel policies in the United States (de Gorter and Just; Ando, Khanna and Taheripour; Lee and Sumner; and Lasco and Khanna) and in Europe (Bureau, Guyomard, Jacquet, et al.) and the motivations behind them. These chapters also discuss the economic, environmental, and social welfare effects of these policies. The impact of a biofuel mandate on fuel consumption and fuel prices is shown to depend on the supply elasticities of gasoline and biofuels, the elasticity of substitution between gasoline and ethanol, and the costs of biofuel production. Ando, Khanna, and Taheripour show that the impact of a biofuel mandate on GHG emissions is ambiguous and decreases as the gasoline supply curve becomes more inelastic. While mandates result in higher biofuel production levels, particularly cellulosic biofuels, than a carbon tax, they can result in significantly lower welfare levels and higher GHG emissions. A tax credit can offset some or all of the GHG mitigation benefits of biofuels, while further lowering social welfare. Lee and Sumner and Lasco and Khanna describe the US import policies governing trade in biofuels. While Lee and Sumner show that the excess supply of ethanol from Brazil is upward sloping indicating that US biofuel policies do influence the world price of ethanol, Lasco and Khanna analyze the effect of those policies on social welfare, GHG emissions, and imports of ethanol from Brazil. Their analysis suggests that removal of the import tariff could lead to a significant increase in imports of sugarcane ethanol and in social welfare while also leading to a reduction in GHG emissions.

In contrast to the United States, biofuels policies in the EU are strongly oriented toward promoting biodiesel production. Concerns about their impacts on food prices and indirect land use changes as well as uncertainty about their economic and environmental sustainability have led policy makers to take a cautionary view toward future biofuel expansion. Taheripour and Tyner describe their assessment of the impact of US and EU biofuel mandates on the world economy over the 2006–2015 period and the way these policies interact with each other in their impact on land use changes across the world. They find that EU mandates can be expected to have larger land use changes in Brazil than US policies primarily because the EU needs to import soybeans and soybean oil from Brazil to meet its biodiesel mandate. Biofuel production in developed countries also has an impact on consumers and farmers in developing countries because they affect the prices of energy and food crops. Kahrl and Roland-Holst find that the poverty gap is fairly elastic with respect to food prices and that food expenditure shares are an order of magnitude higher than energy shares among the poor making low-income countries much more vulnerable to food price increases than to energy price increases.

Expansion of biofuels imposes logistical challenges for the design, capacity, and location of biorefineries. Decisions about the location and capacity of biorefineries are expected to depend on the trade-offs between large facilities that take advantage of economies of size and decentralized but networked production nodes closer to producers of feedstocks, consumers of the biofuels, and consumers of the coproducts from biorefineries. Kang et al., Onal, Ouyang, et al. show the importance of undertaking a spatial and temporal analysis of the factors that influence the location of refineries and their development over time to cost-effectively meet the goals of the RFS. Goldsmith et al., Rasmussen, Signorini, et al. show the importance of developing biorefineries that are flexible in the feedstocks they utilize. Their analysis of the utilization efficiency of sugarcane mills in Brazil shows that efficiency of asset utilization improves by including corn as a complementary feedstock along with sugarcane. The capacity to process more than one feedstock reduces the disadvantage that larger biorefineries face in the form of high costs of transporting feedstock from farm to the processing plant. It also reduces the risks that arise from asset specificity for refineries that are dedicated to particular feedstocks.

The analyses presented in this handbook are based on a diverse set of modeling tools and methods and show the importance of using behavioral models to understand the economic, environmental, and social implications of biofuel production and policies. Several common themes emerge from these chapters: the need to distinguish between biofuels from different feedstocks because they differ in their land requirements, economic implications, and environmental impacts; the importance of technological innovation to increase crop productivity and capacity to convert feedstocks (cellulosic biofuels) to fuels; and the importance of carefully designing policies, incorporating their multiple impacts on food production, land use and the environment.

The biofuel industry is in its infancy. While it is already having an impact on food and fuel markets, its viability (with the exception of sugarcane ethanol in Brazil) is dependent on government policies, technological change, and growing pressures on land availability. Early observations suggest that new feedstocks are likely to be introduced and with technological breakthroughs in fuel conversion technologies could significantly alter the dynamics of the biofuel industry. This book considers a subgroup of feedstocks, the feedstocks that are more prominent today, but the considerations presented here also apply to some extent to other feedstocks like algae and forest products.

Biofuels should be considered as one option in a portfolio of renewable energy technologies to address climate change and reduce dependence on fossil fuels. Conservation and enhancements in energy efficiency are also options that may offer low-cost approaches to reducing greenhouse gas emissions and increasing energy security. Lastly, a holistic policy approach that coordinates climate, energy, and resource conservation policies is needed to induce effective changes in the energy sector that lead to economically, environmentally, and socially sustainable use of natural resources.

There are several emerging issues that will affect the economic viability of biofuels that have not been considered in this book. Uncertainty about biomass feedstock prices and demand for them, risks associated with investments in dedicated energy crops that are perennials, and lack of cellulosic biorefineries have created a "chicken and egg" syndrome that could influence the timing and level of investment. Innovative contracts are likely to be needed between energy companies and farmers to share these risks and reduce the incentives to wait and delay investment in the infrastructure needed for large-scale biofuel production. Future research also needs to explore strategies for mitigating the "blending wall" constraint that limits demand for biofuels with existing vehicle technology.

Index

M. Khanna et al. (eds.), *Handbook of Bioenergy Economics and Policy*,
Natural Resource Management and Policy 33, DOI 10.1007/978-1-4419-0369-3,
© Springer Science+Business Media, LLC 2010

Breinigsville, PA USA
10 December 2009
228975BV00004B/49/P